Notified in Army Orders for February, 1936

TEXT BOOK

OF

AMMUNITION

1936

The Naval & Military Press Ltd

Published by
The Naval & Military Press Ltd
5 Riverside, Brambleside, Bellbrook
Industrial Estate, Uckfield, East Sussex,
TN22 1QQ England
Tel: +44 (0) 1825 749494
Fax: +44 (0) 1825 765701
www.naval-military-press.com

TABLE OF CONTENTS

TABLE OF CONTENTS—*continued*

TABLE OF CONTENTS—*continued*

TABLE OF CONTENTS—*continued*

TABLE OF CONTENTS—*continued*

TABLE OF CONTENTS—*continued*

TABLE OF CONTENTS—*continued*

TABLES

LIST OF FIGURES

LIST OF FIGURES—*continued*

LIST OF FIGURES—*continued*

LIST OF FIGURES—*continued*

APPENDICES

The coloured plates in this reprint are placed after this page

Fig. 1·14.

PICTORIAL REPRESENTATIONS OF CORDITE.

SHEET 1.

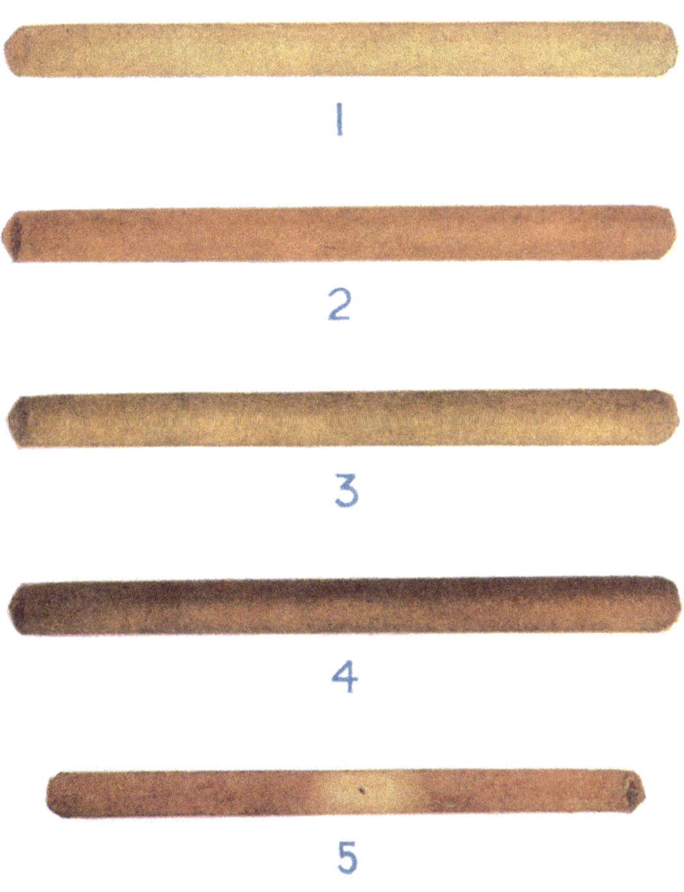

1. }
2. } NEWLY MADE CORDITE.
3. } SHOWING RANGE IN COLOUR.
4. }

5. EARLY STAGE OF CORROSION OF CORDITE MD SIZE 16.

Q 51945. A/D9356/2141. 2/35. W. & Co., Ltd.

Fig. 1·14.

PICTORIAL REPRESENTATIONS OF CORDITE.

SHEET 2.

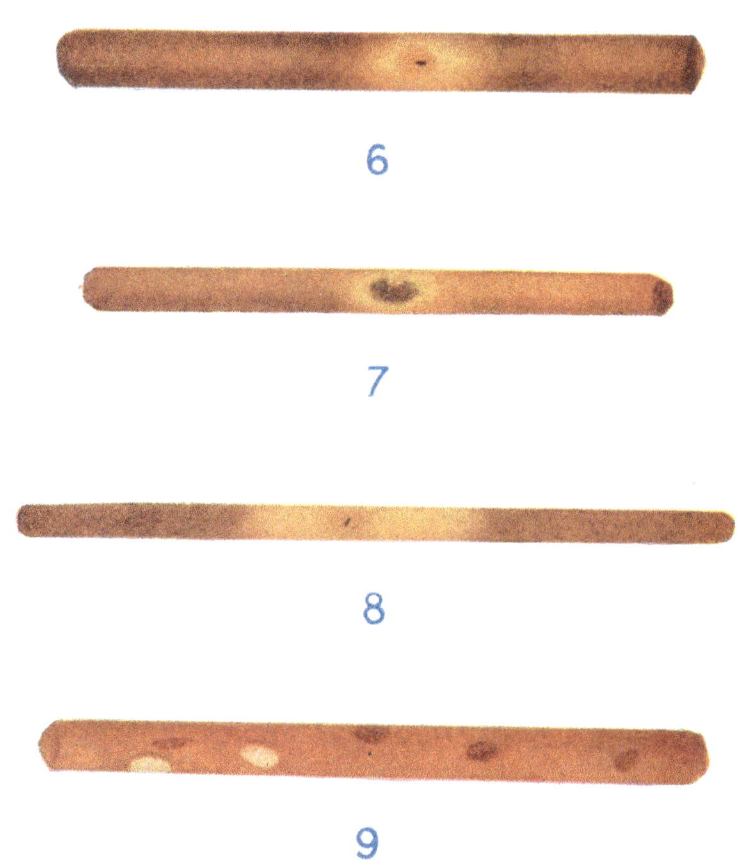

6. INTERMEDIATE STAGE OF CORROSION OF CORDITE MD. SIZE 19.
7. ADVANCED STAGE OF CORROSION OF CORDITE MD. SIZE 16.
8. ADVANCED STAGE OF CORROSION OF CORDITE MD. SIZE 8.
9. CORDITE CONTAINING AIR BUBBLES.

Fig. 1·19.

PICTORIAL
REPRESENTATIONS OF CORDITE S.C.

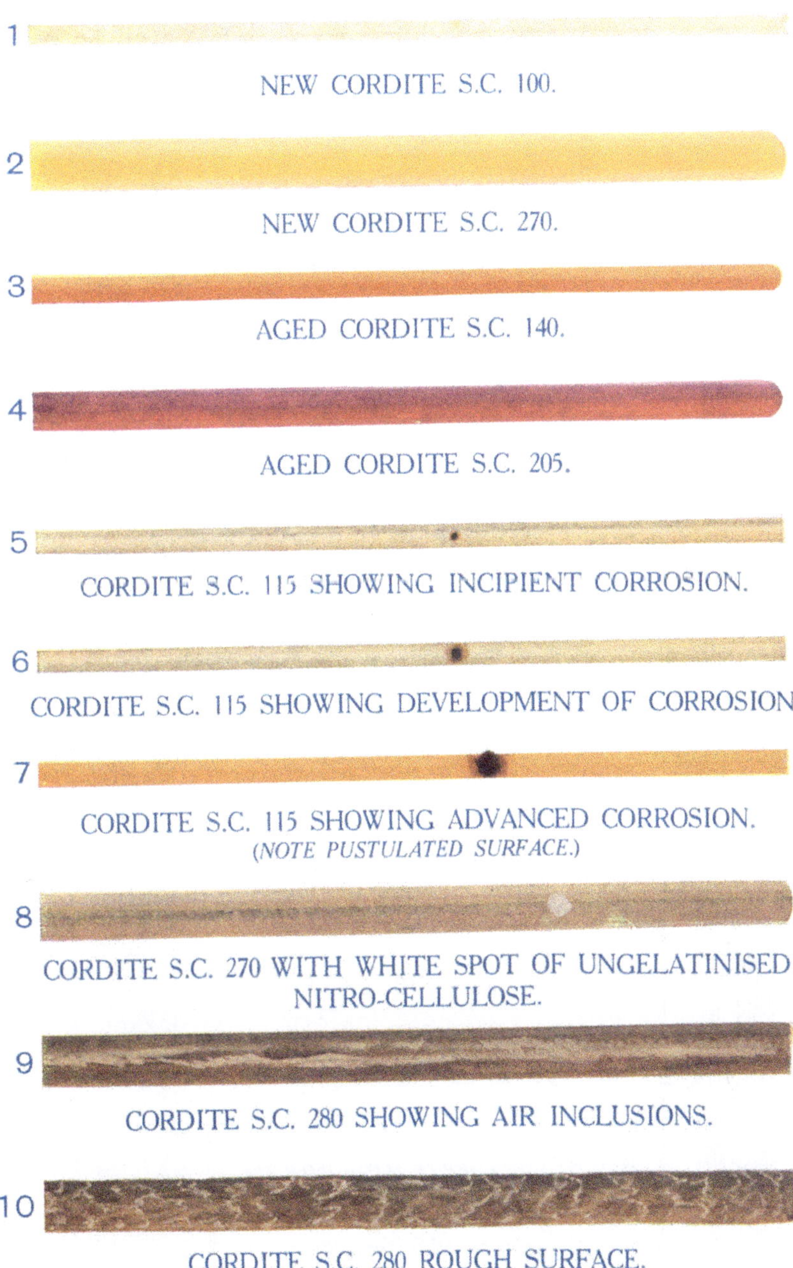

NOTE.—Nos. 1—7 represent Cordite as seen when held up to daylight.

Q 51945. A/D9356/2141. 2/35. W. & Co., Ltd.

Fig. 2·35(b).

CARTRIDGES, Q. F. FIXED AMMUNITION.

TYPICAL FOR FULL CHARGE.

Scale – 1/3.

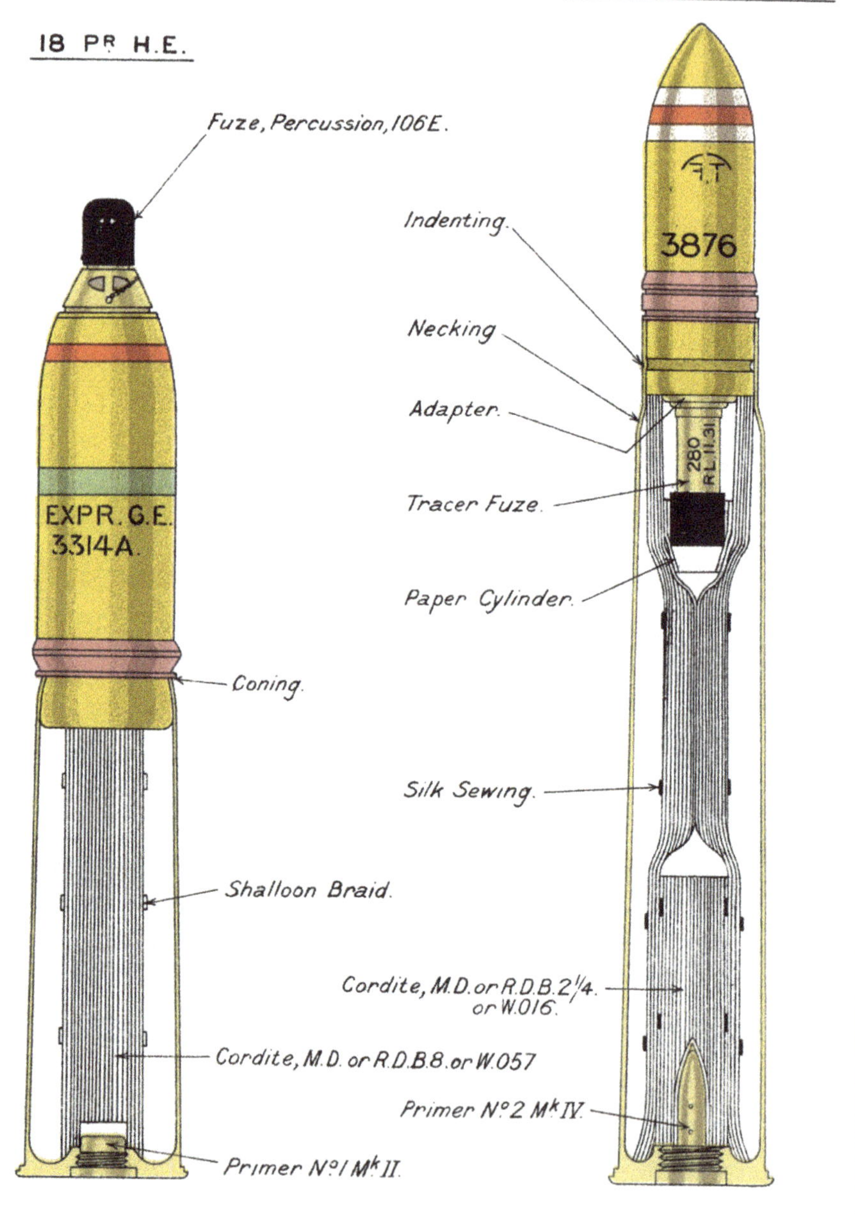

958

Fig. 2·36(b).

STENCILLINGS ON THE SIDE AND BASE OF Q. F. CARTRIDGES, SEPARATE AMM:-

(TYPICAL).

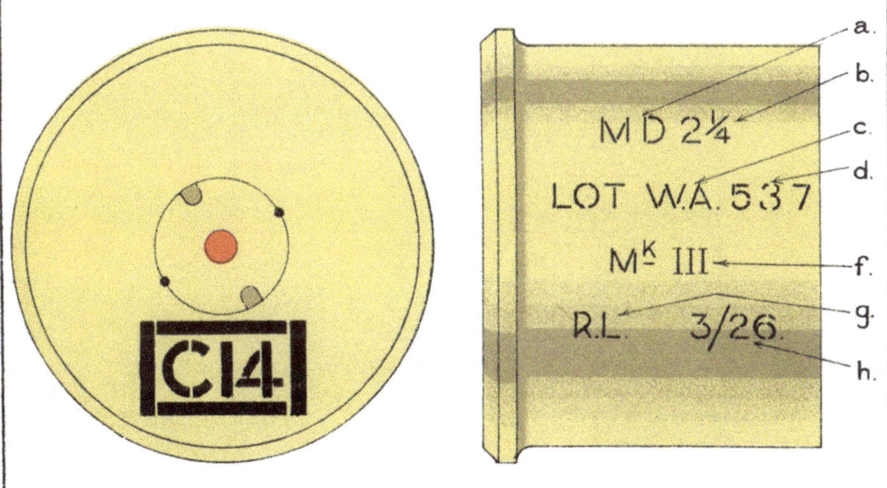

Fig. 3.08.

TUBE, PERCUSSION, S.A. CARTRIDGE, Mk. II L.

SCALE -- 2/1.

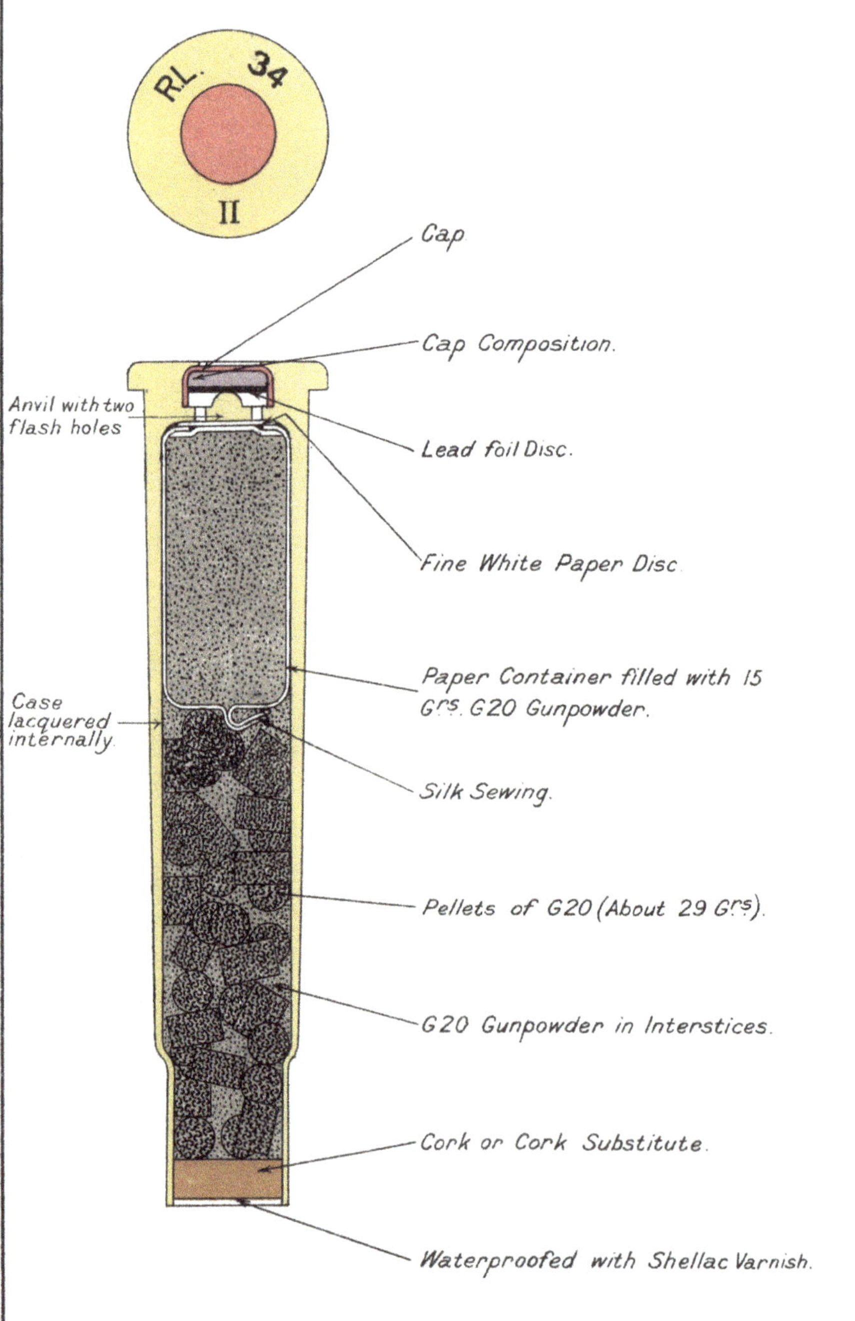

958

Fig 3·10.

TUBE, VENT, PERCUSSION, ·4" Mk. VIII/L/.

SCALE – 2/1.

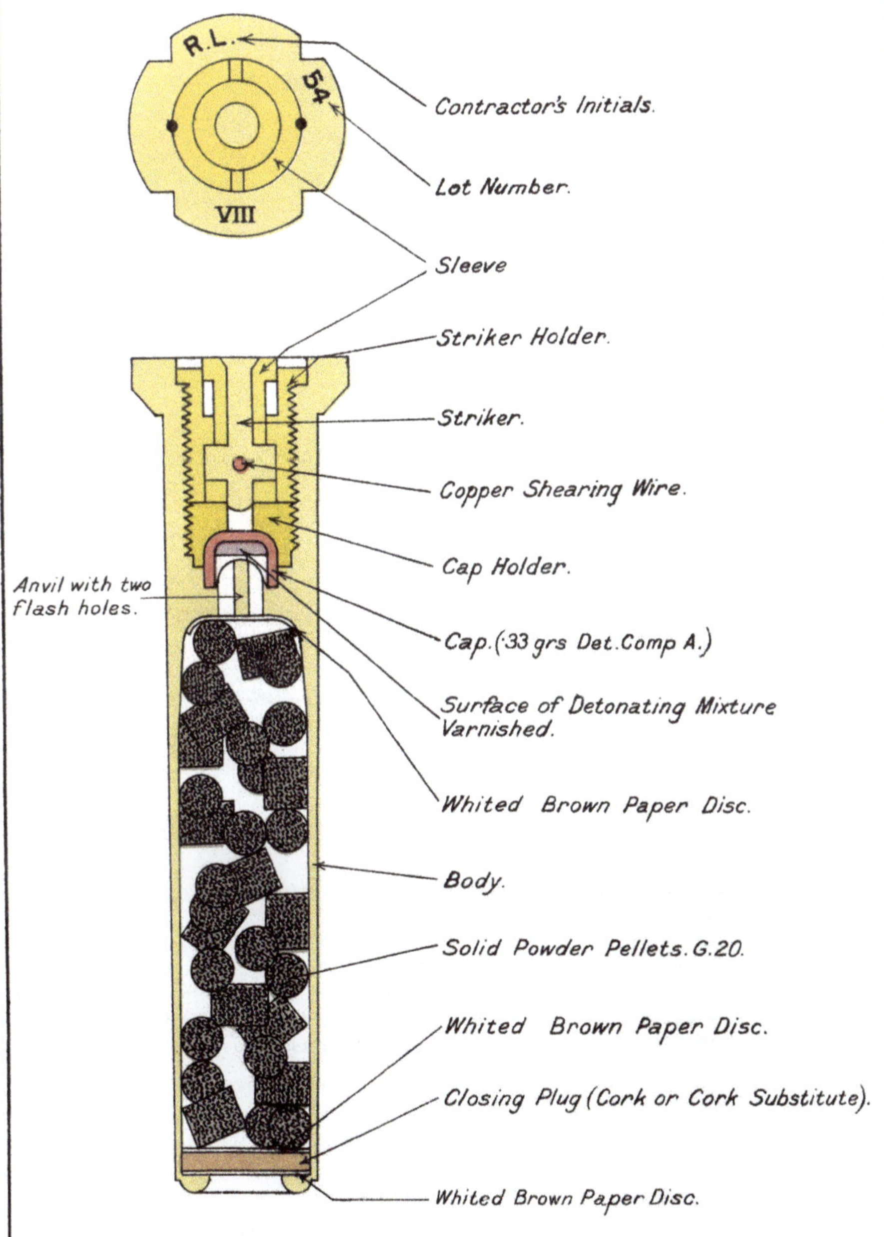

Fig. 3·12.

TUBE, VENT, ELECTRIC ·4" IN MK. X|L|.

SCALE – 2/1.

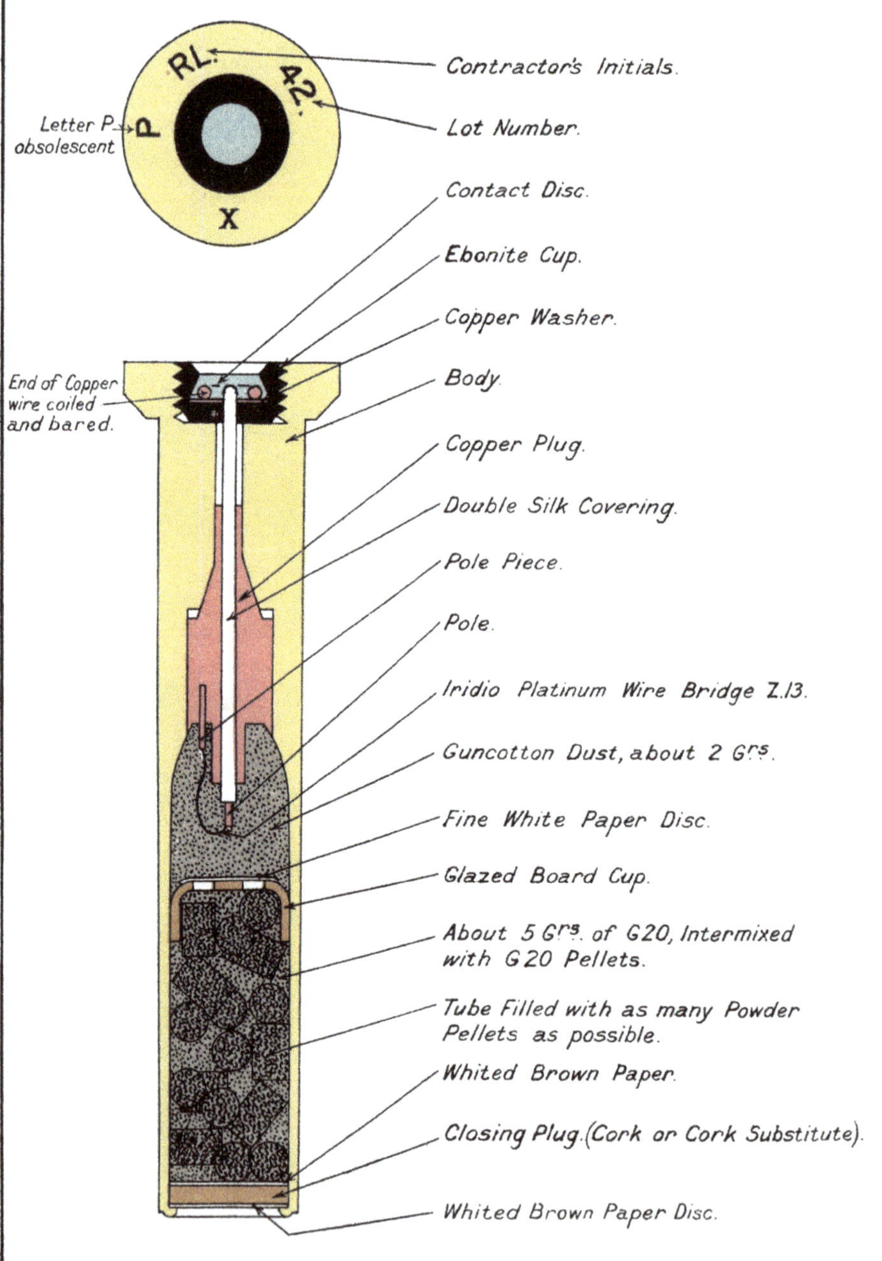

958

Fig 4·05.

PRIMER, PERCUSSION, Q.F. CARTRIDGES, No. I Mk. II [C].

SCALE – 2/1.

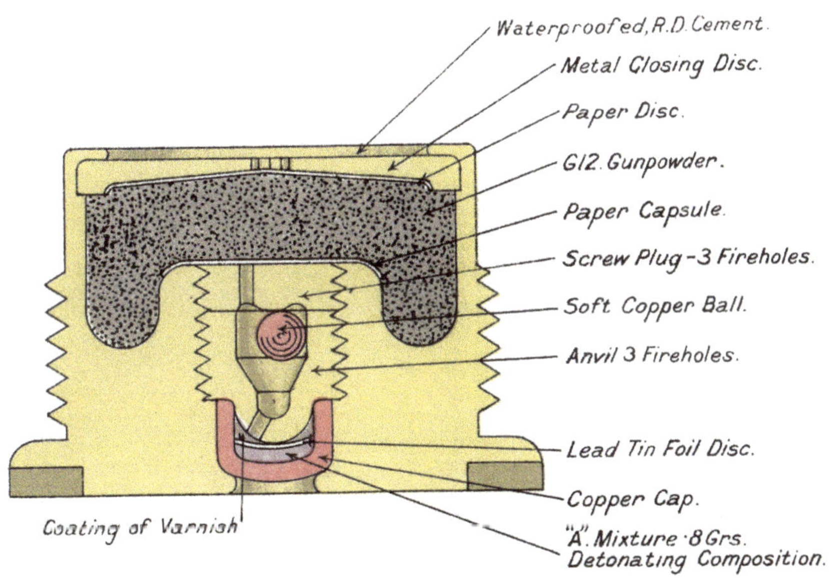

Fig. 4·05(a).

PRIMER, PERCUSSION Q.F. BLANK 6 OR 3 PR Mk. III [L].

SCALE 2/1.

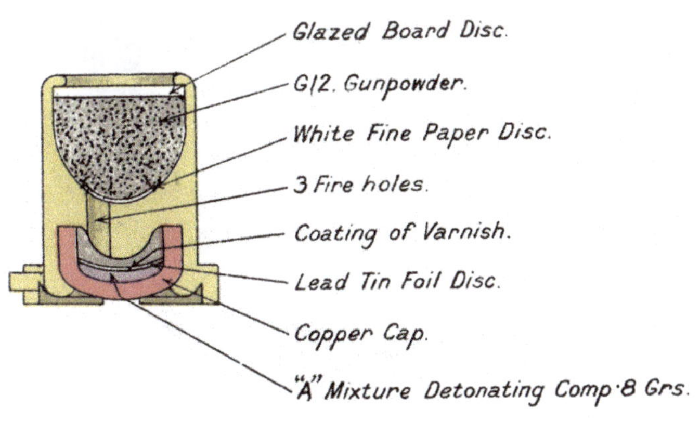

Fig. 4·06.

PRIMER, PERCUSSION, Q.F. CARTRIDGE Nº 2.

Mk III.

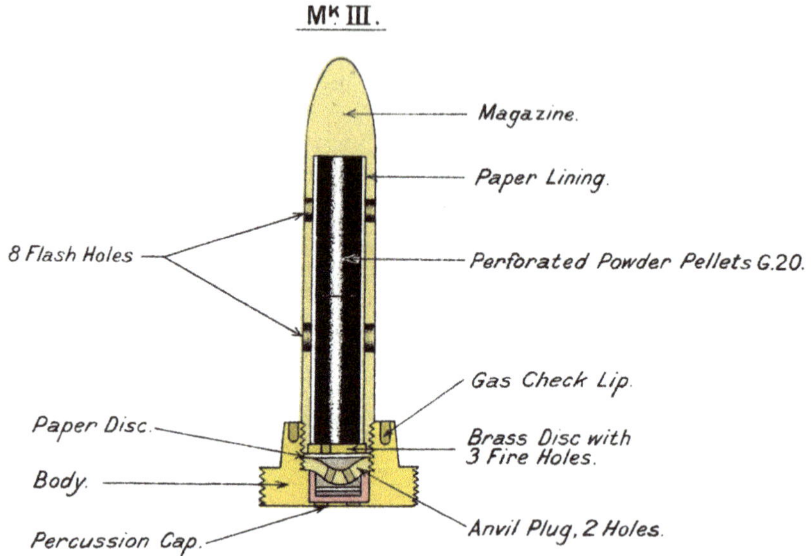

Mk IV. Mk VII.

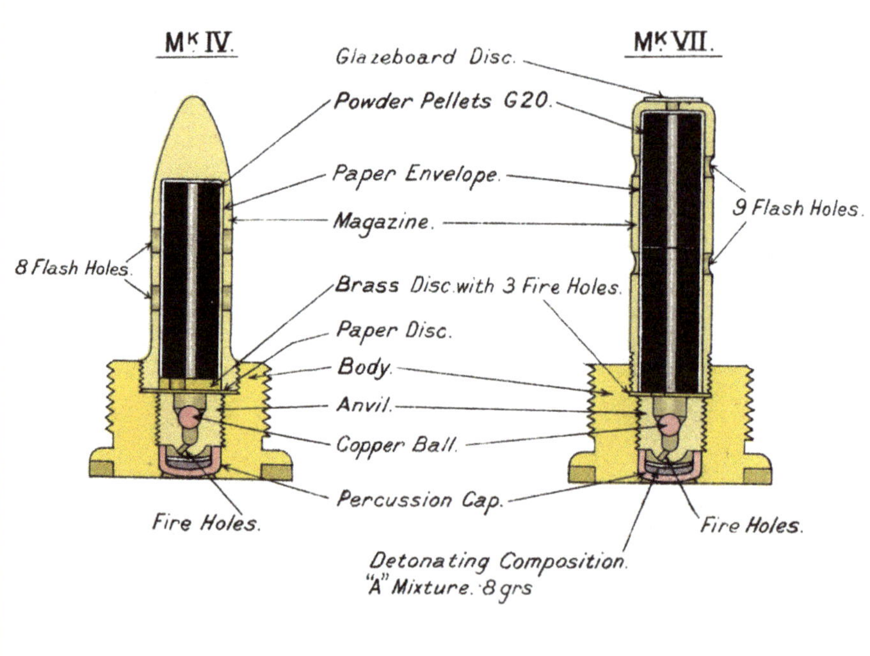

958

Fig. 4.07.
PRIMER PERCUSSION, Q.F. CARTRIDGE No 11 Mk I/L/.
Scale – 2/1.
White Metal Dome.
8 Flash Holes
Shellacked Paper Envelope.
Magazine.
Paper Disc.
Body.
Pea Ball.
Screw Plug 3 Fireholes.
Detonating Comp. ("A" Mixture 8 grs.)
Anvil. 3 Fireholes.
Cap.
26
(vi)
W 9-33
(iv)
R.L. 1-33
(ii)
(v)
(iii)
M
(vii)
(viii)
(i)
I 11oN

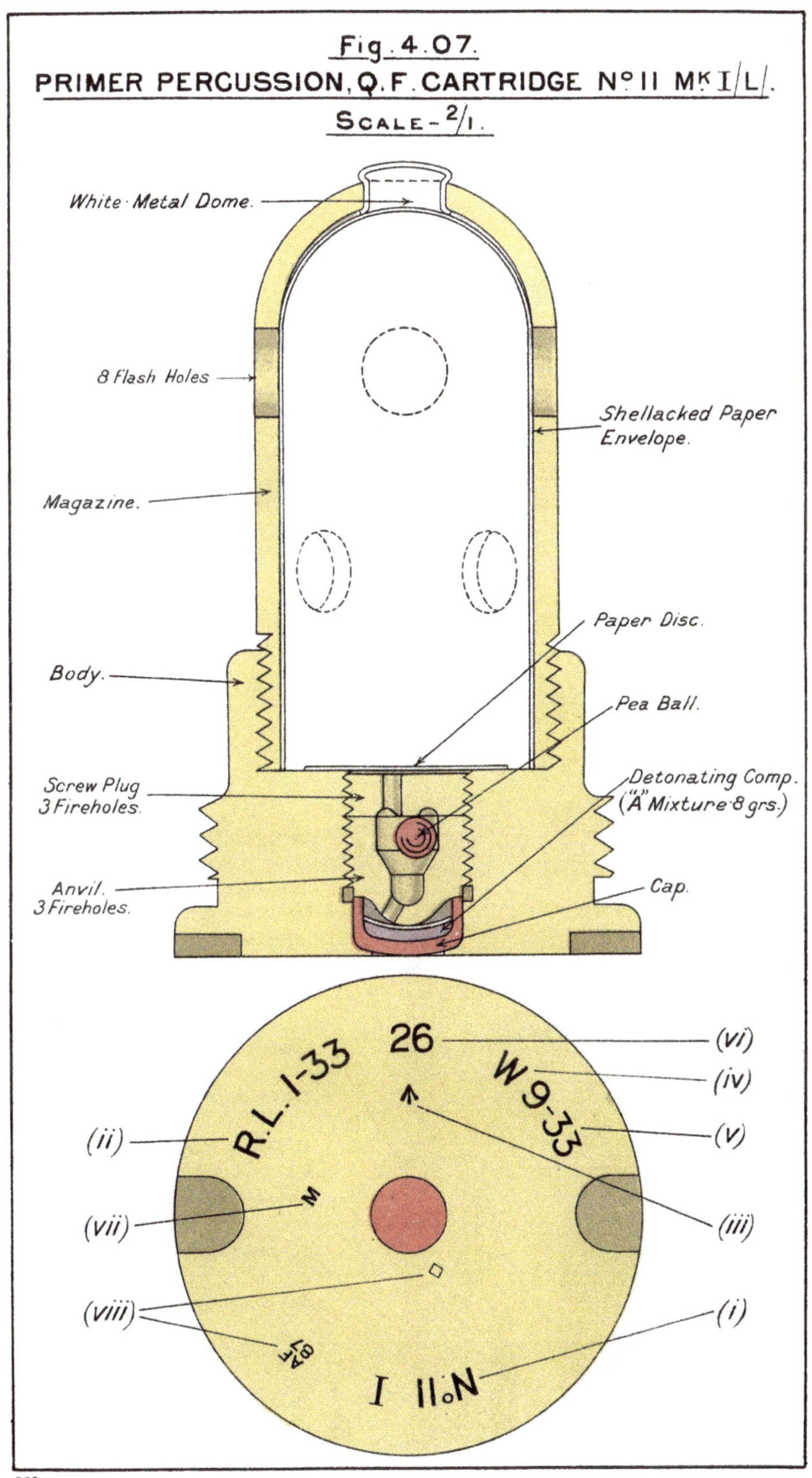

958.

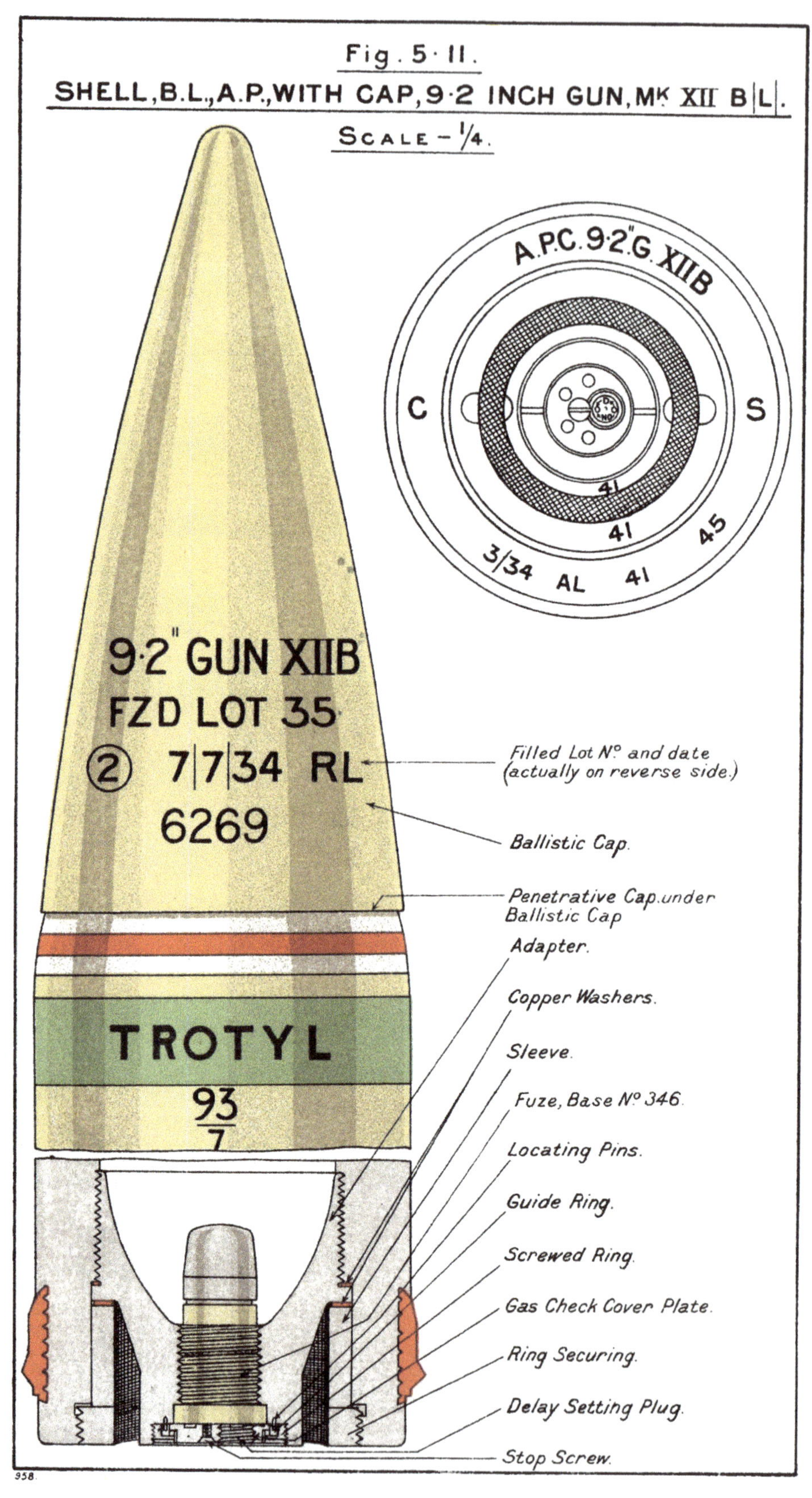
Fig. 5·11.
SHELL, B.L., A.P., WITH CAP, 9·2 INCH GUN, Mk. XII B|L|.
Scale - 1/4.
A.P.C. 9·2"G. XIIB
C
S
41
41
45
3/34
AL
41
9·2" GUN XIIB
FZD LOT 35
② 7|7|34 RL
6269
TROTYL
93
7
Filled Lot No. and date (actually on reverse side.)
Ballistic Cap.
Penetrative Cap under Ballistic Cap
Adapter.
Copper Washers.
Sleeve.
Fuze, Base No. 346.
Locating Pins.
Guide Ring.
Screwed Ring.
Gas Check Cover Plate.
Ring Securing.
Delay Setting Plug.
Stop Screw.
958.

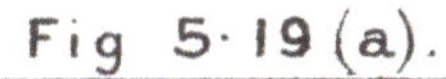
Fig 5·19 (a).

METHOD OF FILLING:-SHELL Q.F. COMMON POINTED 12 PR NON-BURSTER BAG TYPE.

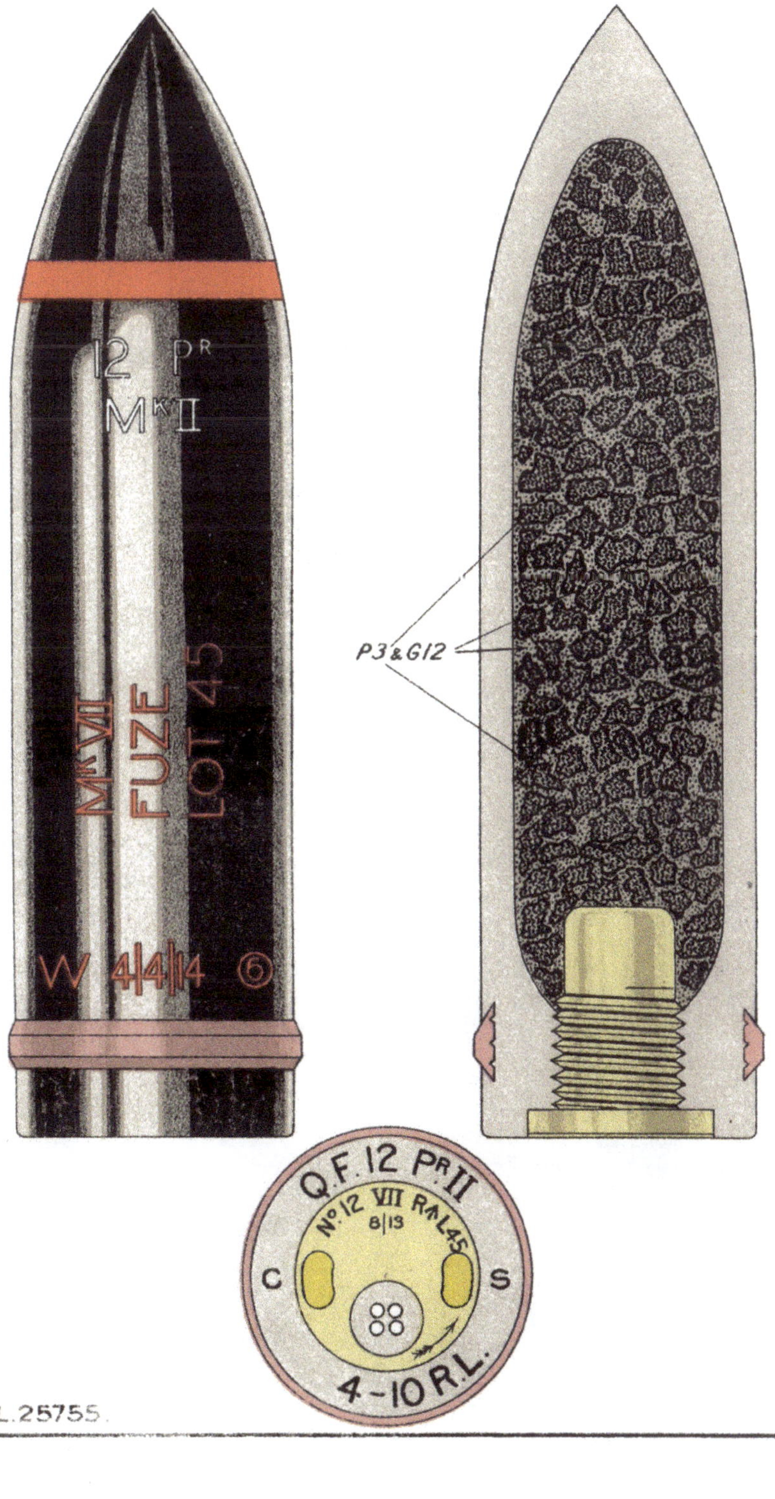

12 PR
MK II
MK VII
FUZE
LOT 45
W 4|4|14 6

P3 & G12

Q.F. 12 PR II
No 12 VII R.L.45
8|13
C
S
4-10 R.L.

R.L. 25755.

Fig. 5. 19 (b).

COMMON POINTED SHELL WITH CAP FILLED POWDER.

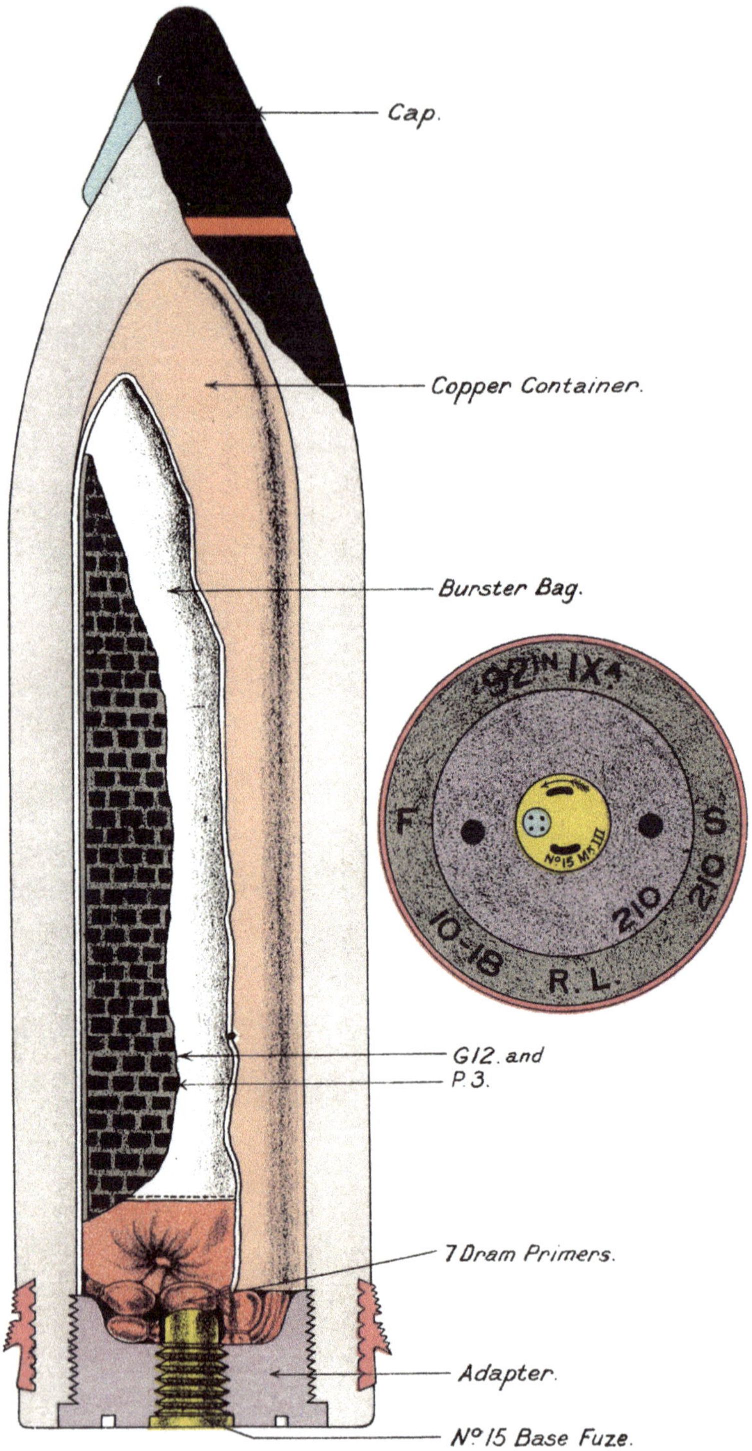
Cap.
Copper Container.
Burster Bag.
G12. and
P.3.
7 Dram Primers.
Adapter.
No. 15 Base Fuze.
9.2 IN IX 4
F
S
210
210
10-18
R.L.
No 15 Mk III

958

Fig 5·19(c).

METHOD OF FILLING:- SHELL B.L. C.P.

(TYPICAL)

FOR 6" GUN, 6" HOWITZERS & ABOVE.

FIELD SERVICE.

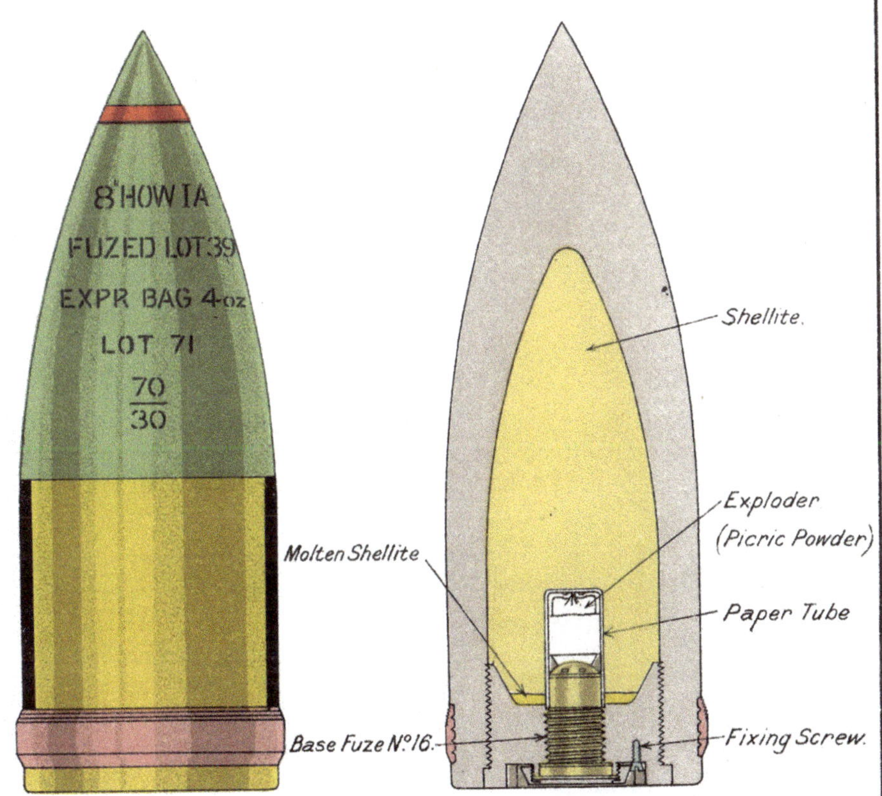

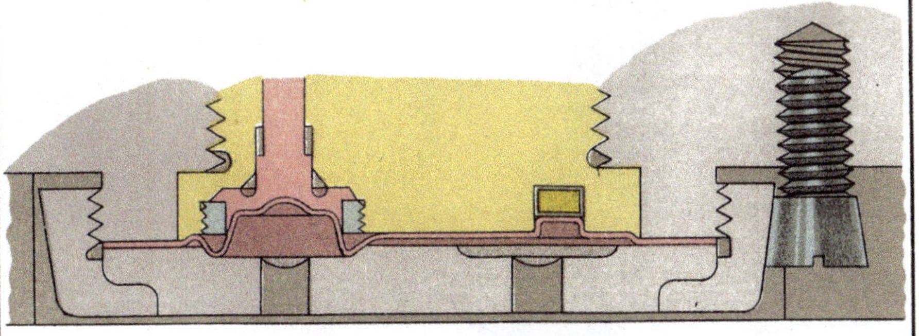

ENLARGED VIEW OF BASE,
SHOWING DETAILS FOR SEALING GAS PRESSURE.

Fig. 5.20(a).

ARMOUR PIERCING SHELL FILLED POWDER.

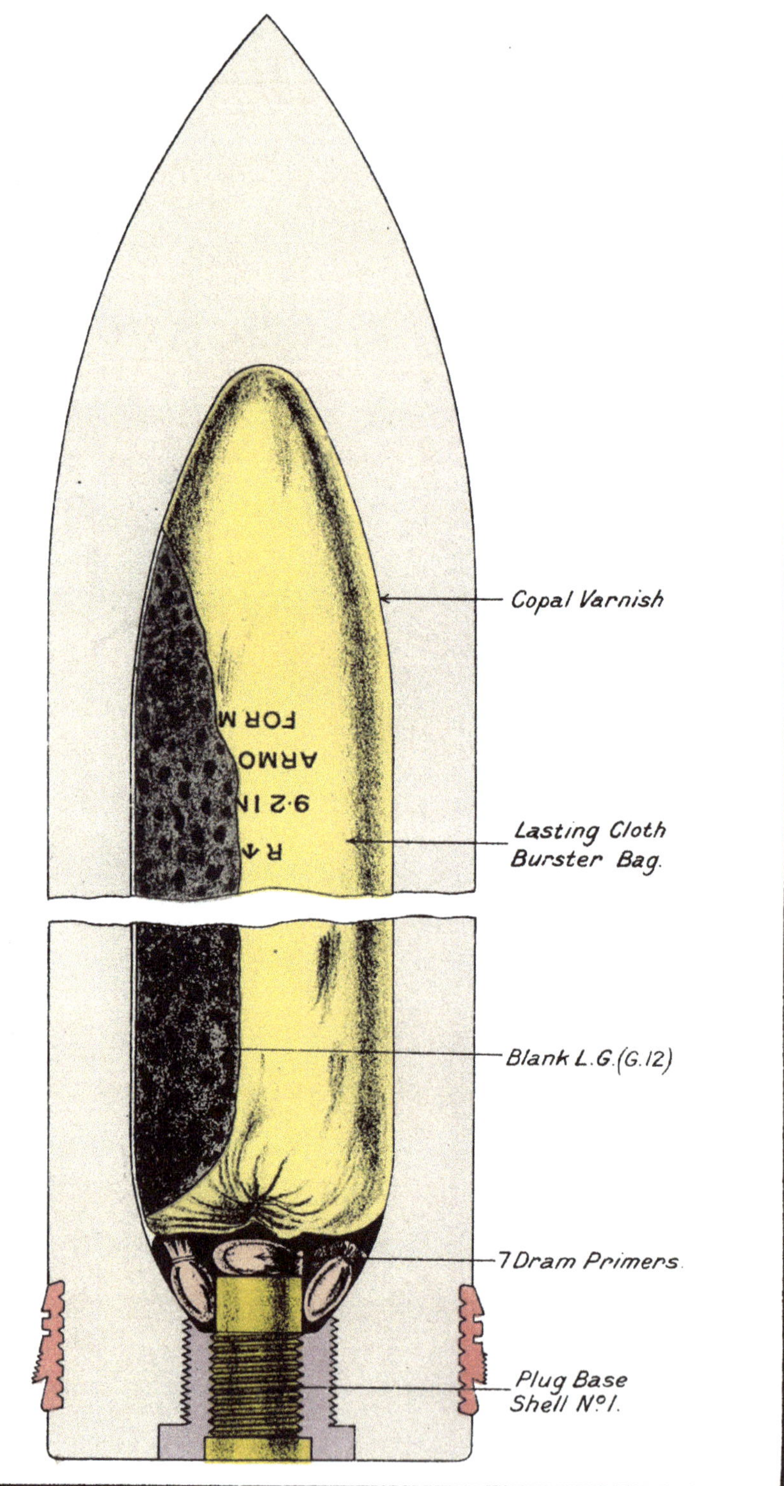

958.

Fig. 5.20(b).

ARMOUR PIERCING SHELL WITH CAP FILLED H.E.

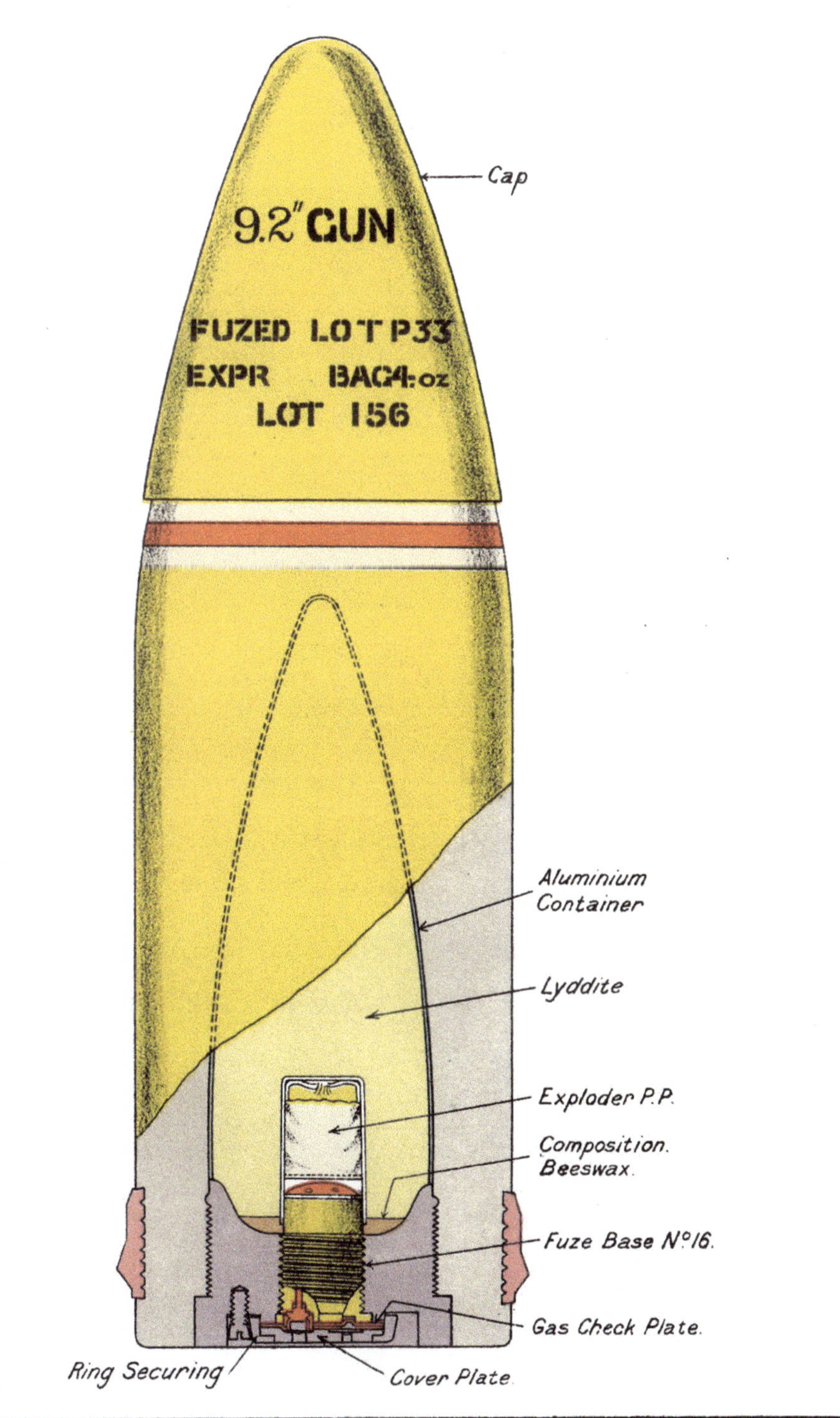

958.

Fig. 5·24 (a).

METHOD OF FILLING :- SHELL, B.L., OR Q.F., H.E., FOR COAST DEFENCE.

12 PR. & ABOVE.

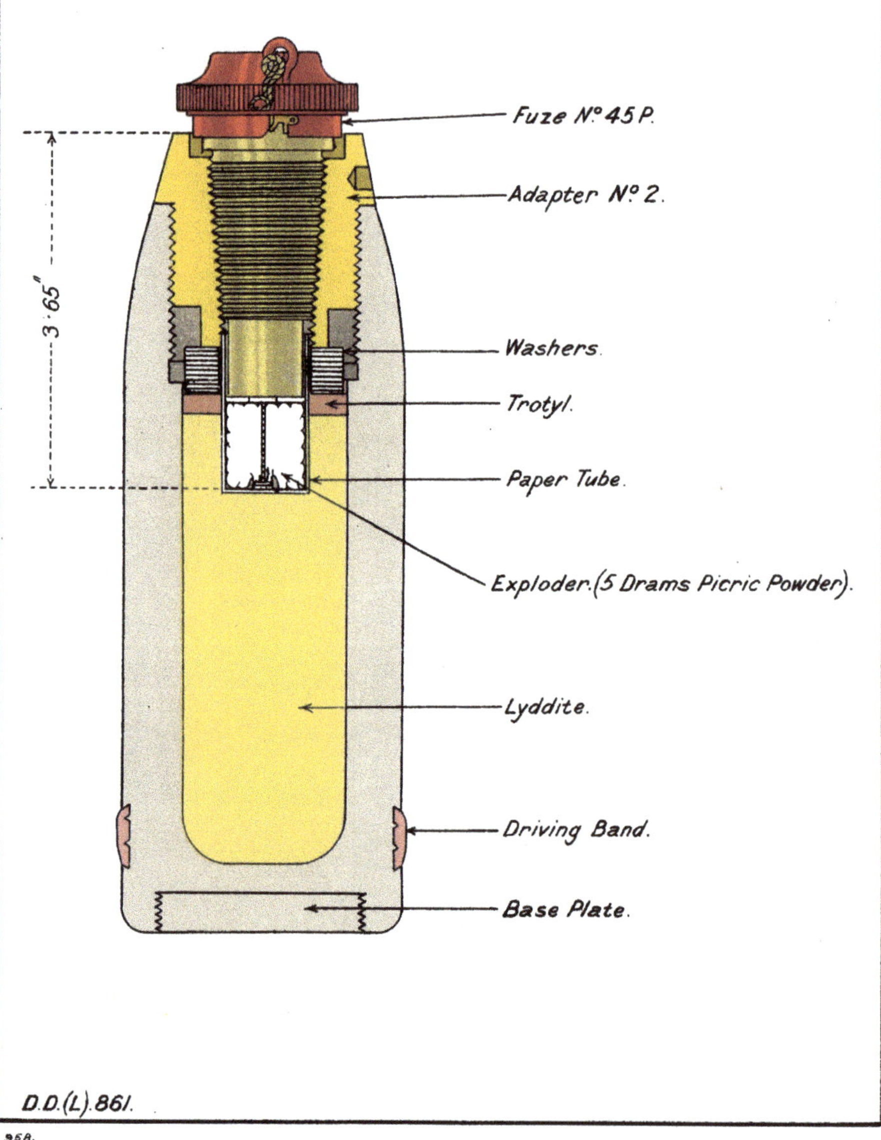

D.D.(L).861.

958.

Fig. 5·24(b).

METHOD OF FILLING:-SHELL, B.L., H.E., 6" GUN Mk XXVI/XXB. COAST DEFENCE.

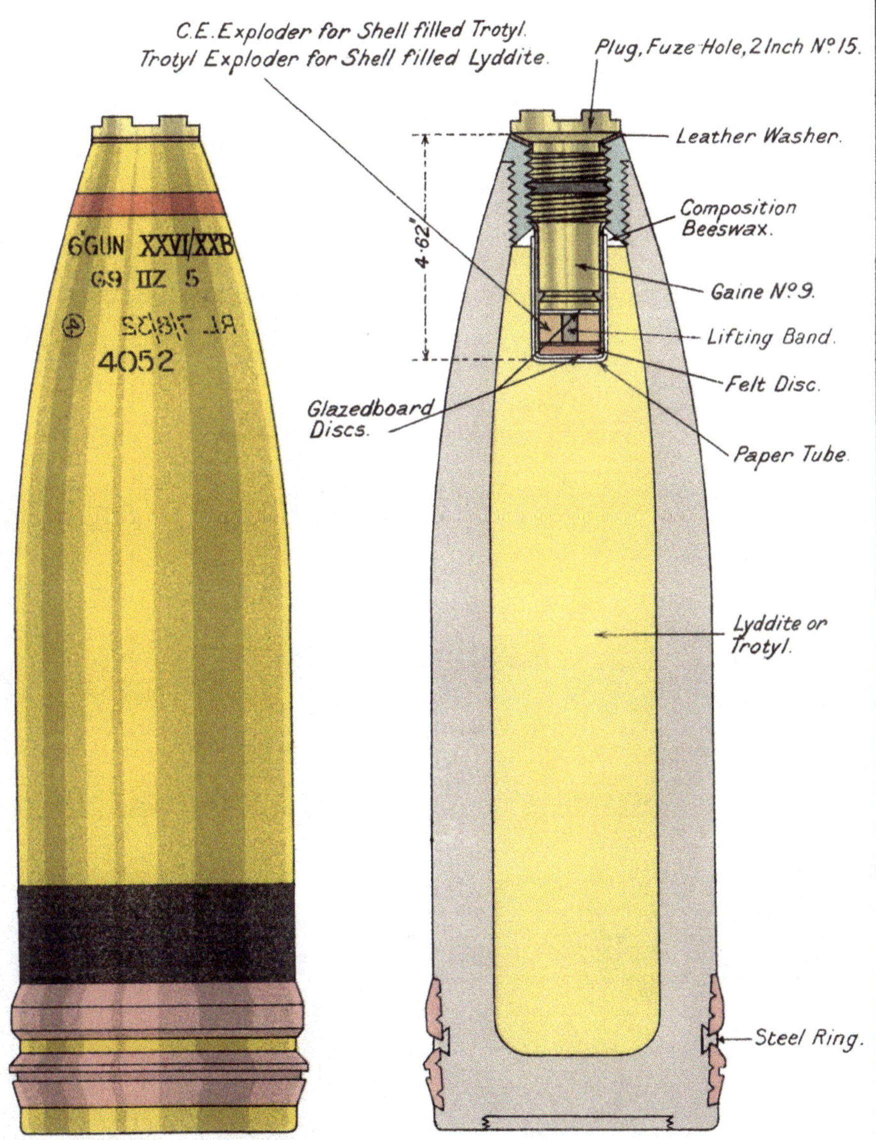

ALLOCATION OF FUZES.

(I). Seawards - Fuze D.A.I. No. 45P, in No. 23 Adapter.

(II). Landwards - Fuze D.A.I. No. 45, in No. 23 Adapter, or Fuze D.A. No. 230. (No Adapter).

(III). Anti-Beach Landings. - Fuze Time No. 188. (No Adapter).

D.D.(L).4052.

958.

Fig. 5·24 (c).

SHELL B.L. OR Q.F. H.E. (TYPICAL) FOR SEPARATE LOADING AMMUNITION. POURED FILLINGS.

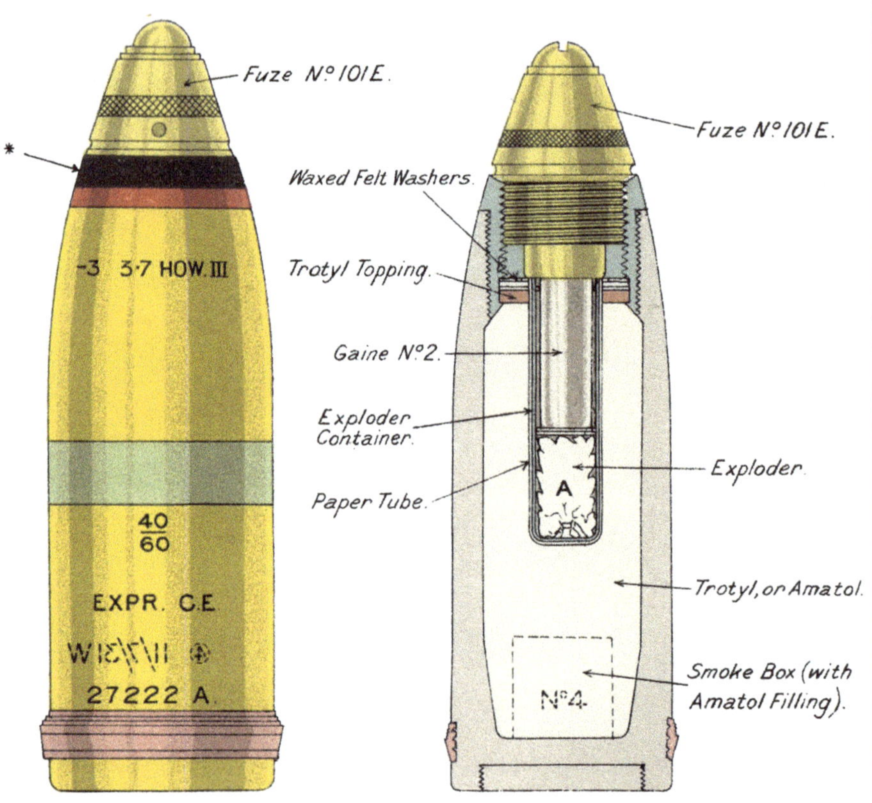

* With Amatol, black tip denotes absence of Smoke Composition.

R.L. 27222 A.

958.

Fig. 5·24(d).

METHOD OF FILLING:- H.E. SHELL 3·7" TO 9·2"

AMATOL $\frac{80}{20}$ HOT MIXED.

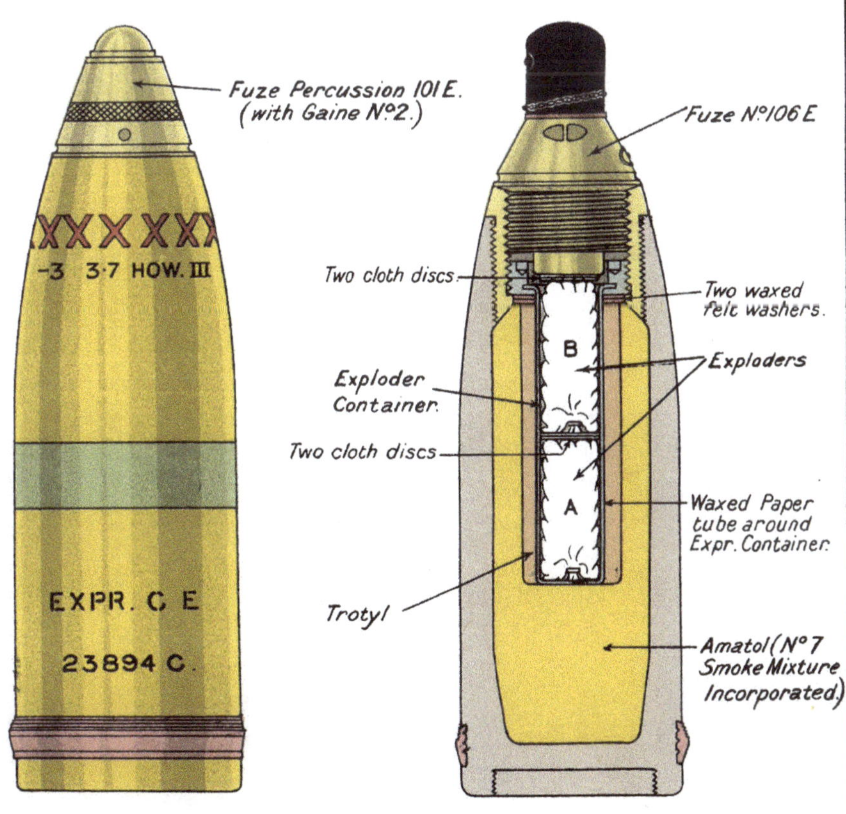

R.L. 23894 C.

Fig. 5·24(e).

METHOD OF FILLING:- 13 PR & 18 PR (TYPICAL).

TROTYL WITH SMOKE BOX (2¼ oz MK III.)

AMATOL $\frac{80}{20}$ CENTRE THIRD TROTYL. (Cold Pressed).

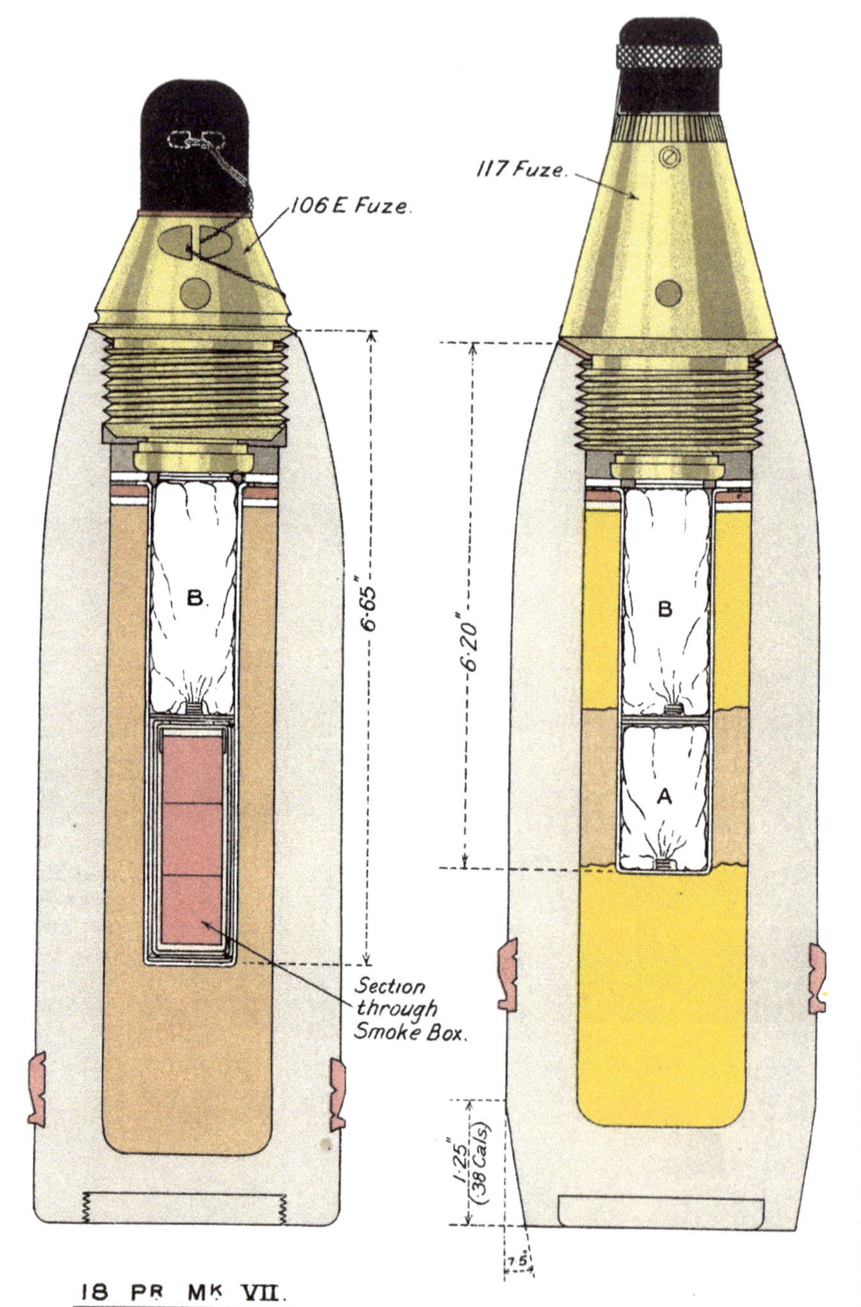

18 PR MK VII.

18 PR MK I.C. (Streamline).

D.D.(L).4837.

D.D.(L).4197.

Fig 5·24 (f).

METHOD OF FILLING:- SHELL, Q.F., H.E., 3 INCH 20 CWT 16 LB., Mk IIB.

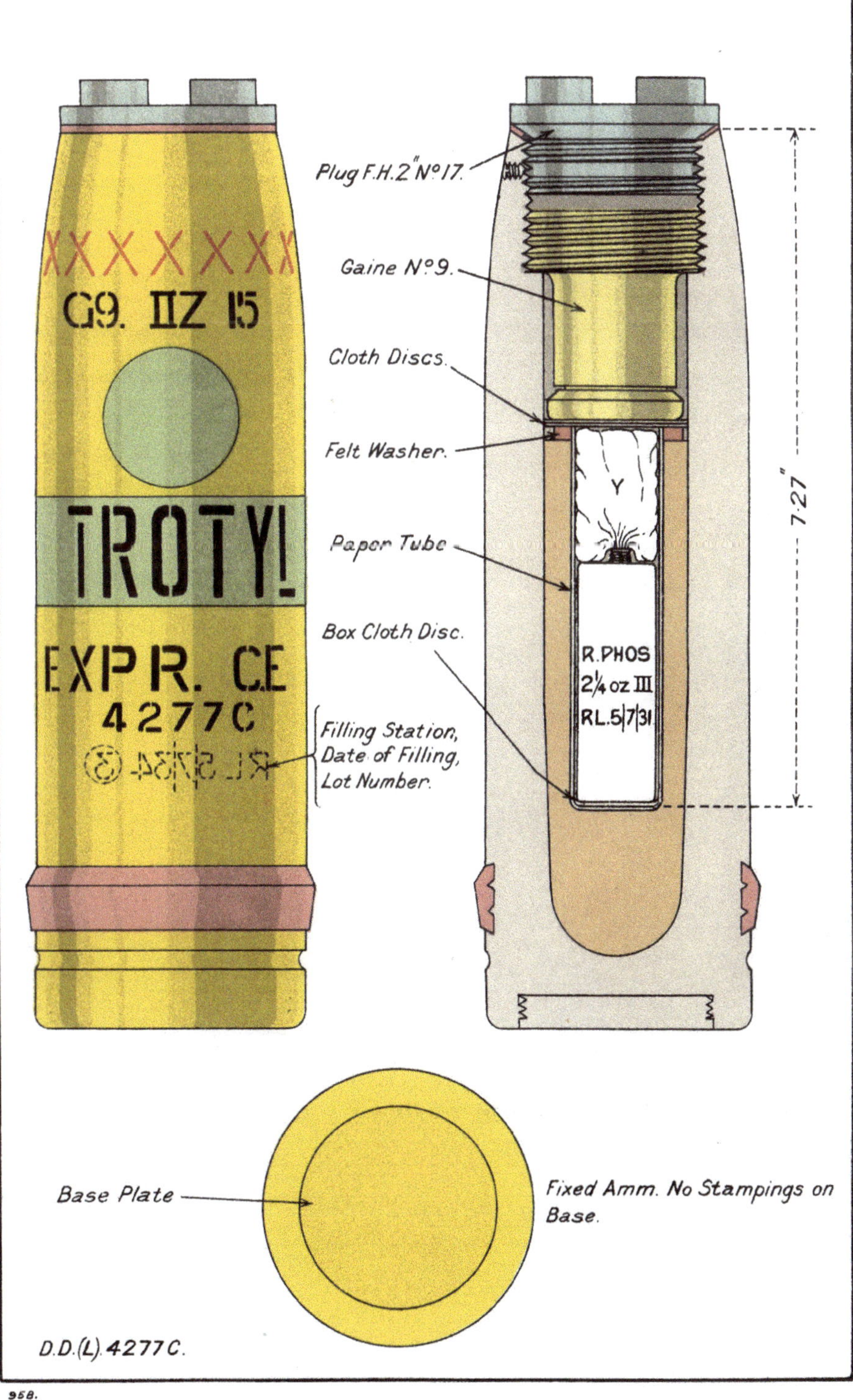

Fig. 5·24 (g).

METHOD OF FILLING:-SHELL, B.L., H.E., STREAM LINE, 6 INCH HOWITZER. AMATOL $\frac{80}{20}$ HOT MIXED.

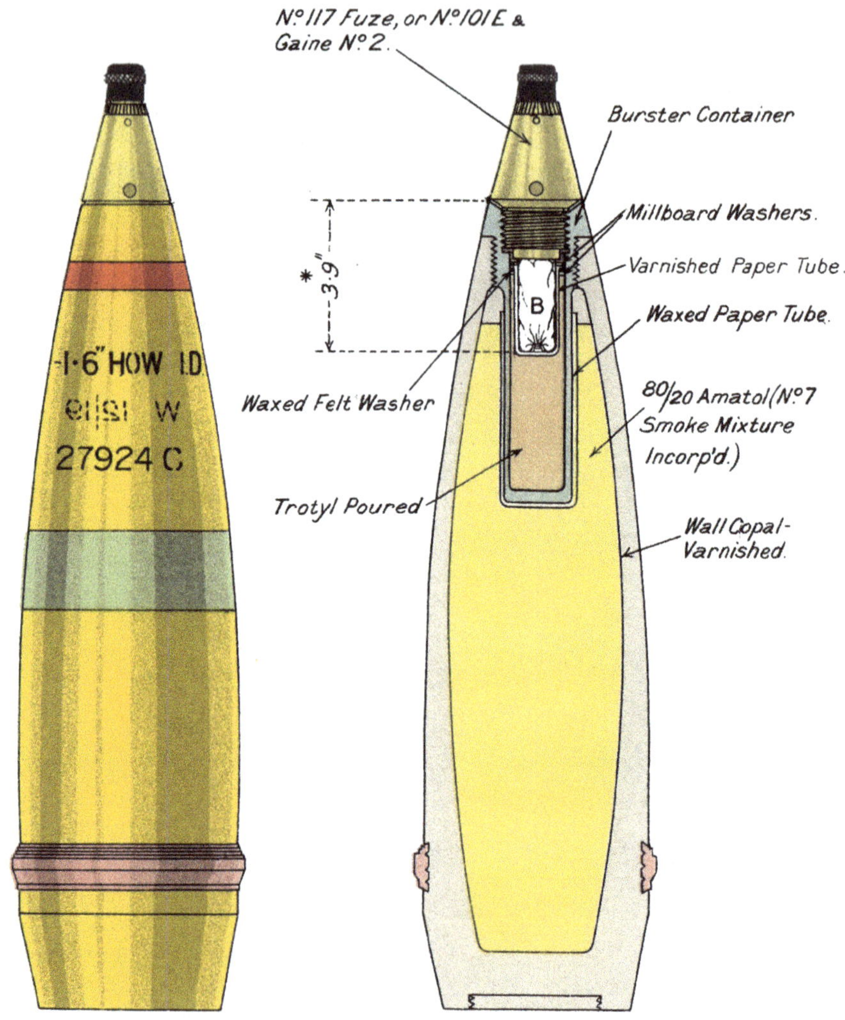

* NOTE:- In future Design, Cavity will be 6·2".

SHELL MAY ALSO BE FILLED:

(I) Amatol $\frac{80}{20}$ Cold Pressed with Nº 4 Smoke Mixture in Bag, (27924.A) or Nº 5 Incorporated (27924 B)

(II) Trotyl. (27924.D)

(III) Lyddite (29194)

27924.C.

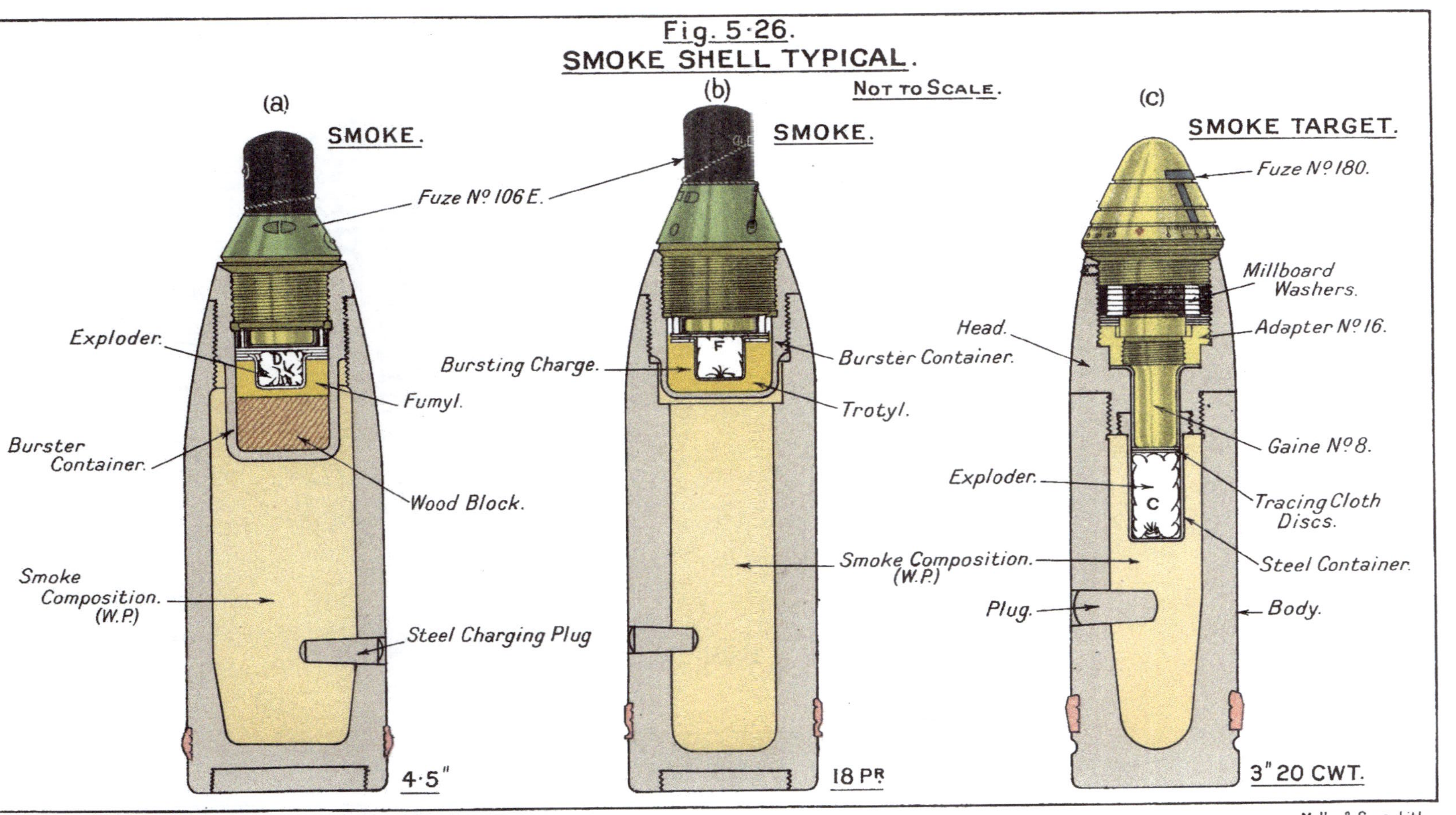
Fig. 5·26.
SMOKE SHELL TYPICAL.
NOT TO SCALE.
(a)
SMOKE.
Fuze Nº 106 E.
Exploder.
Fumyl.
Burster Container.
Wood Block.
Smoke Composition. (W.P.)
Steel Charging Plug
4·5"
(b)
SMOKE.
Bursting Charge.
Burster Container.
Trotyl.
Smoke Composition. (W.P.)
18 PR.
(c)
SMOKE TARGET.
Fuze Nº 180.
Millboard Washers.
Head.
Adapter Nº 16.
Gaine Nº 8.
Exploder.
Tracing Cloth Discs.
Steel Container.
Plug.
Body.
3" 20 CWT.
958
Malby & Sons, Lith.

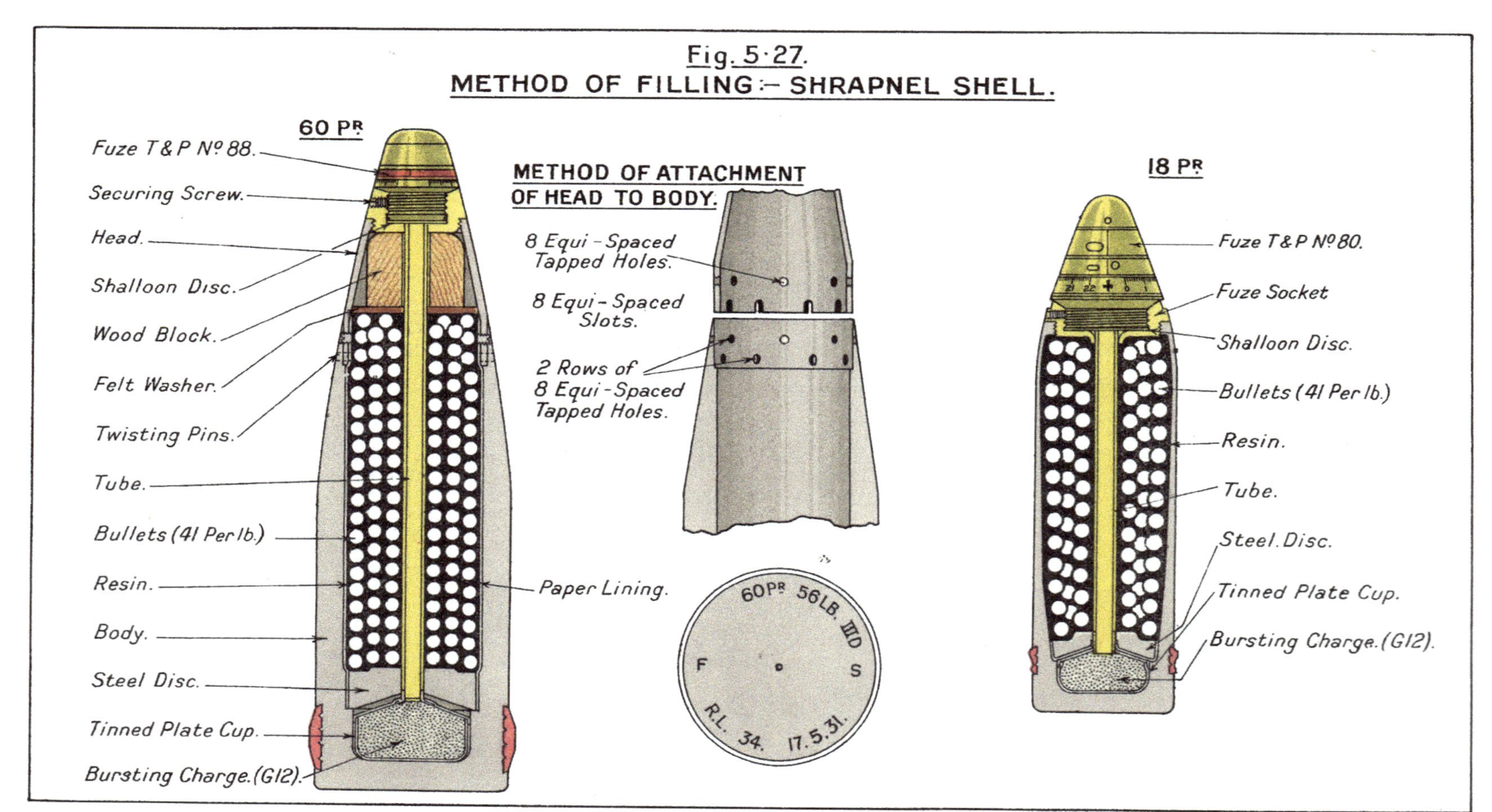

Fig. 5·27.
METHOD OF FILLING:– SHRAPNEL SHELL.

Fig. 5·28.

METHOD OF FILLING:– SHELL, B.L. OR Q.F., STAR, TYPICAL.

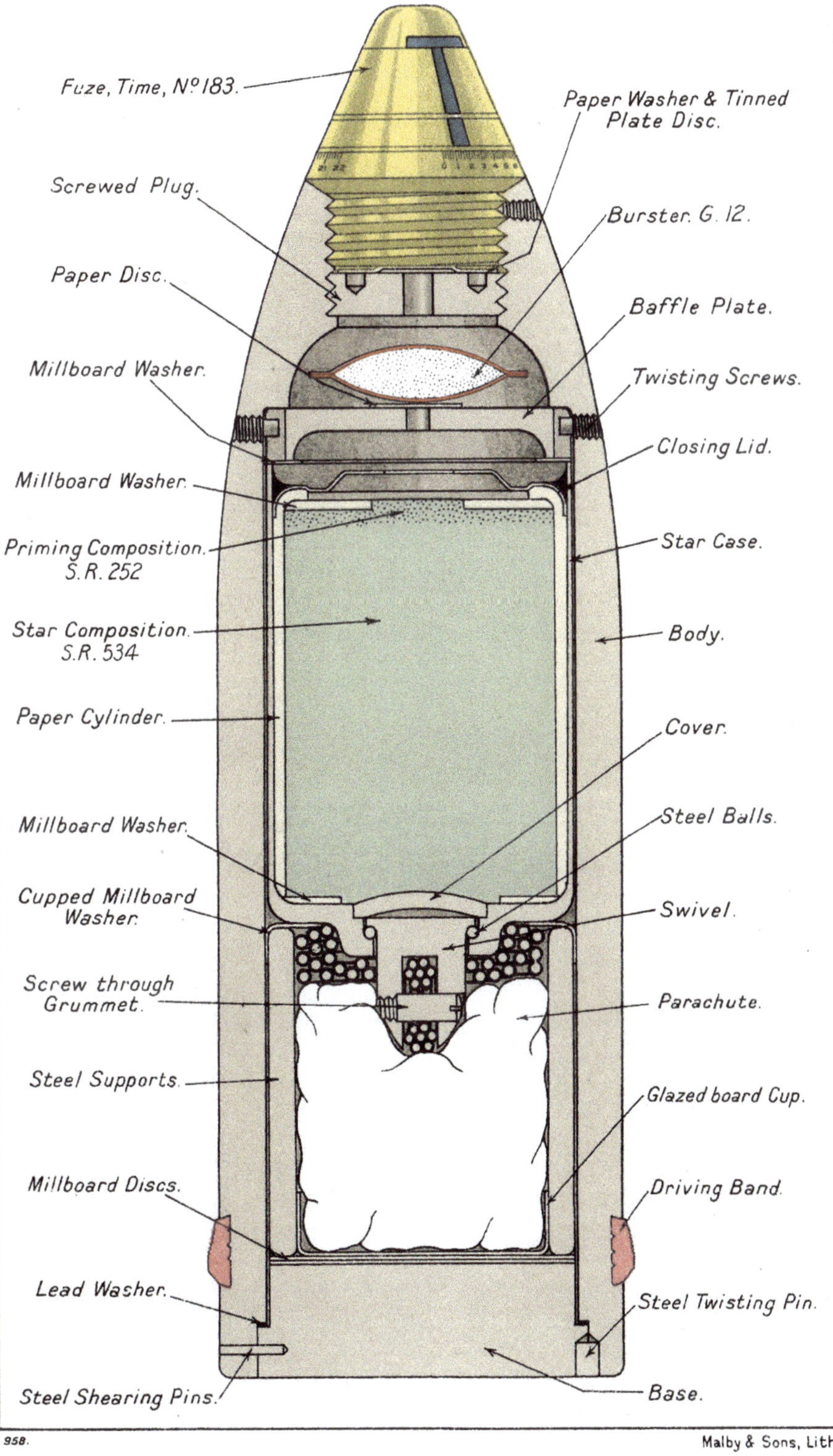

Fuze, Time, Nº 183.
Paper Washer & Tinned Plate Disc.
Screwed Plug.
Burster. G. 12.
Paper Disc.
Baffle Plate.
Millboard Washer.
Twisting Screws.
Closing Lid.
Millboard Washer.
Priming Composition. S.R. 252
Star Case.
Star Composition. S.R. 534
Body.
Paper Cylinder.
Cover.
Millboard Washer.
Steel Balls.
Cupped Millboard Washer.
Swivel.
Screw through Grummet.
Parachute.
Steel Supports.
Glazed board Cup.
Millboard Discs.
Driving Band.
Lead Washer.
Steel Twisting Pin.
Steel Shearing Pins.
Base.

958.

Malby & Sons, Lith.

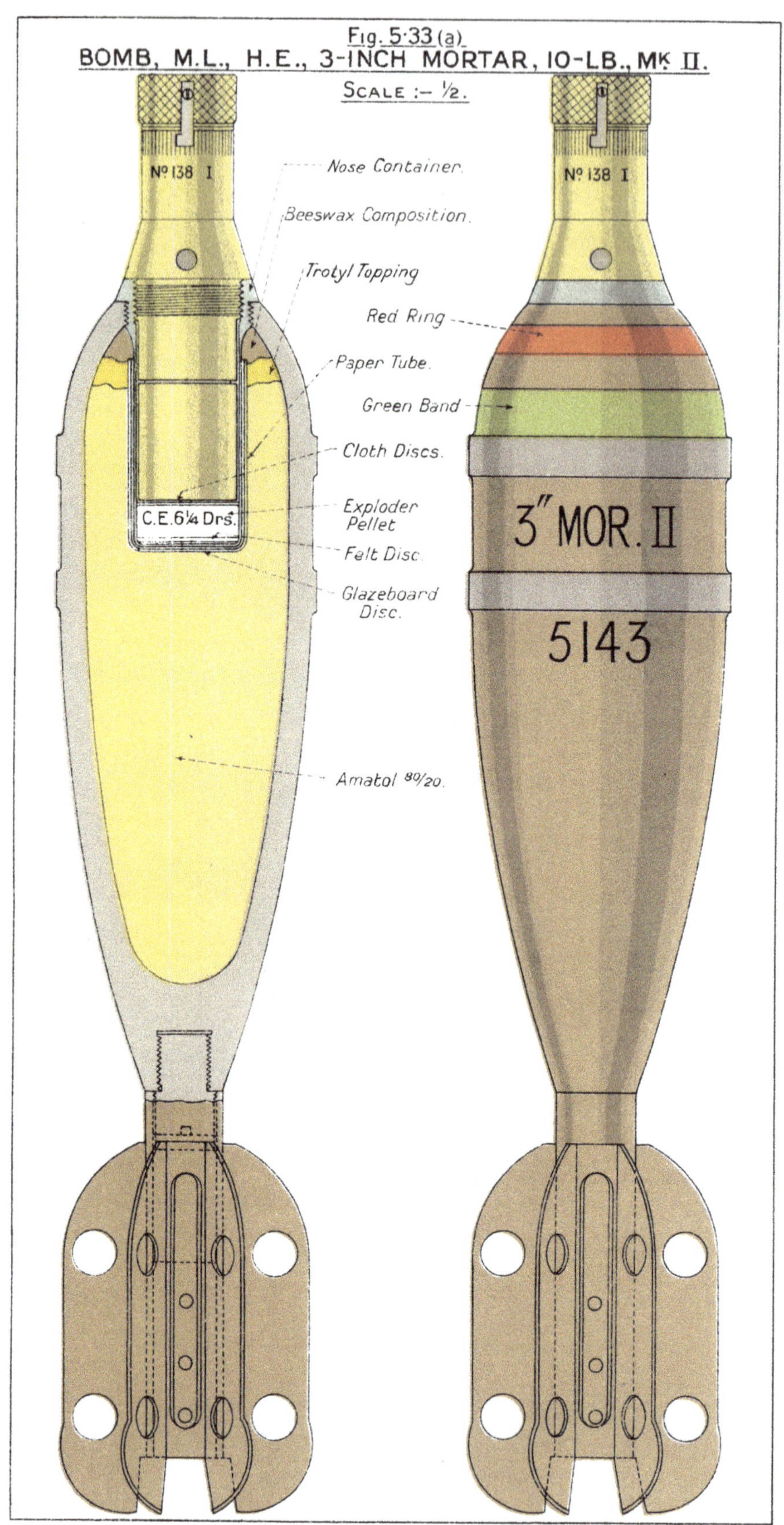

958.
Malby & Sons, Lith.

Fig 5·33(b).

CARTRIDGE, M.L., 3 INCH MORTAR, AUGMENTING, 100 GRAINS N.C.(Y) CYLINDRICAL Mk I.

Scale:- 1/1.

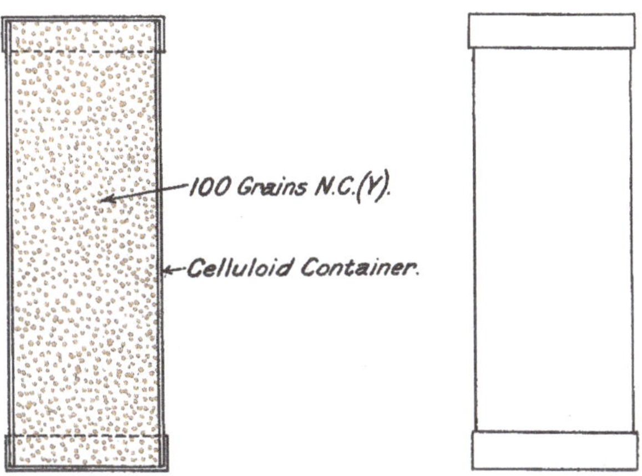

CARTRIDGE, M.L. MORTAR, 95 GRAINS BALLISTITE, WITH STRIKER CLIP Mk. VI.

Scale:- 1/1.

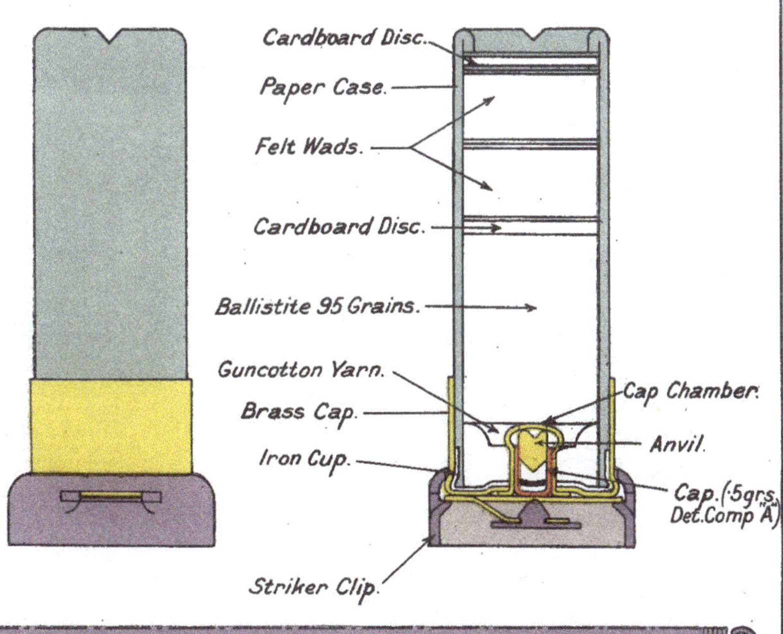

Spring, Retaining, Cartridges, Augmenting.

358

Fig. 5·36

DISTINGUISHING MARKINGS FOR GU

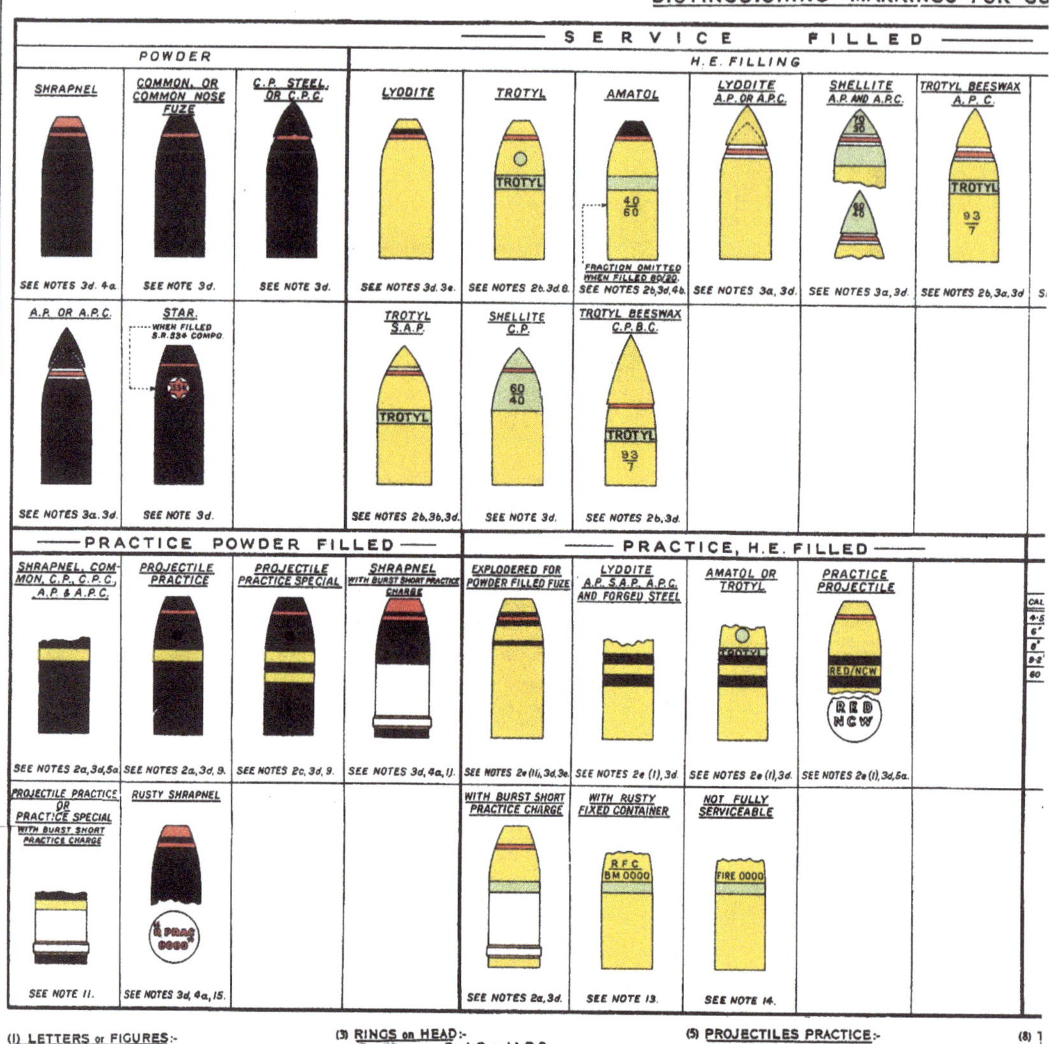

(1) LETTERS or FIGURES:-
Black on Coloured, White on Black.

(2) BANDS ROUND BODY:-

(a) Yellow:- For Projectile Practice.

(b) Green:- For High Explosive Filling (except Lyddite)

(c) Two Yellow:- For Projectile Practice Special.

(d) Black:- For High Explosive Drill.

(e) Two BLACK (i) For High Explosive Proj. Practice
(ii) For High Explosive Practice Shell.—

(3) RINGS on HEAD:-

(a) Two White:- For A.P. and A.P.C.
For A.P.C. the White Rings are below Cap.

(b) One White:- For S.A.P.

(c) Light Brown:- For Cast Iron or Semi-Steel Smoke.

(d) Red:- Denoting Filled Shell.

(e) Black:- Immediately above red ring denotes a Powder Filled Fuze should be used as the shell so distinguished are fitted with P.P. exploders.

(4) TIPS:-

(a) Red:- Denotes Shrapnel

(b) Black:- Denotes absence of Smoke producing mixture.

(c) White:- For Shot Practice (pointed)

(5) PROJECTILES PRACTICE:-

(a) Projectiles restricted for use with reduced charge are stencilled "RED" in black on yellow band, and in white on base. On calibres 6 inch and above, the body stencilling is in two places diametrically opposite. If not suitable for use over crowded waterways they are stencilled with the letters NCW in addition, thus:- RED NCW

(b) Projectiles weighted with SAND or SALT are stencilled "SAND" or "SALT" in white, as applicable.

(c) 6Pr. and 3Pr. Steel Shell which have been annealed are stencilled "Ā" in white above driving band on opposite side to the stamping.

(6) Empty or Weighted Projectiles for use in "apparatus practising loading" will be stencilled "LOADING TEACHER" on body in addition to black band.

(7) If prepared for Tracer but not fitted, the "T" is to be omitted.

(8) T

(9) T

(10)

(11) I

(12) S

(13) A

1066 FORMS, G. 910/14.

R GUN AND HOWITZER PROJECTILES

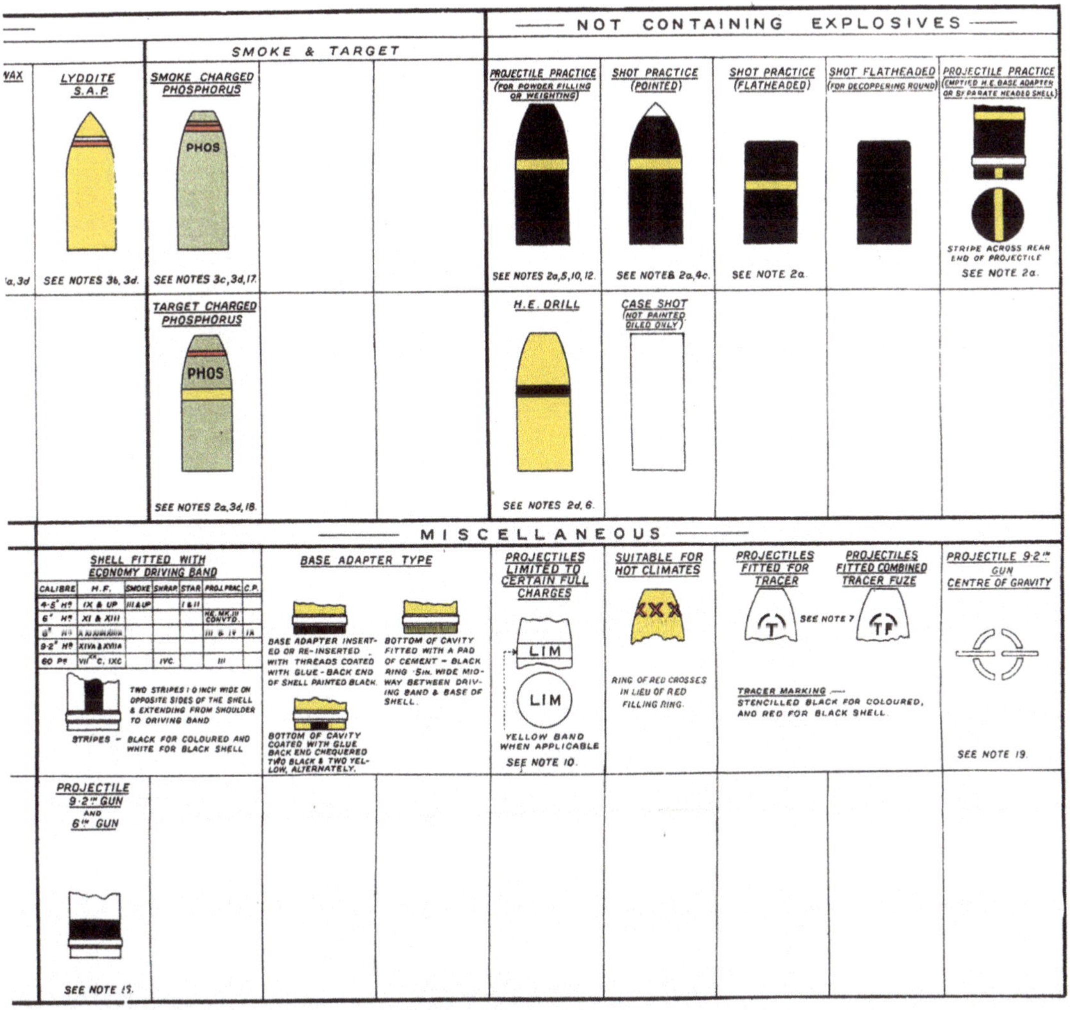

SMOKE & TARGET

WAX — a, 3d

LYDDITE S.A.P. — SEE NOTES 3b, 3d.

SMOKE CHARGED PHOSPHORUS — PHOS — SEE NOTES 3c, 3d, 17.

TARGET CHARGED PHOSPHORUS — PHOS — SEE NOTES 2a, 3d, 18.

NOT CONTAINING EXPLOSIVES

PROJECTILE PRACTICE (FOR POWDER FILLING OR WEIGHTING) — SEE NOTES 2a, 5, 10, 12.

SHOT PRACTICE (POINTED) — SEE NOTES 2a, 4c.

SHOT PRACTICE (FLATHEADED) — SEE NOTE 2a.

SHOT FLATHEADED (FOR DECOPPERING ROUND)

PROJECTILE PRACTICE (EMPTIED H.E. BASE ADAPTER OR SEPARATE HEADED SHELL) — STRIPE ACROSS REAR END OF PROJECTILE — SEE NOTE 2a.

H.E. DRILL — SEE NOTES 2d, 6.

CASE SHOT (NOT PAINTED OILED ONLY)

MISCELLANEOUS

SHELL FITTED WITH ECONOMY DRIVING BAND

CALIBRE	H.F.	SMOKE	SHRAP.	STAR	PROJ. PRAC.	C.P.
4·5" H^{R}	IX & UP	III & UP		I & II		
6" H^{R}	XI & XIII				HE. MK. III CONVTD.	
8" H^{R}	[illegible]				III & IV	IA
9·2" H^{R}	XIVA & XVIIA					
60 P^{R}	VII^{XX}C, IXC		IVC.		III	

TWO STRIPES 1·0 INCH WIDE ON OPPOSITE SIDES OF THE SHELL & EXTENDING FROM SHOULDER TO DRIVING BAND

STRIPES – BLACK FOR COLOURED AND WHITE FOR BLACK SHELL

BASE ADAPTER TYPE

BASE ADAPTER INSERTED OR RE-INSERTED WITH THREADS COATED WITH GLUE - BACK END OF SHELL PAINTED BLACK.

BOTTOM OF CAVITY FITTED WITH A PAD OF CEMENT – BLACK RING ·5IN. WIDE MIDWAY BETWEEN DRIVING BAND & BASE OF SHELL.

BOTTOM OF CAVITY COATED WITH GLUE BACK END CHEQUERED TWO BLACK & TWO YELLOW, ALTERNATELY.

PROJECTILES LIMITED TO CERTAIN FULL CHARGES

LIM

LIM

YELLOW BAND WHEN APPLICABLE

SEE NOTE 10.

SUITABLE FOR HOT CLIMATES

XXX

RING OF RED CROSSES IN LIEU OF RED FILLING RING.

PROJECTILES FITTED FOR TRACER — T — SEE NOTE 7

PROJECTILES FITTED COMBINED TRACER FUZE — TF

TRACER MARKING — STENCILLED BLACK FOR COLOURED, AND RED FOR BLACK SHELL.

PROJECTILE 9·2" GUN CENTRE OF GRAVITY

SEE NOTE 19.

PROJECTILE 9·2" GUN AND 6" GUN

SEE NOTE 19.

(8) Two Green Discs diametrically opposite denote 3in. 2[illegible] cwt. or 18 Pr. shell fitted with smoke box.

(9) Two Discs of Aluminium Paint diametrically opposite denote 3in. 20 cwt. projectile with flash producing composition.

(10) Projectiles which are limited as to the full charge with which they may be fired to be stencilled "LIM" in two places diametrically opposite as under

PRACTICE PROJECTILES——on yellow band.
OTHER PROJECTILES——on centre of body.

(11) Projectiles for Burst Short Practice Charge have lower part of body above driving band painted white in addition to other markings.

(12) Shell for subcalibre are marked as for practice.

(13) Amatol filled shell with rusty fixed containers authorised for firing at practice are marked :-

R.F.C.
B.M.00 year of examination)

(14) Amatol filled shell other than the above which are not fully serviceable but are sentenced for practice are marked :-

FIRE 0000 (year as applicable)

(15) Unboxed shrapnel shell with central tube affected by rust are stencilled in red on base as follows :-

(i) "R.PRAC. 0000" (year as applicable) when powder is caked (central tube choked or not choked)
(ii) "R" when powder is free.

(16) Where a Gun and Howitzer of the same calibre exist, H.E. and Smoke Shell suitable for the gun have a black band painted. immediately in front of the driving band.

(17) The red ring is painted on when head filling is completed.

(18) The red ring is painted on only when shell is explodered.

(19) This symbol will be stencilled in three places. equally spaced round body, at centre of gravity of the filled shell—(black on coloured shell and white on black shell).

EMPTY PROJECTILES ARE PAINTED AND MARKED TO INDICATE THEIR TYPE AND USE ONLY.

Malby & Sons, Lith.

Fig. 6·11.

FUZE, PERCUSSION, D.A. WITH CAP.

No. 44, Mk. X Z/L/.

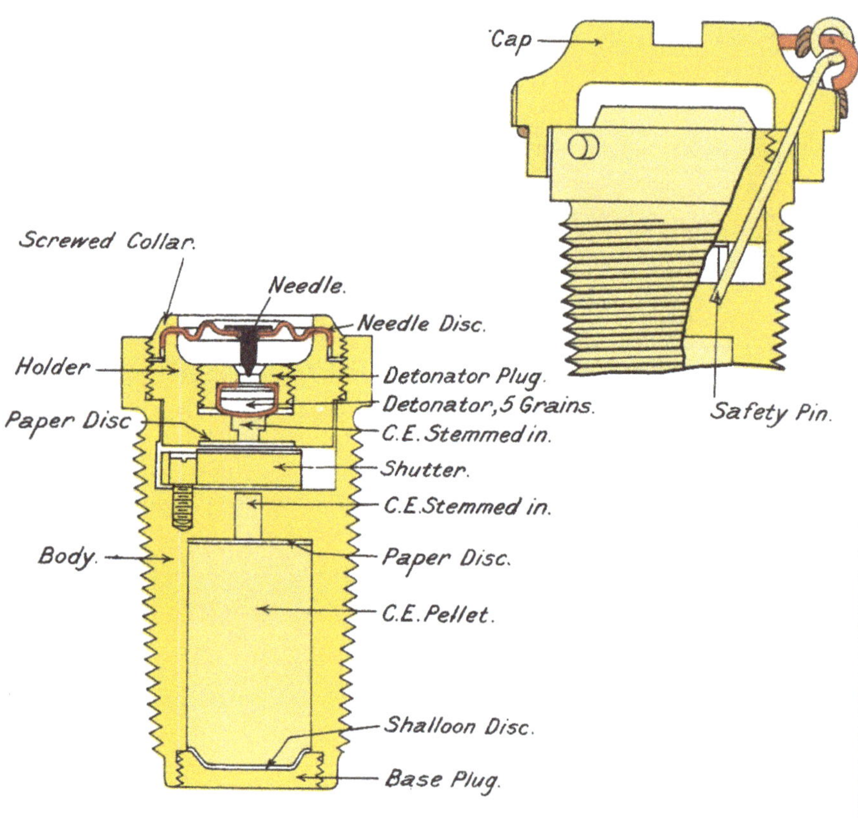

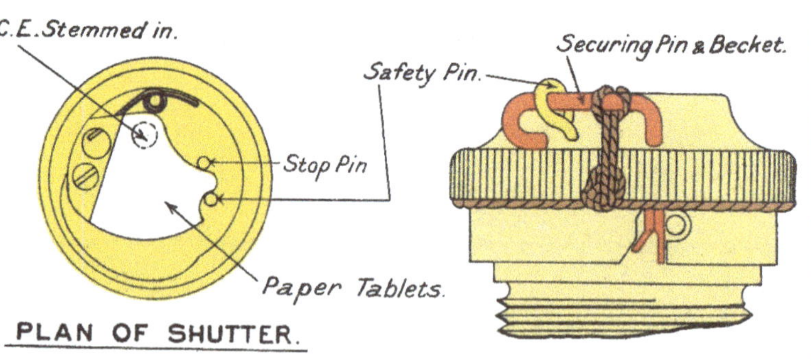

PLAN OF SHUTTER.

958

Fig. 6·12.

FUZE, PERCUSSION, D. A. No. 106 E, Mk. VIII Z /L/.

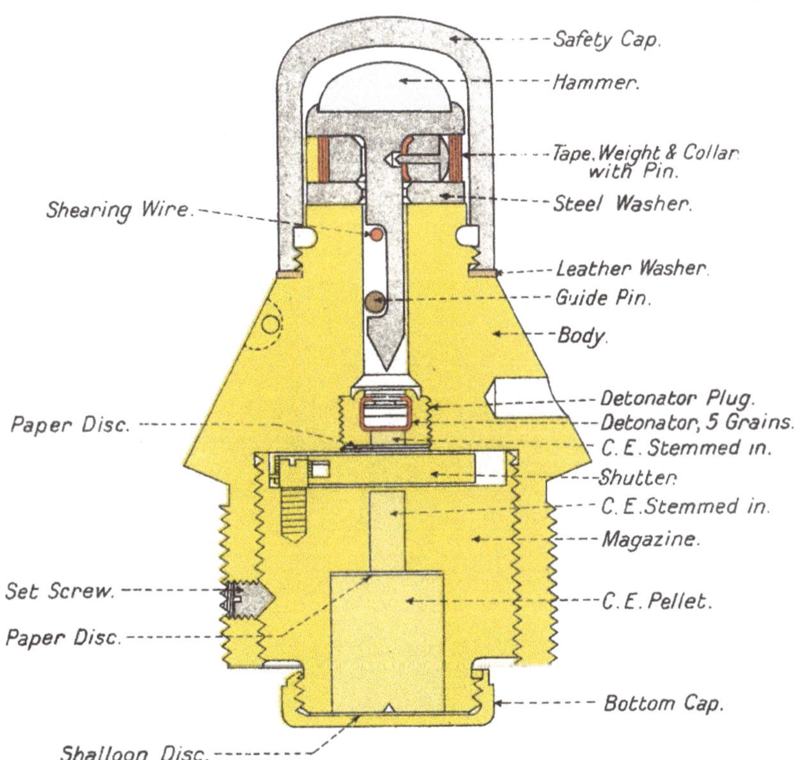

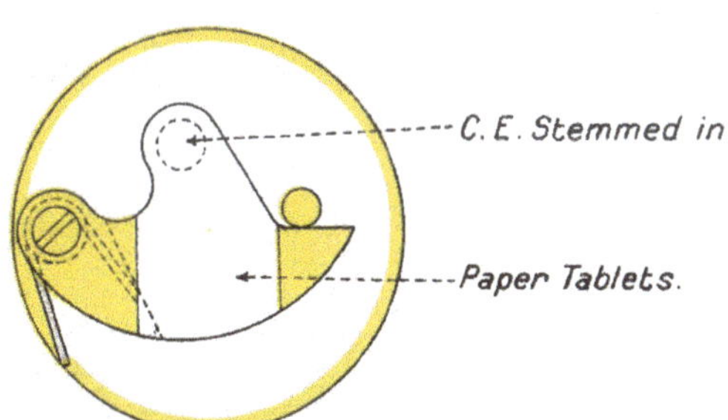

PLAN OF SHUTTER.

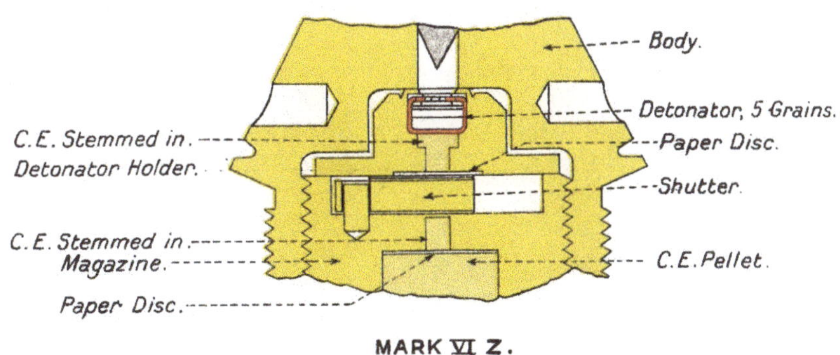

MARK VI Z.

958.

Malby & Sons Lith.

Fig 6:13.

FUZE, PERCUSSION, D.A. No 117, MARK III Z/L/.

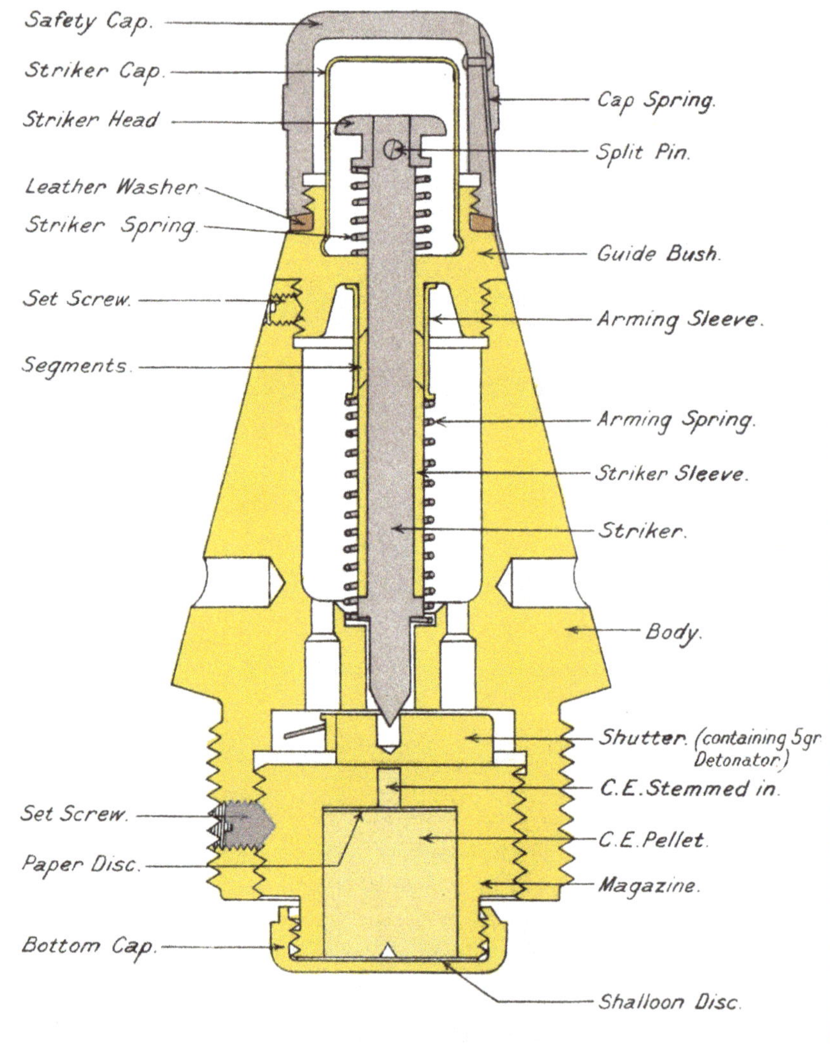

958

Fig. 6·14.

FUZE, PERCUSSION, D. A. IMPACT.

No. 45, Mk. IX Z /L/.

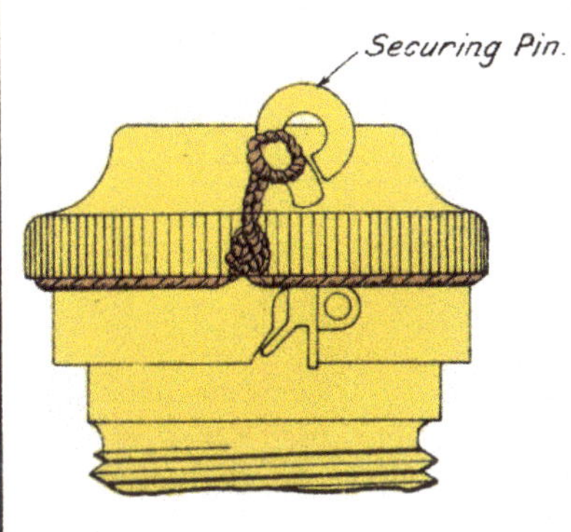

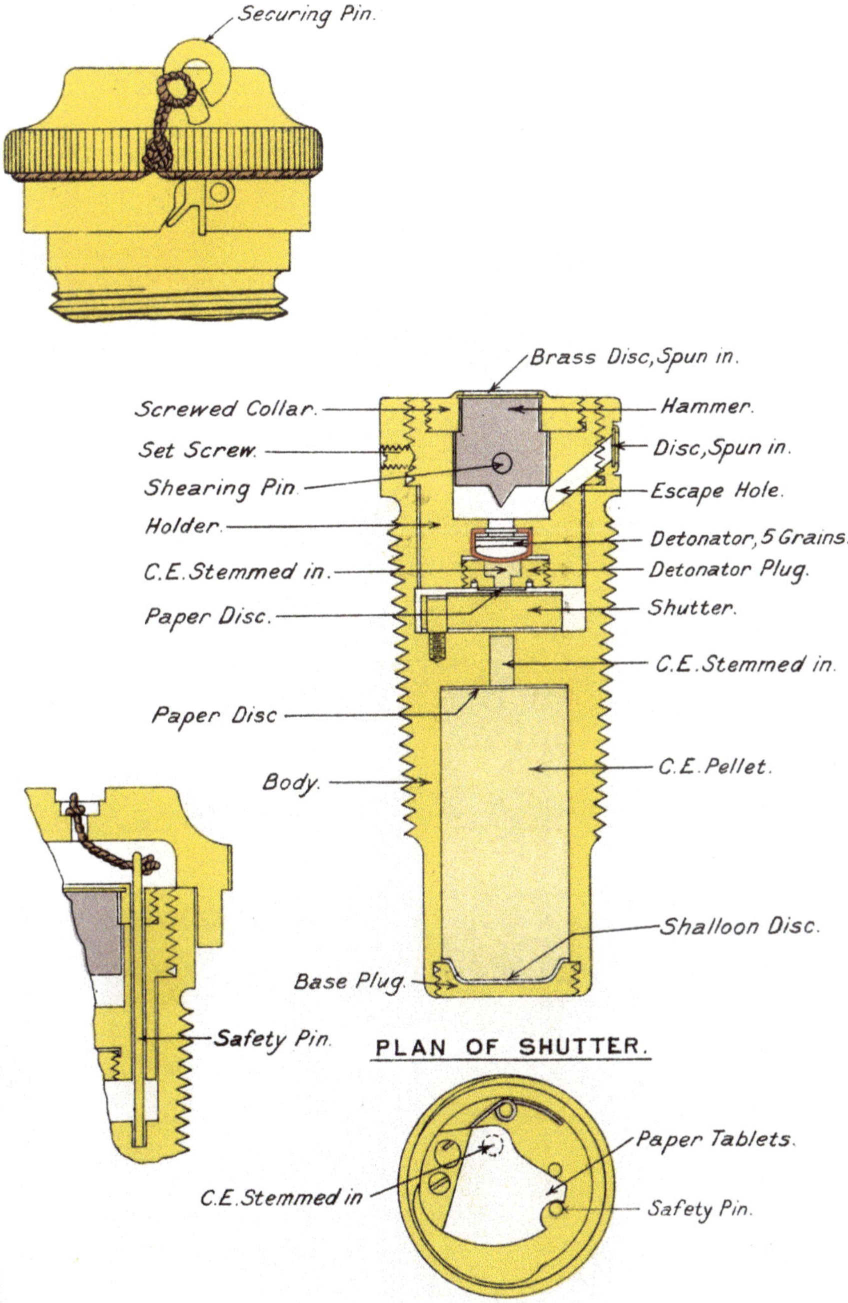

Securing Pin.
Brass Disc, Spun in.
Screwed Collar.
Hammer.
Set Screw.
Disc, Spun in.
Shearing Pin.
Escape Hole.
Holder.
Detonator, 5 Grains.
C.E. Stemmed in.
Detonator Plug.
Paper Disc.
Shutter.
C.E. Stemmed in.
Paper Disc
Body.
C.E. Pellet.
Shalloon Disc.
Base Plug.
Safety Pin.
PLAN OF SHUTTER.
Paper Tablets.
C.E. Stemmed in
Safety Pin.

Fig 6·16.

GAINE Nº 9 MARKS II & II Z.

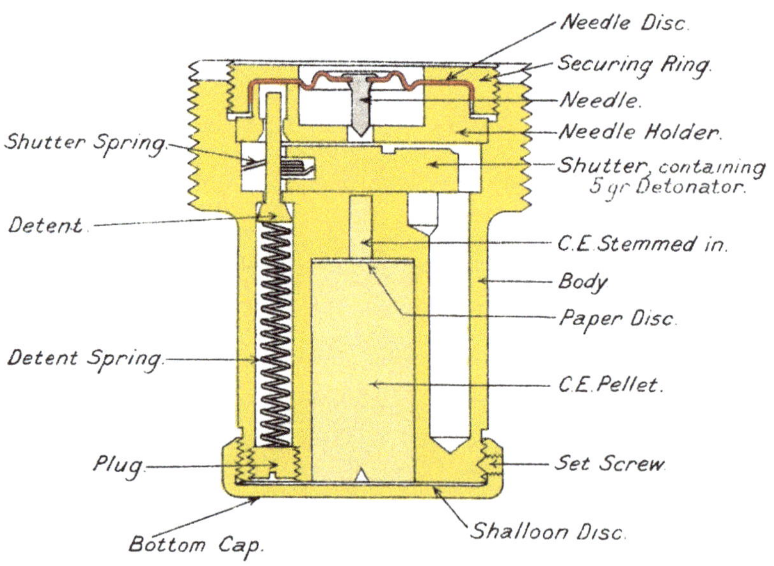

Fig. 6·17.

FUZE, PERCUSSION, Nº 101 E Mk. II M/L/.

WITH GAINE Nº 2 Mk. IV WITH DELAY.

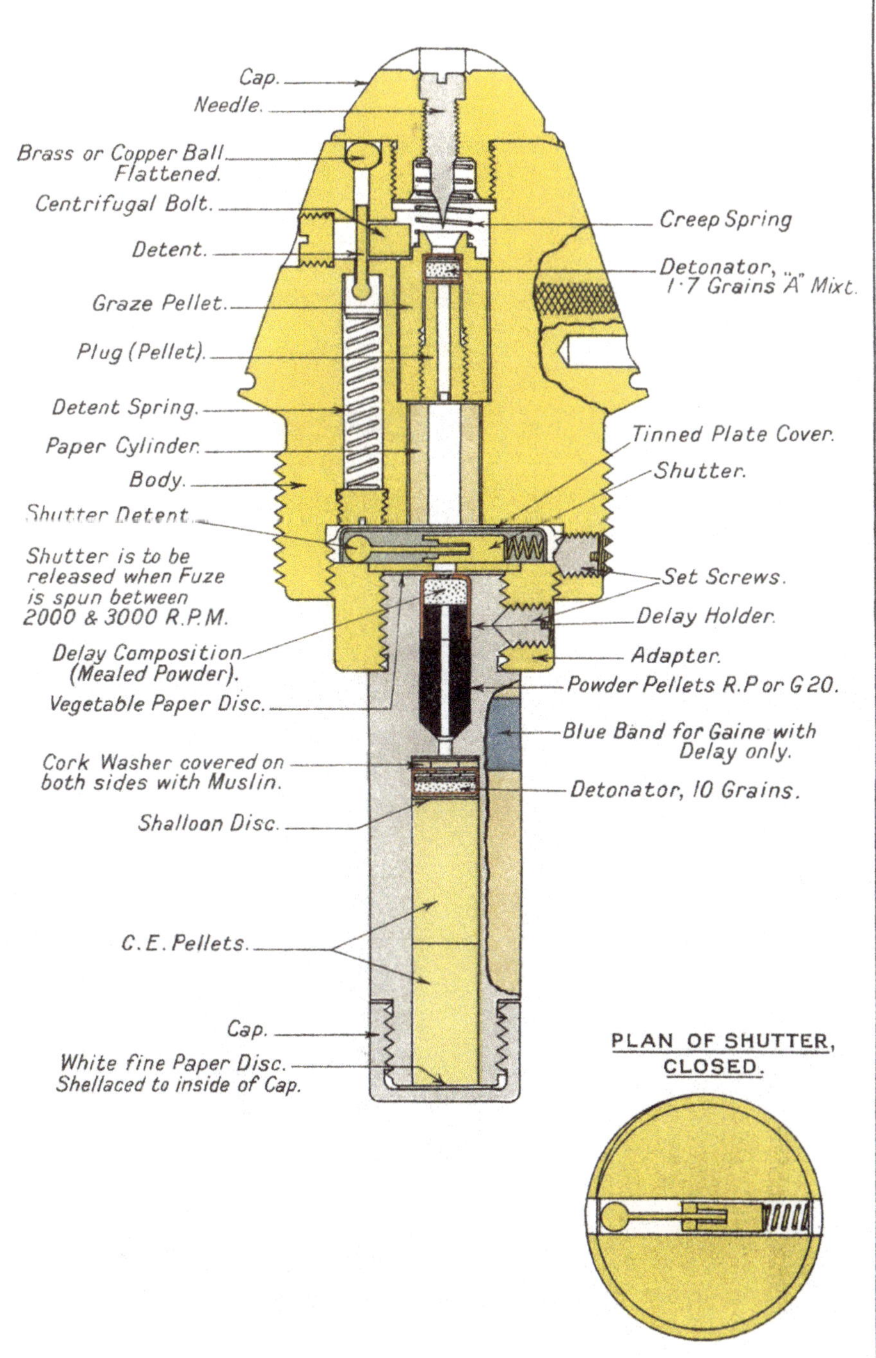

958.

Malby & Sons, Lith.

Fig. 6·19.

FUZE, PERCUSSION, BASE, HOTCHKISS, Mk. X/L.A/.

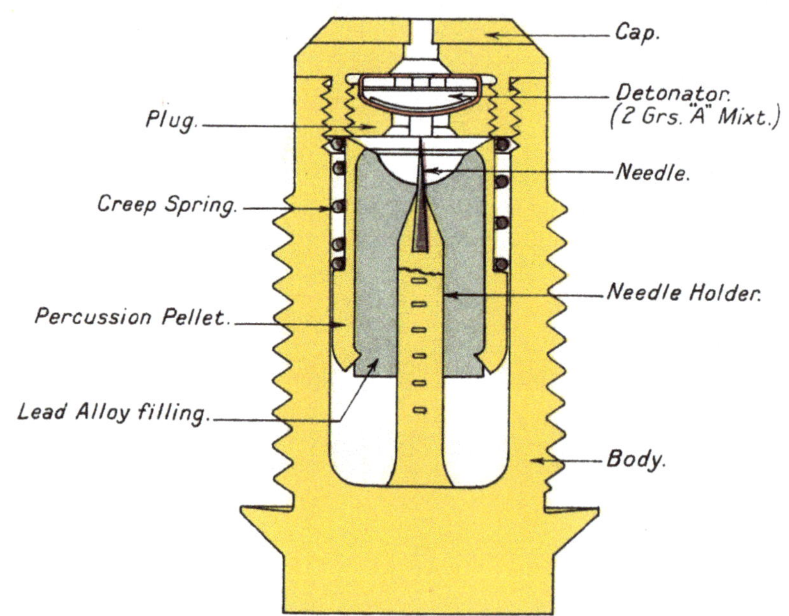

958.

Malby & Sons, Lith.

Fig. 6·20.

FUZE, PERCUSSION, BASE, LARGE. BRONZE, No 16, Mk IV /C/.

FUZES WITH DELAY HAVE LETTER D ADDED TO THE NUMBER.

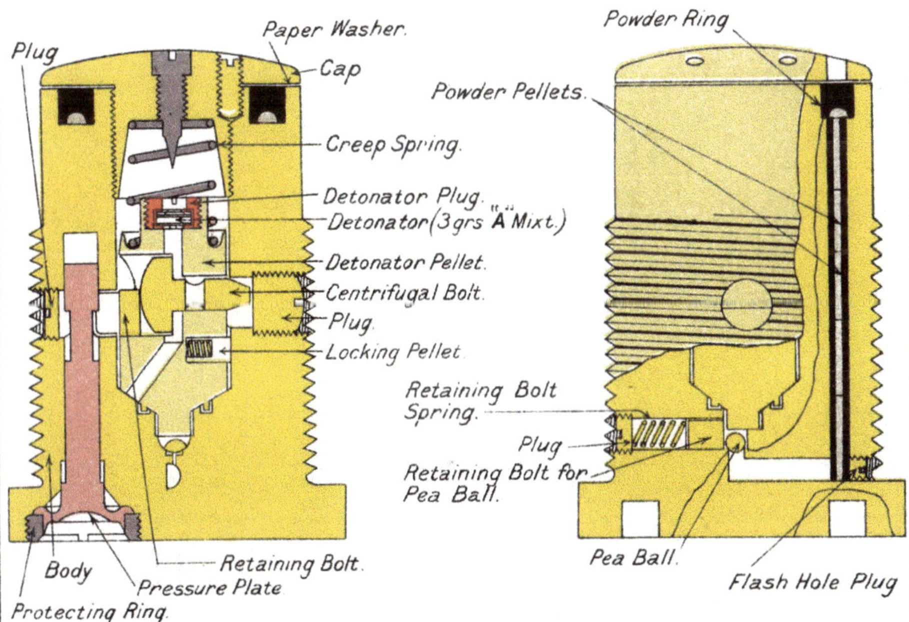

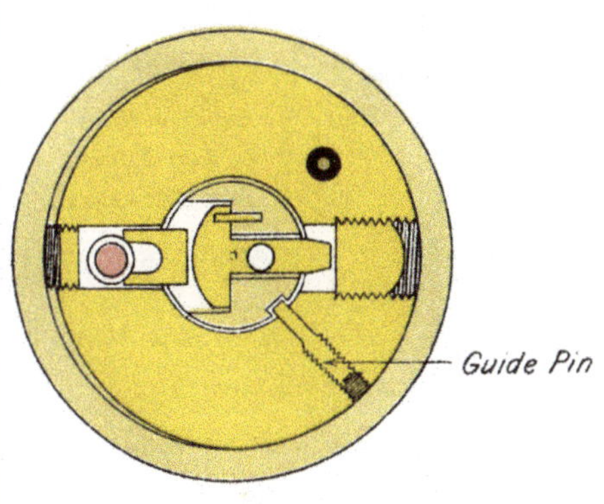

958

Fig. 6·25(a).

FUZE, TIME & PERCUSSION.

No. 80, Mk. XI/L/.

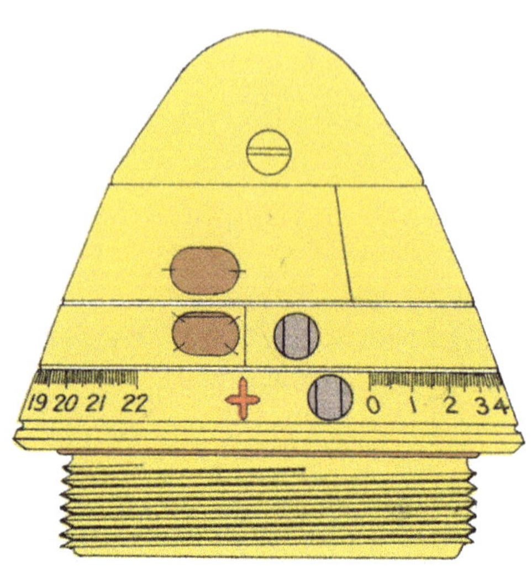

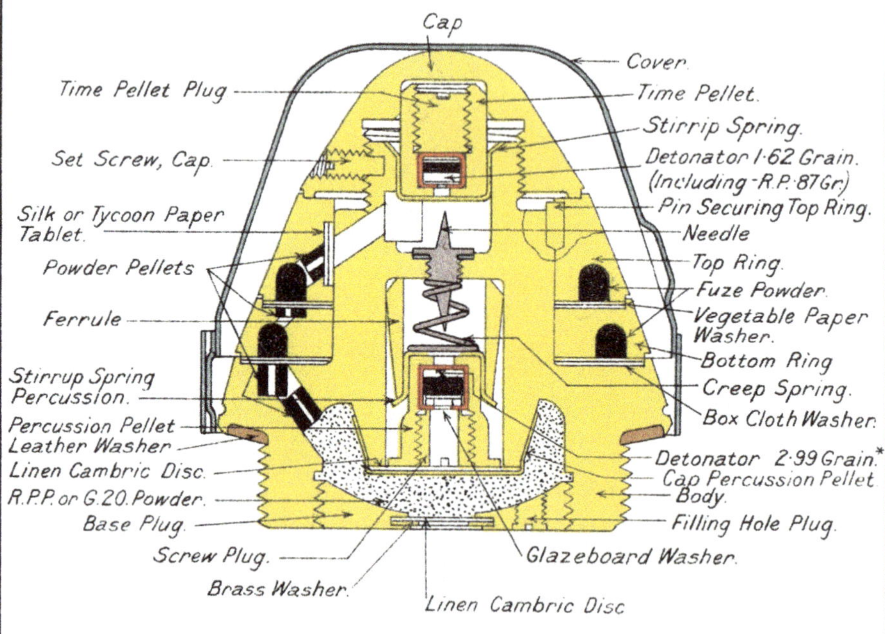

* Including 1·6 Grs. R.P.

958

Fig. 6·25(b).

FUZE, TIME No 80/44 Mk V/L/.

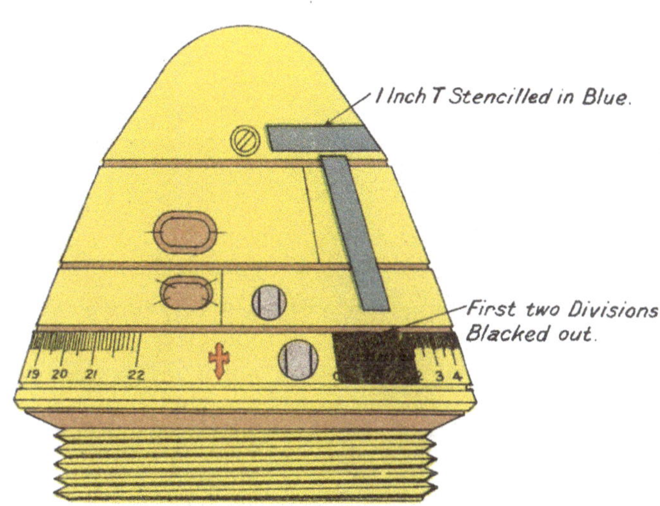

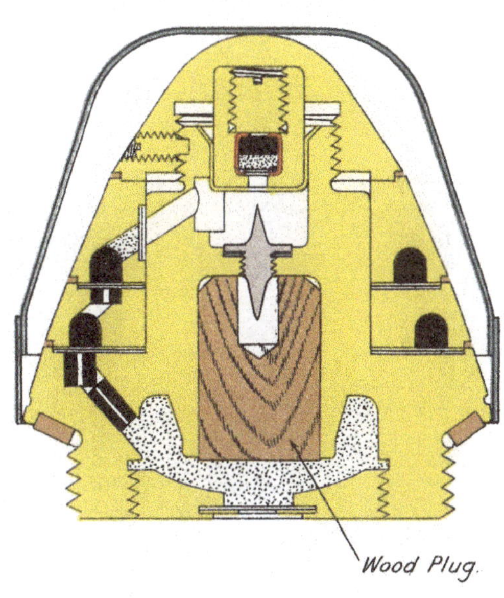

Fig. 6·26.

FUZE, TIME & PERCUSSION,

No. 88 Mk. VI/L/.

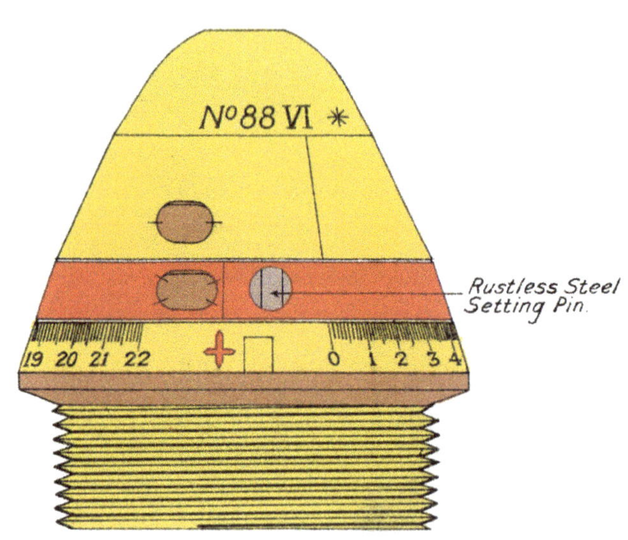

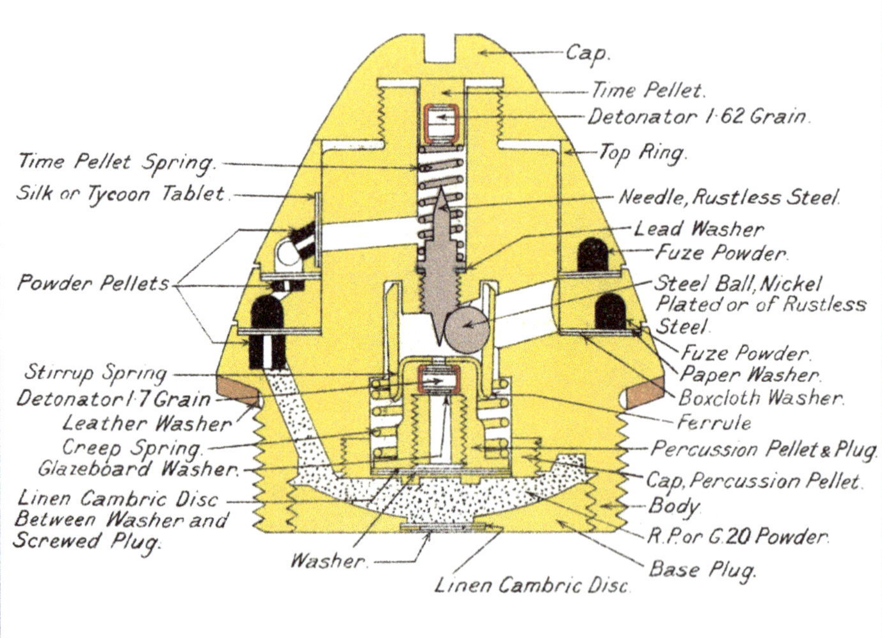

958.

Fig. 6·26(a).

FUZE TIME AND PERCUSSION, Nº 220 MARK I./L/.

Full Size.

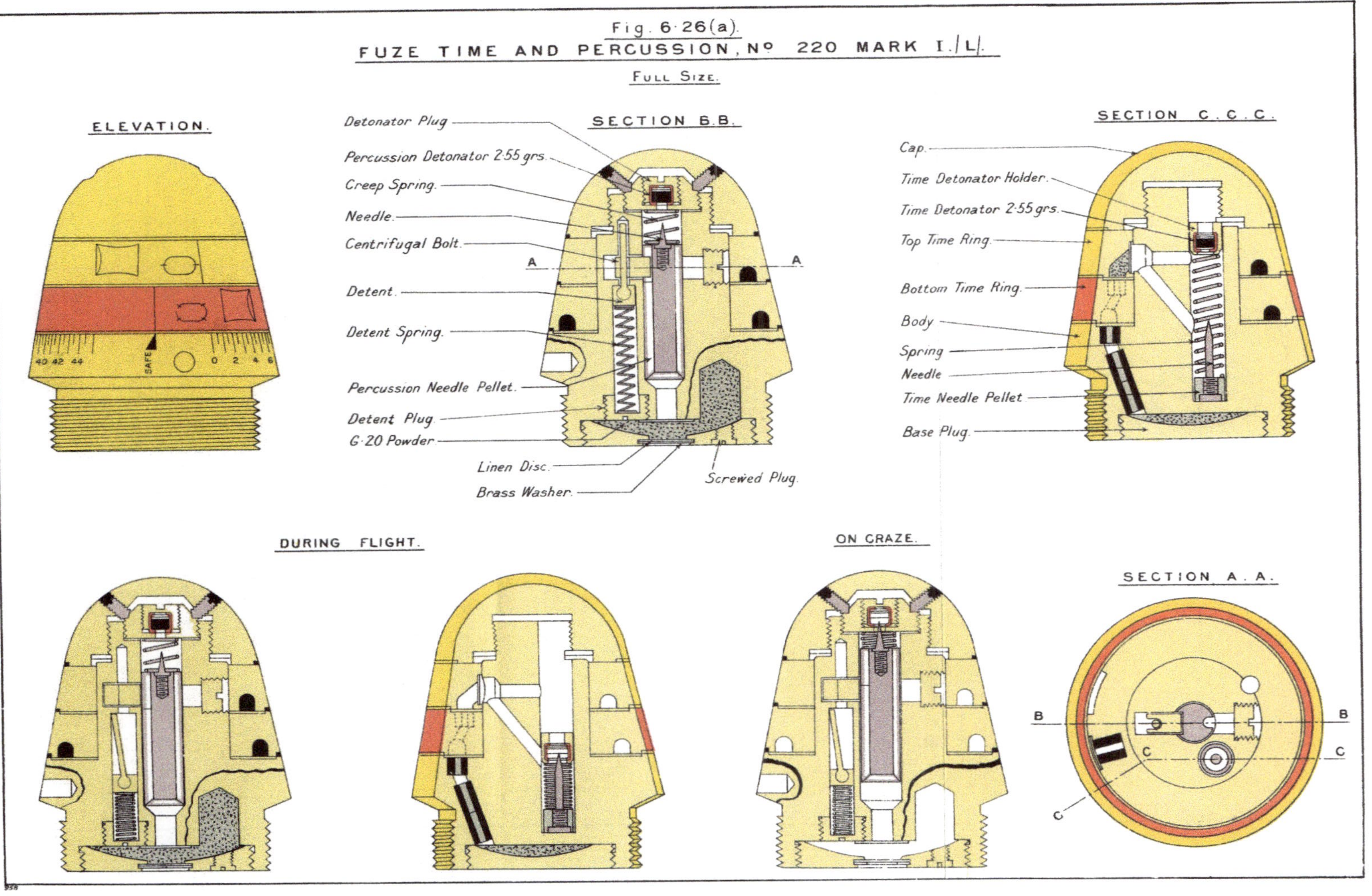

Fig. 6·27.

FUZE, TIME, No. 199 MK. III /L/.
WITH COVER.

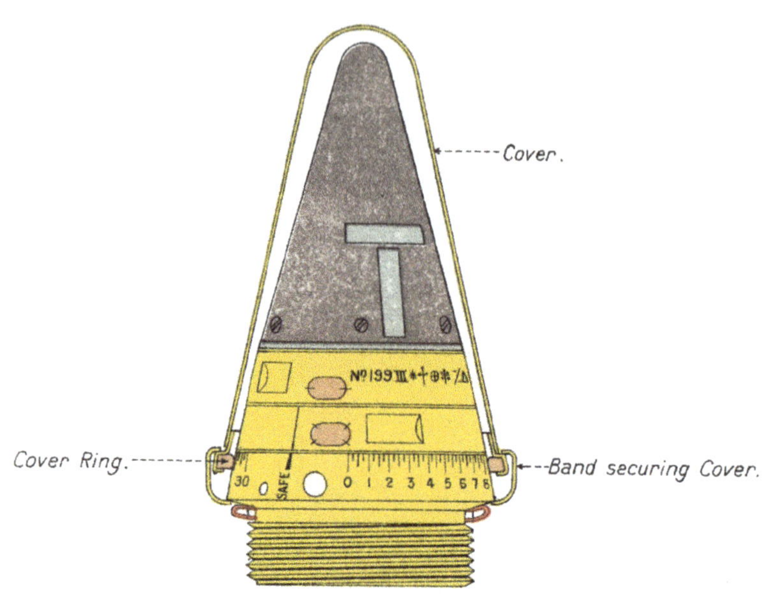

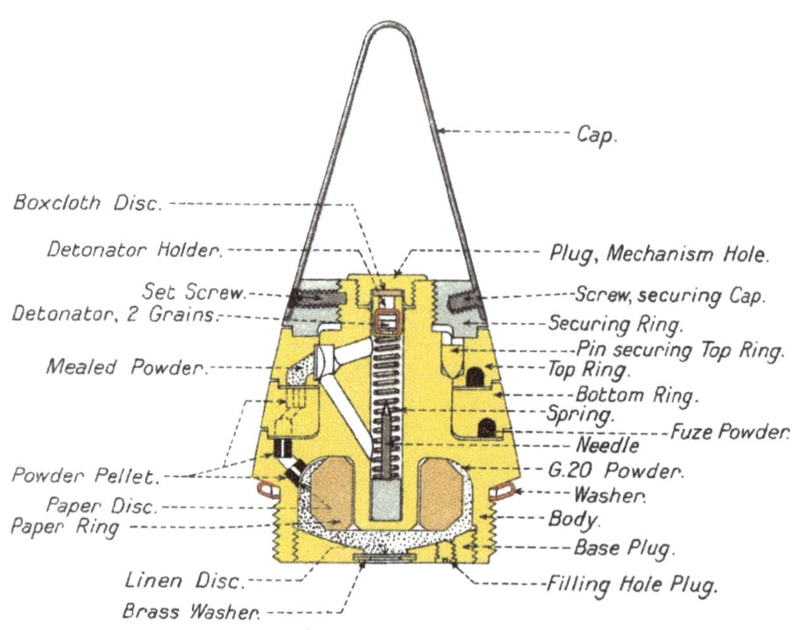

958.

Malby & Sons, Lith.

Fig. 6·30 (a).

FUZE, TIME, No 203, Mk I.

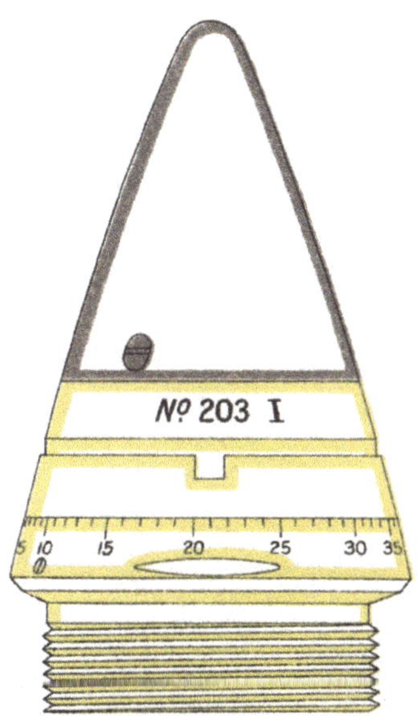

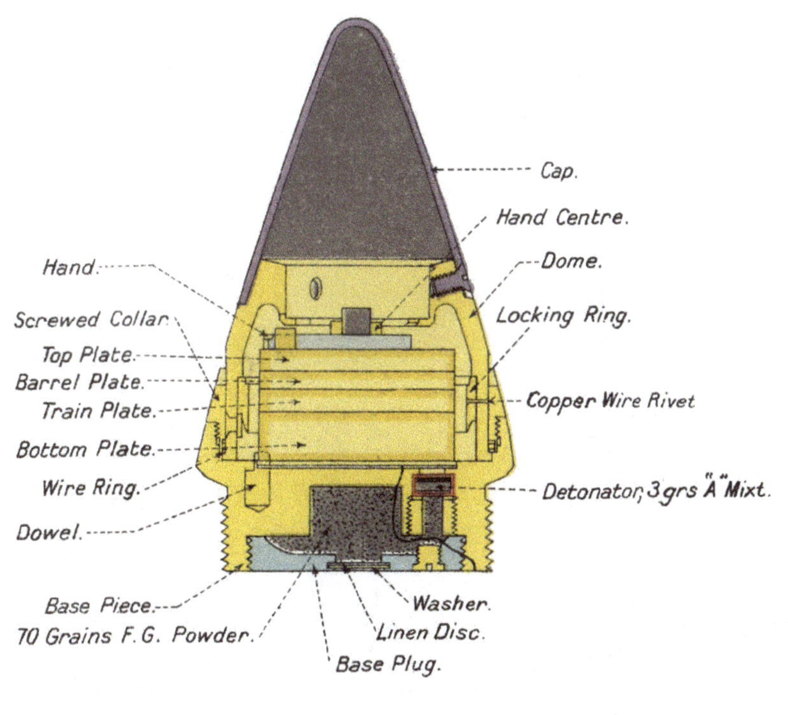

958.

Malby & Sons, Lith.

Fig. 6·30 (b).

FUZE, TIME, No 203, Mk I.

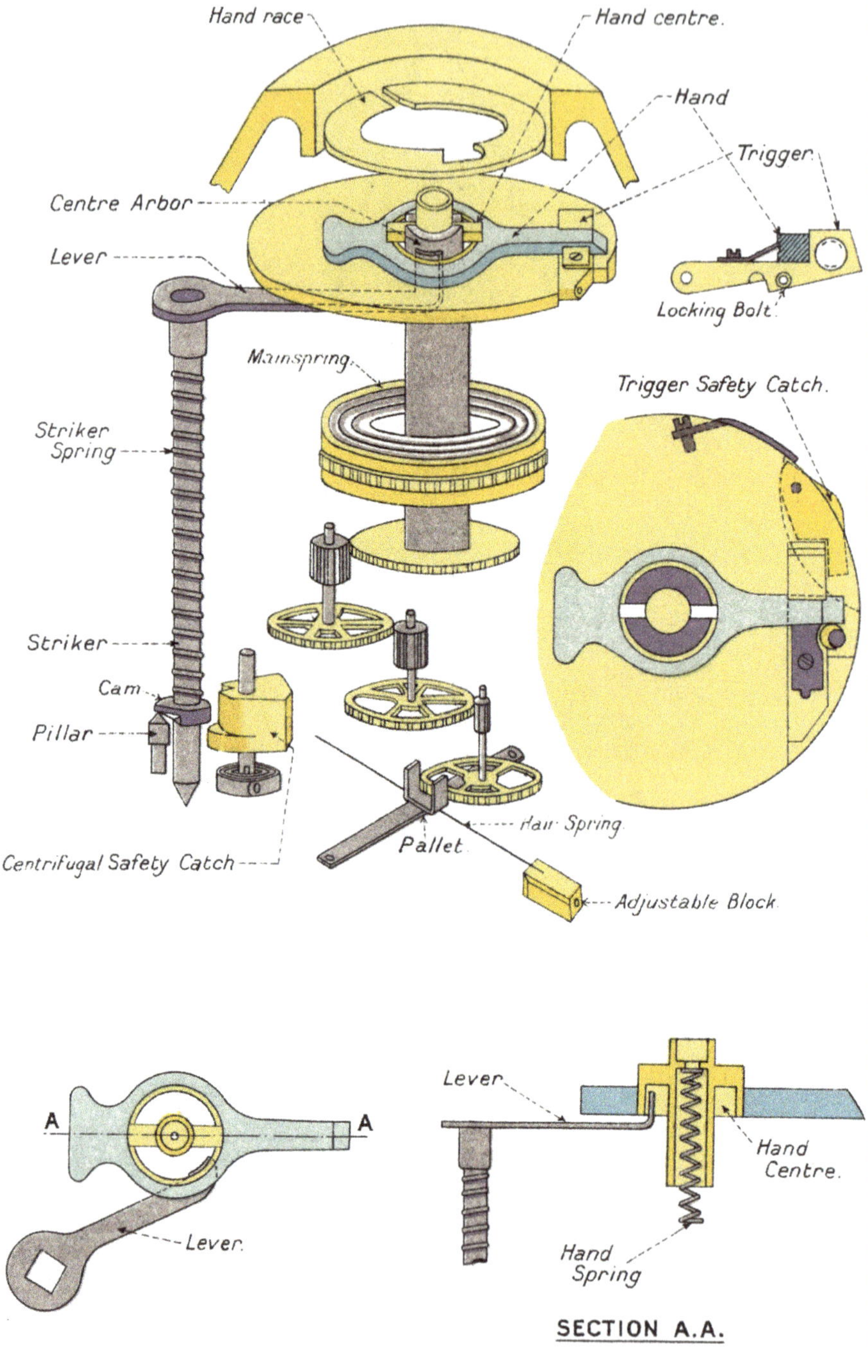

Hand race
Hand centre.
Hand
Trigger
Centre Arbor
Lever
Locking Bolt
Mainspring
Trigger Safety Catch.
Striker Spring
Striker
Cam
Pillar
Hair Spring
Pallet
Centrifugal Safety Catch
Adjustable Block

A
A
Lever
Lever
Hand Centre.
Hand Spring

SECTION A.A.

Malby & Sons, Lith

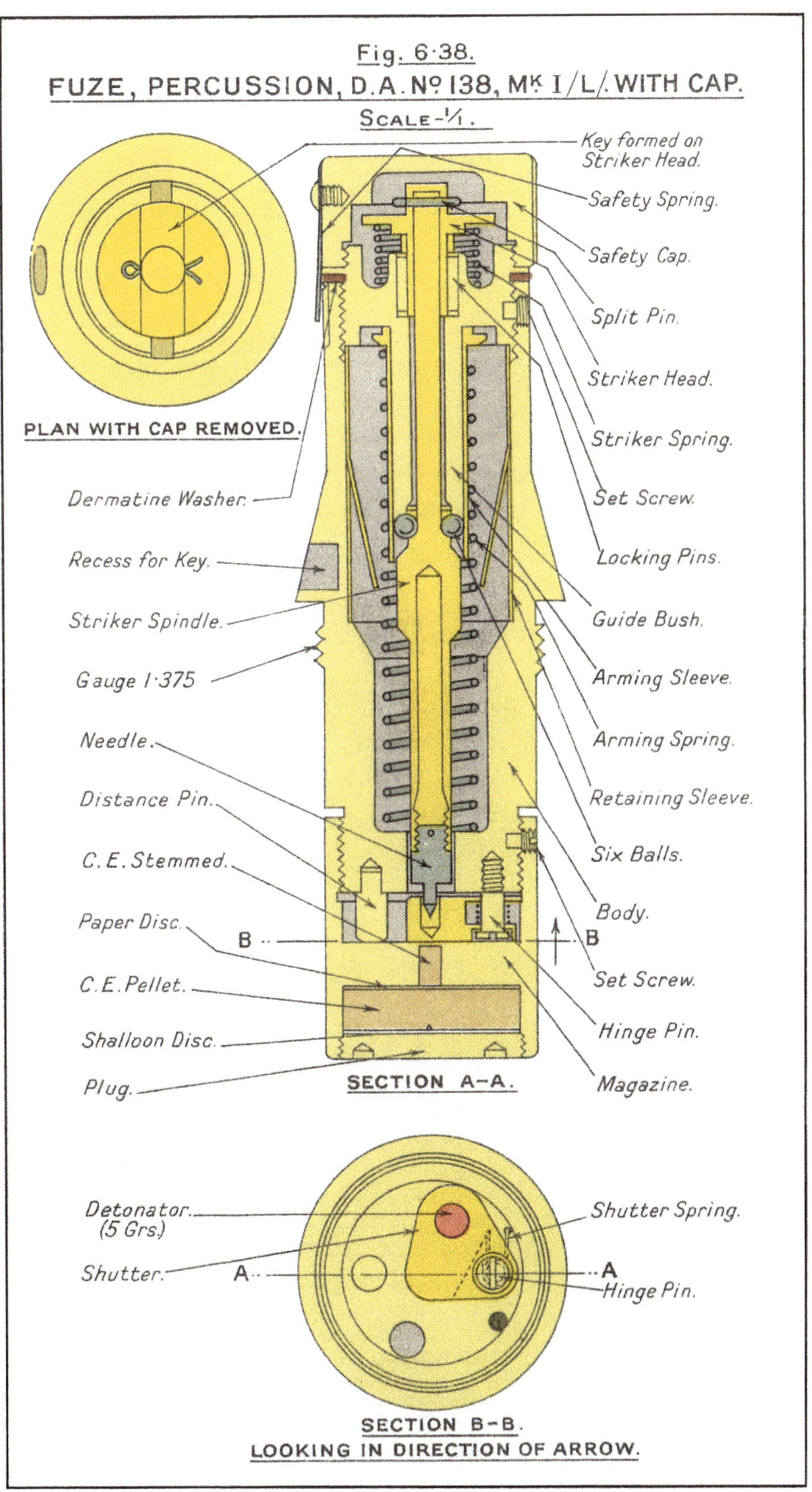
Fig. 6·38.
FUZE, PERCUSSION, D.A. No 138, Mk I/L/. WITH CAP.
Scale-1/1.
PLAN WITH CAP REMOVED.
Key formed on Striker Head.
Safety Spring.
Safety Cap.
Split Pin.
Striker Head.
Striker Spring.
Set Screw.
Locking Pins.
Guide Bush.
Arming Sleeve.
Arming Spring.
Retaining Sleeve.
Six Balls.
Body.
Set Screw.
Hinge Pin.
Magazine.
Dermatine Washer.
Recess for Key.
Striker Spindle.
Gauge 1·375
Needle.
Distance Pin.
C.E. Stemmed.
Paper Disc.
C.E. Pellet.
Shalloon Disc.
Plug.
B
B
SECTION A-A.
Detonator. (5 Grs.)
Shutter.
A
A
Shutter Spring.
Hinge Pin.
SECTION B-B.
LOOKING IN DIRECTION OF ARROW.

Fig 6·39.

FUZE, PERCUSSION, D. A., Nº 139 Mᴷ II.

SCALE :- 1/1.

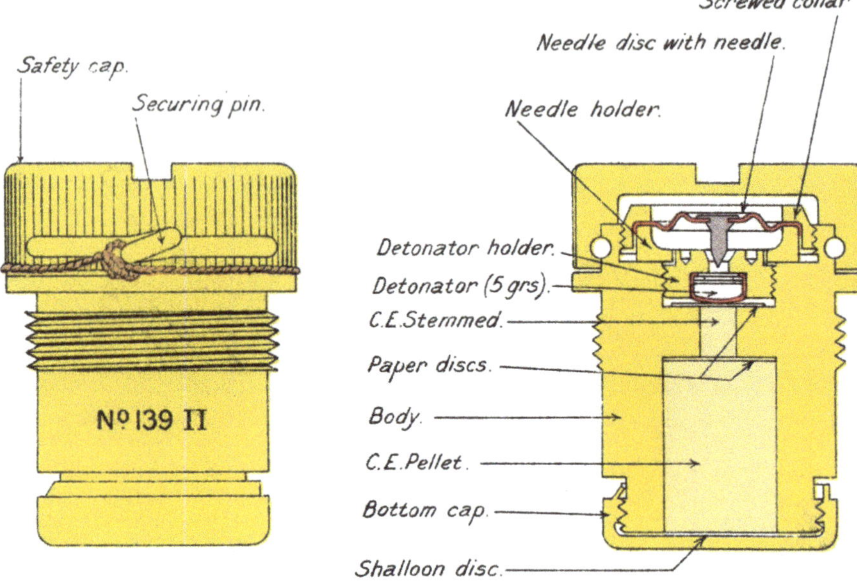

FUZE, PERCUSSION, D. A., Nº 139 P. Mᴷ II.

SCALE :- 1/1.

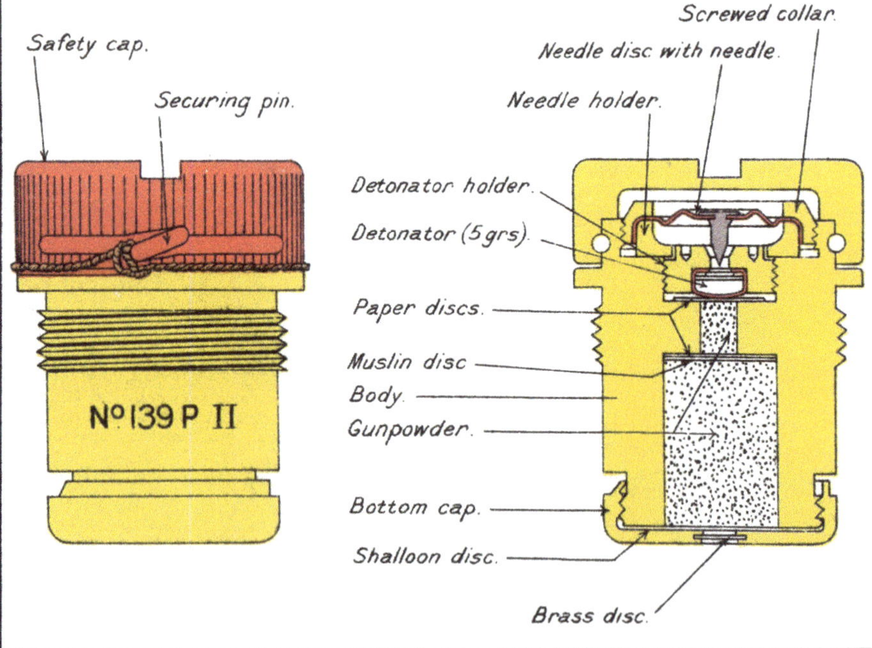

968

Fig. 8·02

CARTRIDGE S. A. BALL, 303 IN., Mk VII. /C/.

Scale = 2/1.

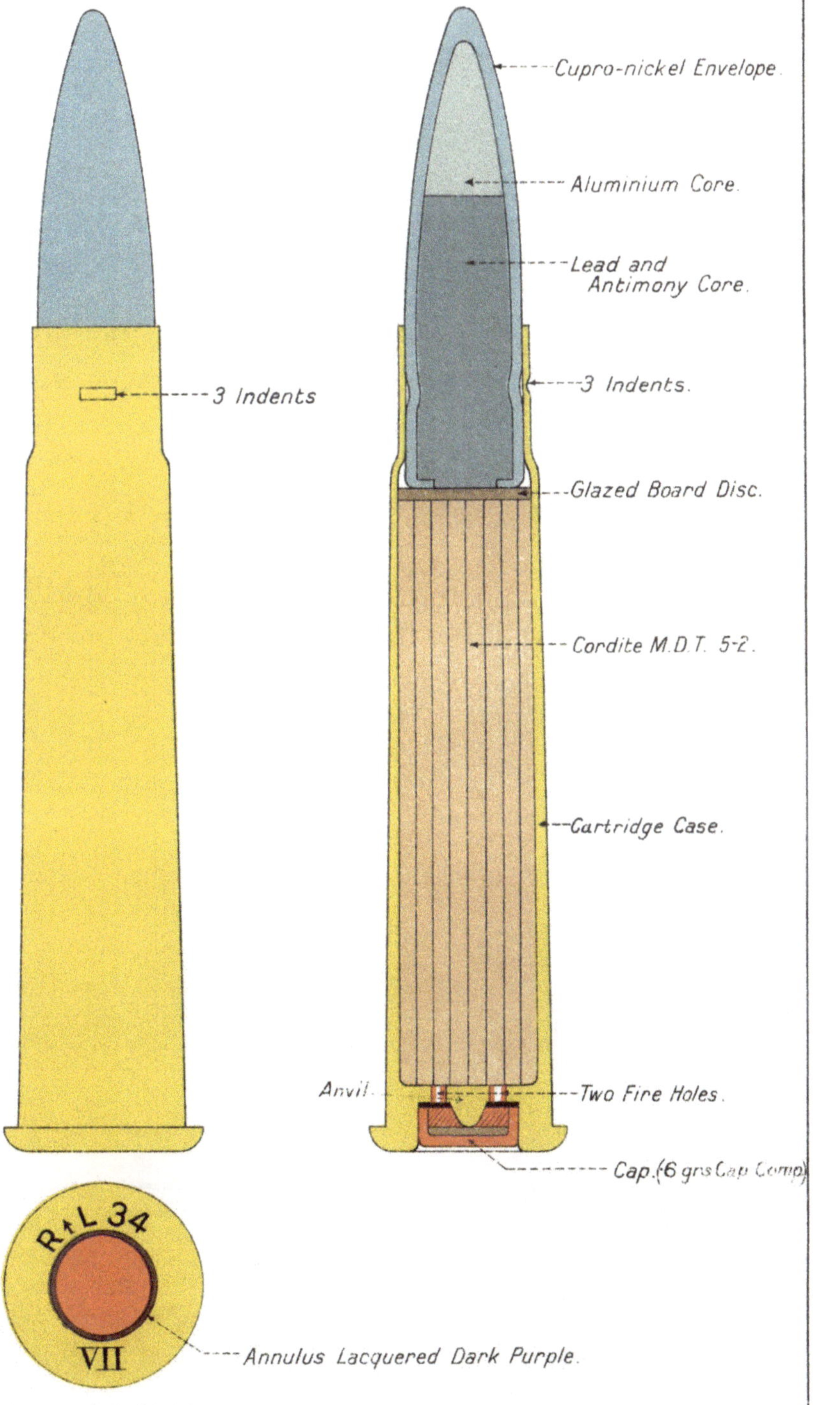

3 Indents

Cupro-nickel Envelope.

Aluminium Core.

Lead and Antimony Core.

3 Indents.

Glazed Board Disc.

Cordite M.D.T. 5-2.

Cartridge Case.

Anvil.

Two Fire Holes.

Cap.(6 grs Cap Comp)

R ↑ L 34

VII

Annulus Lacquered Dark Purple.

PLAN OF BASE.

958

Malby & Sons, Lith.

Fig. 8·03 & 8·04.

CARTRIDGES, S. A., ·303 INCH.

SCALE .. 2/1.

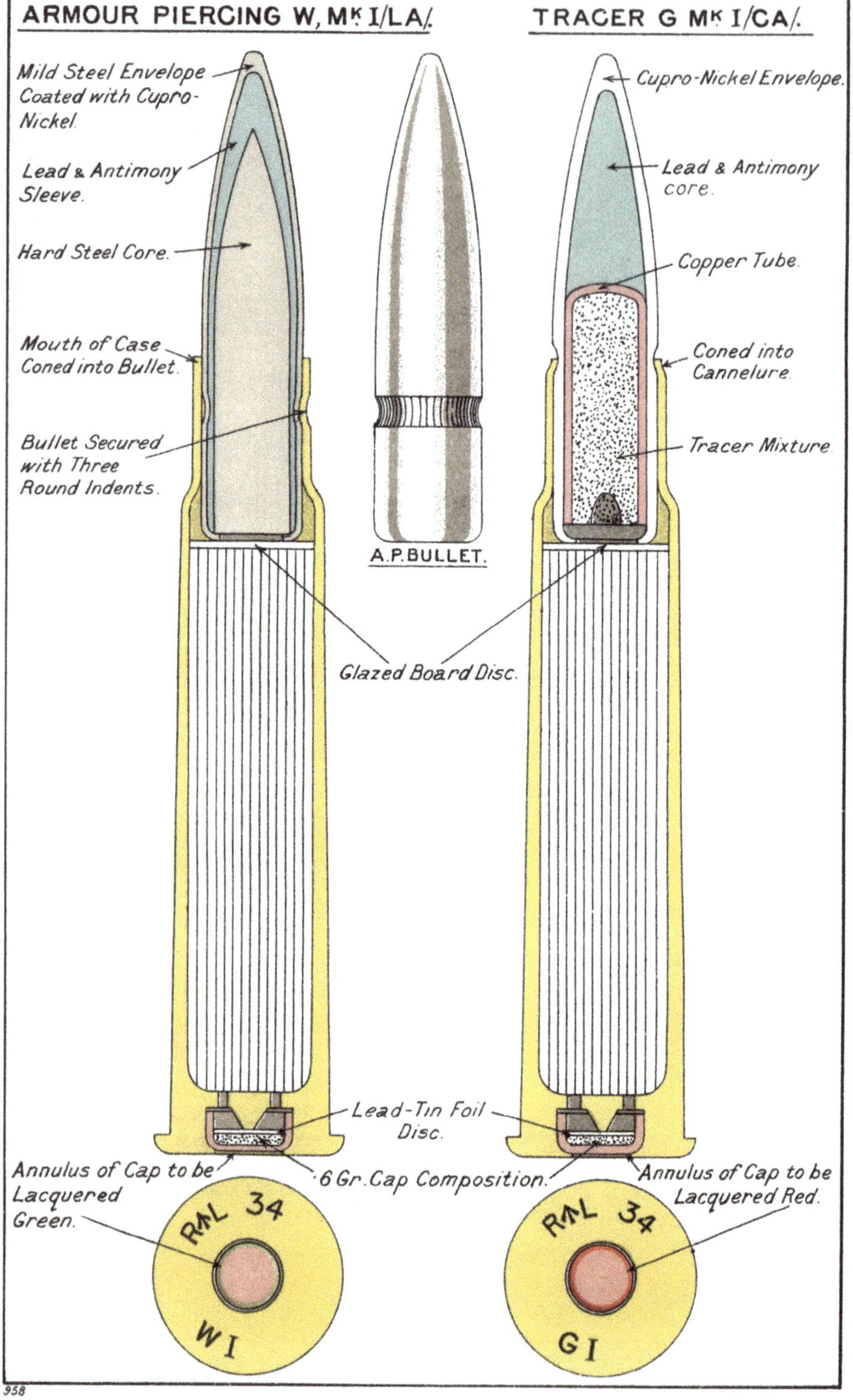

ARMOUR PIERCING W, Mk I/LA/.

Mild Steel Envelope Coated with Cupro-Nickel.

Lead & Antimony Sleeve.

Hard Steel Core.

Mouth of Case Coned into Bullet.

Bullet Secured with Three Round Indents.

A.P. BULLET.

TRACER G Mk I/CA/.

Cupro-Nickel Envelope.

Lead & Antimony core.

Copper Tube.

Coned into Cannelure.

Tracer Mixture.

Glazed Board Disc.

Lead-Tin Foil Disc.

6 Gr. Cap Composition.

Annulus of Cap to be Lacquered Green.

Annulus of Cap to be Lacquered Red.

R^L 34

W I

R^L 34

G I

958

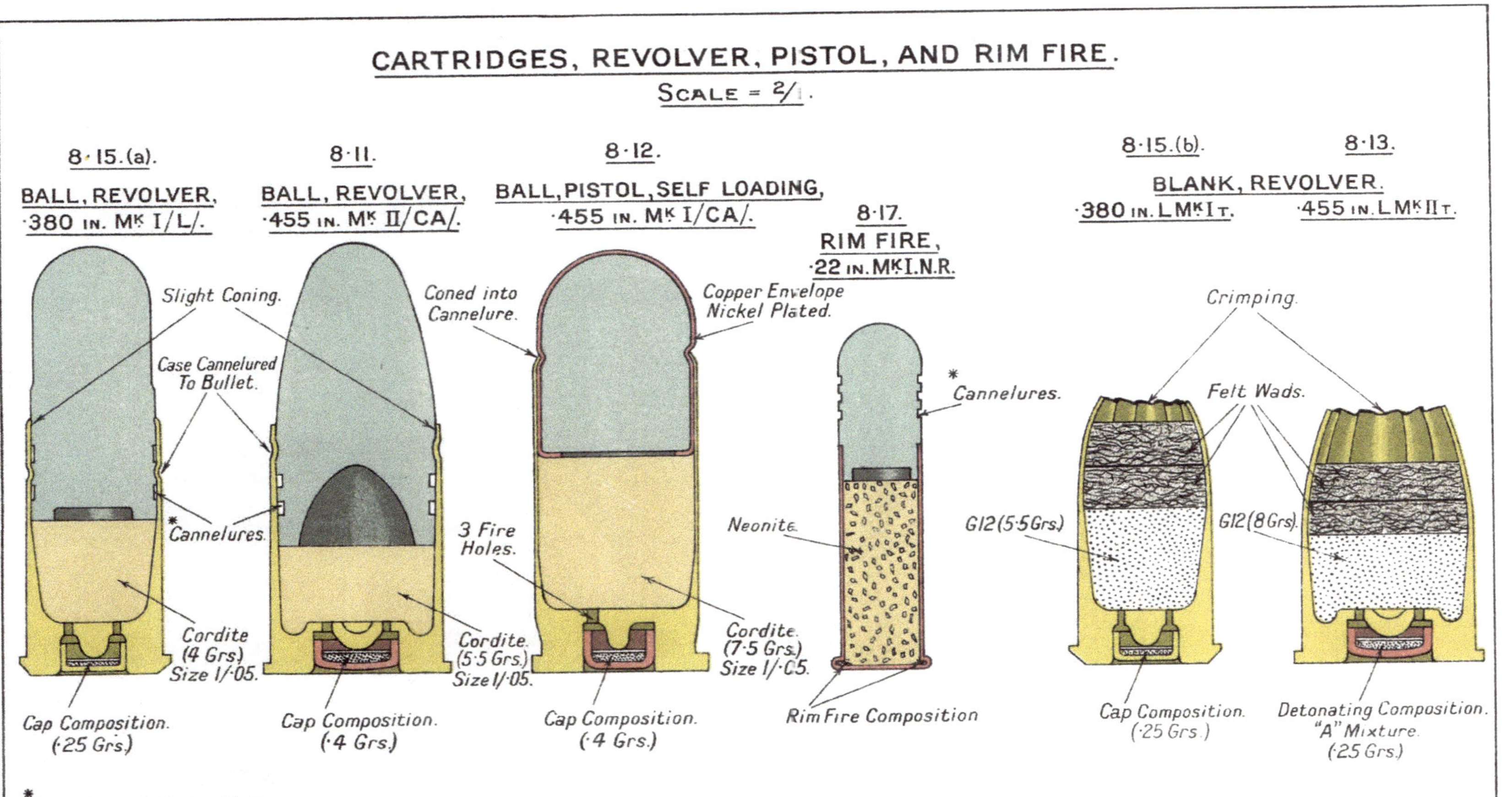

CARTRIDGES, REVOLVER, PISTOL, AND RIM FIRE.
SCALE = 2/1.
8·15.(a).
BALL, REVOLVER,
·380 IN. MK I/L/.
8·11.
BALL, REVOLVER,
·455 IN. MK II/CA/.
8·12.
BALL, PISTOL, SELF LOADING,
·455 IN. MK I/CA/.
8·17.
RIM FIRE,
·22 IN. MK I.N.R.
8·15.(b).
8·13.
BLANK, REVOLVER.
·380 IN. L MK I T.
·455 IN. L MK II T.
Slight Coning.
Case Cannelured To Bullet.
*Cannelures.
Coned into Cannelure.
Copper Envelope Nickel Plated.
*Cannelures.
3 Fire Holes.
Neonite.
Crimping.
Felt Wads.
G12 (5·5 Grs.)
G12 (8 Grs.)
Cordite (4 Grs.) Size 1/·05.
Cordite. (5·5 Grs.) Size 1/·05.
Cordite. (7·5 Grs.) Size 1/·05.
Cap Composition. (·25 Grs.)
Cap Composition. (·4 Grs.)
Cap Composition. (·4 Grs.)
Rim Fire Composition
Cap Composition. (·25 Grs.)
Detonating Composition. "A" Mixture. (·25 Grs.)
*Cannelures Filled with Beeswax.
958.
Malby & Sons, Lith.

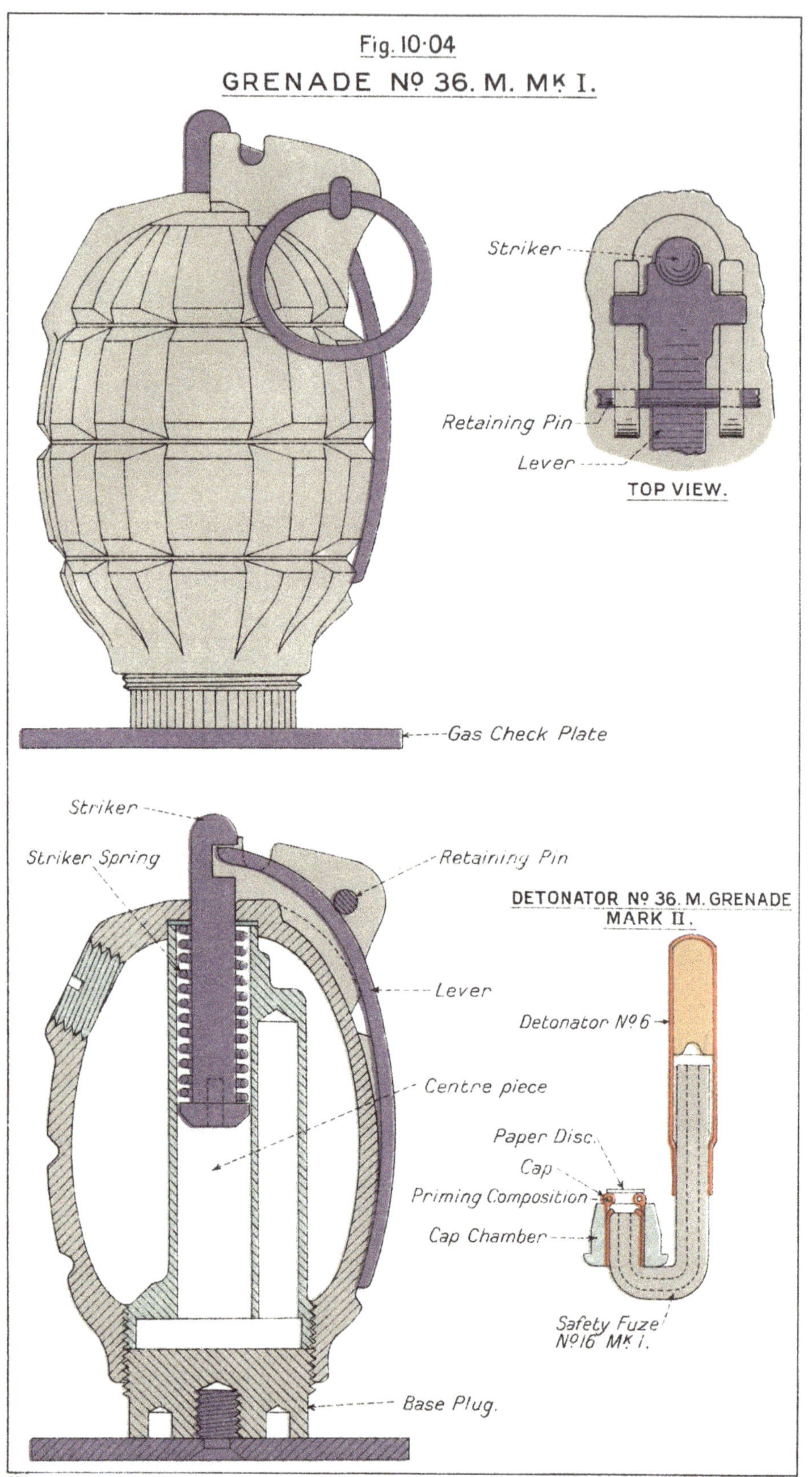
Fig. 10·04
GRENADE Nº 36. M. Mk I.
Striker
Retaining Pin
Lever
TOP VIEW.
Gas Check Plate
Striker
Striker Spring
Retaining Pin
Lever
Centre piece
Base Plug.
DETONATOR Nº 36. M. GRENADE
MARK II.
Detonator Nº 6
Paper Disc
Cap
Priming Composition
Cap Chamber
Safety Fuze
Nº 16 Mk I.
858
Malby & Sons, Lith.

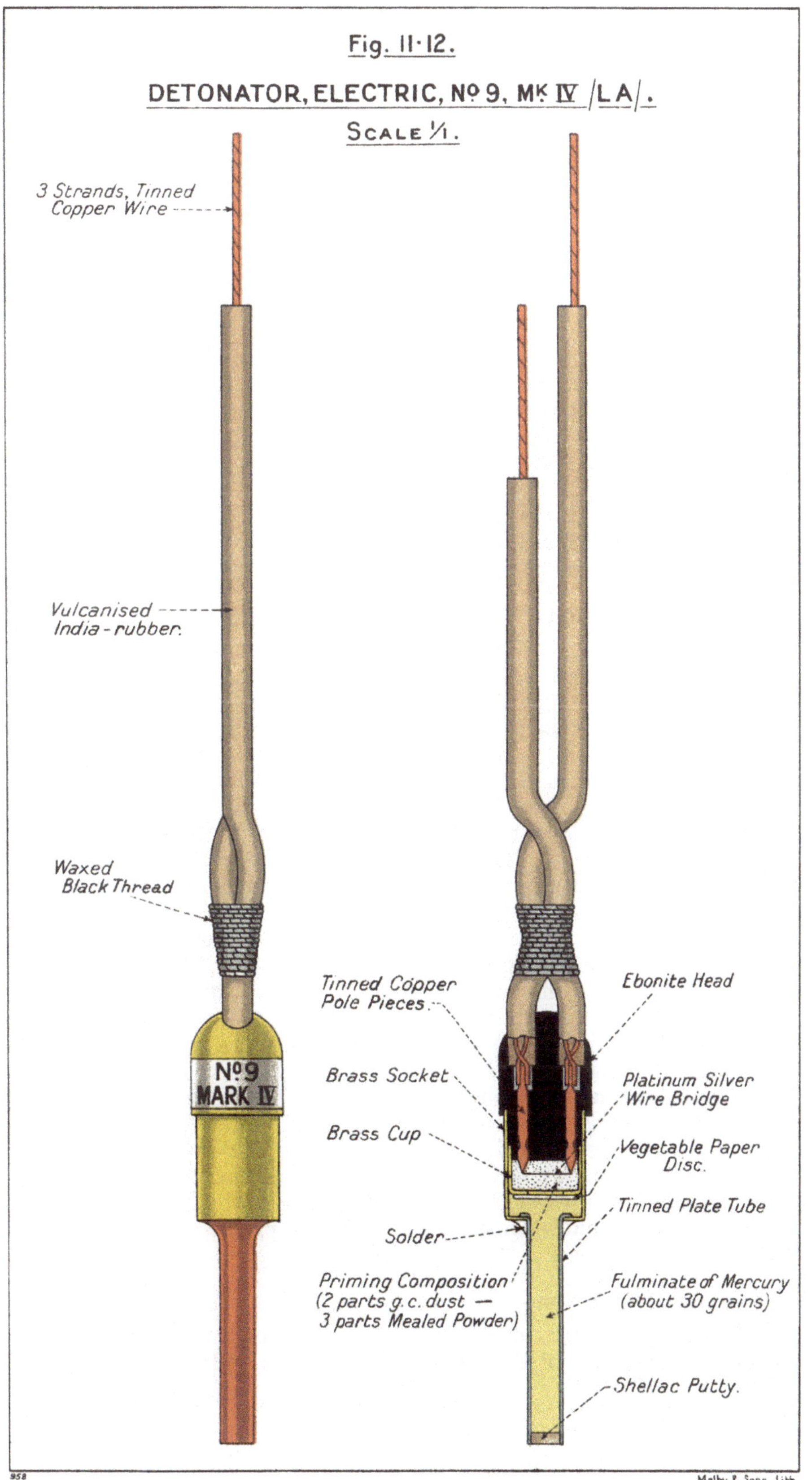

958 Malby & Sons, Lith.

CHAPTER I

EXPLOSIVES USED IN THE PRODUCTION OF AMMUNITION

§1.01. Definition of explosive.

An explosive is a substance which, on being suitably initiated, is capable of exerting a sudden and intense pressure on its surroundings. This pressure is produced by the extremely rapid decomposition of the explosive into gas, the volume of which even at ordinary atmospheric pressure and temperature would be many times greater than that of the original explosive. The volume of gas liberated, and hence the pressure developed, is greatly increased by the heat produced during the explosive decomposition.

An explosive may either explode or detonate.

§1.02. Explosion.

The phenomenon of combustion is familiar, and it is well known that different substances burn at different rates. Thus the rusting of iron is an example of very slow combustion ; while the ignition of petrol vapour mixed with air is followed by extremely rapid combustion.

Ordinary combustion is a process by which substances combine with the oxygen of the air, and therefore the rate of combustion is limited by the amount of oxygen present and the rate at which it can be supplied. If, however, the oxygen required can be closely associated with the combustible, in some other form (as, for instance, in gunpowder, where the charcoal and sulphur are the combustibles, and the oxygen is supplied in the potassium nitrate), combustion can proceed independently of the atmospheric oxygen and can take place within the confines of the mixture itself. This will, therefore, result in a very much more rapid rate of combustion, which can be still further increased by confinement in some vessel. This, then, is what is known as *explosion.*

Explosion is, with a few exceptions, a rapid combustion. Its velocity may vary from about $\frac{1}{3}$ of a metre a second (if controlled) as with modern propellants, up to about 300 metres a second (under the severest conditions as to pressure, etc.). It is, however, constant for a given substance, provided that the latter is homogeneous, and under constant conditions of pressure and temperature, etc.

§1.03. Detonation.

The explanation of the phenomenon known as detonation is outside the scope of this book but it may be said to be an action closely allied to explosion though far more rapid in character. Generally speaking it proceeds through the bulk of the explosive in the form of a wave known as the detonation wave.

The velocity of this detonation wave varies with service explosives from about 3000 to 8000 metres per second, depending on the chemical nature, physical condition and degree of confinement of the explosive.

Detonation is characterized by its powerful shattering effect, which is the result of extremely high local pressures due to an almost instantaneous action.

§1.04. Classification of Explosives.

For military purposes explosives may be divided into :—

1. Propellants.
2. High explosives.
3. Miscellaneous, including gunpowder, pyrotechnic and other compositions that cannot usually be detonated.

Propellants are explosives that are not normally intended to do more than explode.

High explosives are those that are capable of being detonated, and are normally used for that purpose. They may be sub-divided for gun ammunition purposes, into :—

(A) **Shell fillings,** which are very insensitive and difficult to detonate ;

(B) **Initiating agents,** which are very sensitive, and detonate quite easily ; and

(C) **Intermediaries,** which are midway between (*a*) and (*b*), and which are used to pick up the small but concentrated shock given by the initiator and transform it into a sufficiently violent wave to detonate the main filling.

§1.05. Variations in the velocity of an explosion wave.

The changes in velocity observed during the progress of explosion of a solid explosive rising to the maximum velocity known as detonation may be divided into three periods :—

1. Ignition.—During which the velocity of decomposition increases from zero to that of ignition, ignition being the point at which the heat evolved by decomposition exceeds the heat lost by radiation and conduction.

2. Explosion.—The explosion then proceeds at a uniform velocity, provided that the external pressure remains constant, but increasing rapidly with pressure. This pressure rise may be sufficiently rapid to cause detonation.

3. Detonation.—At a maximum velocity of several thousand metres a second, which is roughly unaffected by external pressure.

Hence the service classification of explosives may be defined according to the duration of the above periods.

Propellants are ignited and explode at a velocity depending on the surrounding pressure, which should not rise at a rate sufficiently rapid to cause detonation.

High explosives of the initiator type, when ignited under conditions of confinement, rapidly reach the detonation stage.

High explosives of the shell filling and intermediary types detonate by picking up a detonating wave from an adjacent detonating explosive.

§1.06. The effect of an air gap in the detonating train.

If an air gap is introduced in a detonating explosive system, the detonation wave, in jumping that gap, loses energy and the amplitude of the wave falls off to an extent depending on the size of the gap. If the falling off is sufficient, there may be a failure to detonate. Even if the gap is so small that there remains sufficient energy in the wave on the other side of it to carry detonation over, it may not thereafter be sufficient to pass right through the explosive, which, under the best conditions of filling, can never be perfectly homogeneous. The result will then be a partial detonation, a portion of the explosive being either consumed by burning, or else scattered unconsumed. Such an effect can usually be distinguished by the eye, a detonation generally giving a dense black cloud, while in the case of partial detonation, or explosion, the cloud is grey or yellowish (with picric acid explosives) owing to unconsumed particles scattered in the burst.

It is therefore essential that an explosive system should possess continuity, and every effort is made to ensure this.

§1.07. The properties of a propellant.

A propellant is an explosive which, by its regularity of burning, produces moderately high and sustained gas pressure in the bore of a gun, thereby imparting an acceleration to the projectile. It may be made of ingredients some or all of which are capable of being detonated, but it is essential that there should be no tendency to detonate on the part of a propellant when used under the conditions for which it is intended. The following are the chief properties which an ideal propellant should possess :—

(a) *Its rate of burning should be regular and readily controllable.*

This is usually attained by using a gelatinized (*i.e.* colloidal) explosive of a uniform composition, which can only burn from the surface, layer by layer. For most guns the propellant is made in the form of cords, and the rate of burning can thus be controlled by altering the size of the cord, since the ratio of the surface area to the diameter diminishes as the diameter increases. In this way the same propellant can be used for natures of weapons varying from a revolver to a large gun.

It must not be liable to break up, so giving a sudden change of surface area.

(*b*) *It should be smokeless and leave no residue.*

The former of these is important from a tactical standpoint, as smoke will divulge the position of the active gun. The latter has a practical significance. Hot smouldering fragments remaining in the bore of a gun are obviously dangerous, and any solid residue fouls the bore and increases erosion.

(*c*) *It should be free from flash.*

Freedom from muzzle flash is as important tactically as freedom from smoke.

Freedom from backflash is important, as there is danger to the detachments, and a risk of igniting cartridges or other inflammable material in the vicinity of the breech. The cause of muzzle and backflash is the formation of incompletely burnt gases, which are at a high temperature. These gases ignite at the muzzle on coming into contact with the outside air, after the shell has departed, and on opening the breech.

(*d*) *It should not cause erosion of the bore.*

Erosion is due to the washing action of the hot gases; being caused partly by their high velocity, but mainly by their high temperature. The rapid heating and cooling to which the bore of a gun is subjected further tends to disintegrate the surface metal.

(*e*) *It should be easy to ignite.*

Most propellants are not easy to ignite, and the use of gunpowder igniters is almost universal.

(*f*) *It should be stable in storage and transport.*

The explosives used in propellants differ from those used for other classes, in that they undergo a continuous though slow process of decomposition. Direct sunlight, heat and damp accelerate the rate of decomposition. The progress of decomposition of propellants in storage is ascertained by periodical tests.

Decomposition is accompanied by an evolution of heat and the formation of free acids. If the former is not dissipated and the latter are not neutralized, decomposition is accelerated and may eventually become so pronounced as to result in spontaneous ignition. Cool, dry storage and the addition of stabilizers, which neutralize the free acids, are the best means of prolonging the "life" of a propellant.

Propellants should be insensitive to shock and friction.

(*g*) *It should be unaffected by moisture and temperature.*

The general effect of moisture, if absorbed by a propellant, is to alter the density of loading, which causes erratic shooting. In addition, damp may cause interaction between the explosive and its surroundings, leading to instability.

Exposure, *for short periods*, to extreme temperatures may lead to physical changes in the propellant, which are usually only temporary.

(*h*) *It should not be too bulky.* The propellant must be sufficiently powerful.

TABLE 1.07.

Composition and Explosion Constants of Different Propellants.

Composition	Ballistite 40/60	Cordite Mk. I.	Cordite M.D.	Cordite R.D.B.	Cordite W.	Cordite S.C.	N.C.(Z.)
Nitroglycerine	40	58	30	42	29	41	—
Nitrocellulose	60	37	65	52	65	50	92·4
Mineral jelly	—	5	5	6	—	—	—
Carbamite	—	—	—	—	6	9	—
Diphenylamine	—	—	—	—	—	—	0·6
Dinitrotoluene	—	—	—	—	—	—	6·5
Graphite ...	—	—	—	—	—	—	0·5
Volatile solvent	—	acetone	acetone	ether-alcohol	acetone and water	none	ether-alcohol
Other particulars:—							
N. in N/C (average)	12·6	13·1	13·1	12·2	13·1	12·2	13·0 35 per cent of 12·6 65 per cent of 13·2
Specific gravity	—	1·57	1·58	1·54	1·60	1·57	1·6
Heat of explosion* (Calories per gram)	1,135	1,125	940	915	945	900	815
Vol. of gases* including aqueous vapour (c.c. per gram) ...	830	885	935	960	935	940	930
Ignition temp. (degrees C.)	166	151	160	155	153	154	165

* The water formed is reckoned to be in the gaseous condition.

§1.08. The properties of a high explosive.

High explosives are intended to detonate, and in so doing they must exert a violent disruptive effect on their surroundings. Their chief properties are, therefore, the following :—

(*a*) *They must possess power and violence.*

A high explosive must be sufficiently violent to shatter thoroughly the walls of a shell, and sufficiently powerful to impart a high velocity to the fragments.

Power is of importance when these explosives are used for the destruction of material.

(*b*) *They must be insensitive to ordinary shocks.*

A high explosive should be able to withstand transport, handling, rapid acceleration in the gun and shock of impact without exploding or detonating. The last is particularly important in attacking armour.

(*c*) *They must be free from any tendency to detonate sympathetically.*

Some high explosives are found to be more liable than others to detonate when stored in a dump in which an isolated detonation takes place. This property is closely allied to the sensitiveness of the explosive.

(*d*) *They should be stable in storage.*

High explosives, particularly when filled in shell, should be stable to all ordinary conditions of storage, as they cannot then be subjected to periodical inspection so easily as can propellants. Fortunately, most modern high explosives are so stable that inspection after issue to the filling factory is unnecessary.

(*e*) *They should be unaffected by damp or temperature.*

Moisture may affect a high explosive shell-filling by increasing its volume and thereby forcing out some of the filling from the shell, by partially dissolving some constituent of the filling, or by causing chemical action.

Usually low temperatures are harmless, though they may render some high explosives more difficult to detonate. High temperatures may result in the partial liquefaction of the filling.

(*f*) *They should not form unsuitable compounds with metals.*

The importance of this property and the precautions thereby entailed will become more evident in the later parts of this chapter, where picric acid and ammonium nitrate are considered.

(*g*) *They should be easy to fill at a sufficiently high density.*

A high density in a H.E. shell-filling is important for two reasons :

(i) in order to attain the maximum rate of detonation, and

(ii) to maintain the continuity of the explosive system against " set-back " on firing.

(*h*) *They should provide smoke for observation.*

High explosives in which total combustion occurs on detonation provide little or no smoke, rendering the observation of the burst difficult.

TABLE 1.08.

Properties of High Explosives.

Explosive	Specific gravity	Density as applied	Figure of insensitiveness (picric acid = 100)	Heat value (gram-calories per gram, water gaseous)	Volume of gases produced on explosion (c.c. per gram)	Power lead block test, equal weights (picric acid = 100)	Rate of detonation (metres a sec.) at density
Amatol 40/60	1·70	1·54	115	920	890	114	6,470 (1·55)
Amatol 80/20	1·71	1·25 to 1·45	111 to 120	990	900	120	4,620 (1·2) to 5,080 (1·5)
Ammonal ...	—	1·1	over 110	1,400	900	140	4,135 (1·07
Baratol 20/80	1·86	1·75	101	910*	650*	85	5,000 (1·75)
Baratol 70/30	2·57	2·4	100	760*	380*	64	4,100
Cap composition (S.A.A.) ...	—	—	8	490	85	—	—
Detonating composition ("A" mixture) ...	—	—	6	650	115	—	—
Guncotton, dry	1·67	1·15	23	960	850	115	7,300 (1·2)
Guncotton, wet (13 per cent. water) ...	—	1·2	70–100	750	900	111	5,500 (1·1)
Lead azide ...	4·81	4·01	15 to 25	385	230	37	4,500 (3·8)
Lead styphnate	3·09	2·93	18	450	325	39	2,200 (1·1) to 4,900 (2·6)
Mercury fulminate ...	4·43	4·07	10	410	230	39	4,500 (3·3)
Picric acid (lyddite) ...	1·76	1·6	100	990	700	100	7,250 (1·63)
(C.E.) Tetryl ...	1·76	1·45 to 1·55	70	1,100	750	120	7,520 (1·63)
Trotyl (T.N.T.)	1·68	1·56	106 to 115	1,040	640	95	6,950 (1·57)

* Estimated.

§1.09. Initiation.

The means by which the explosion or detonation of a substance is brought about is known as *initiation.*

The particular method of initiation is decided by :—

(i) the nature of the explosive, and

(ii) the circumstances in which initiation is to take place.

The different methods of initiation are :—

1. *By heat.*—This includes ignition by flash or by the incandescence of a fine metallic wire. The former is exemplified in the case of gunpowder, which is usually ignited by the flash from a cap or detonator. An example of the latter is the electric tube, in which a wire in the tube is rendered incandescent by the passage of an electric current through it. The heat so generated suffices to ignite the priming composition surrounding the wire, which in turn ignites the gunpowder in the body of the tube.

   The above are examples of the initiation of explosion ; an instance of the initiation of detonation by heat is the picric powder exploder. Picric powder, when so ignited, burns at a very rapidly increasing rate, passing almost at once from the speed of explosion to that of detonation. The detonation of H.E. filling through the medium of a powder-filled fuze is readily accomplished by this means.

2. *By mechanical shock.*—An example of this is the cap of the cartridge primer. On being struck by the firing pin, the composition in the cap is nipped between the anvil (which is to the front) and the rear wall of the cap. This is sufficient to explode the cap filling.

3. *By friction.*—Detonators in fuzes are usually actuated by this means. A needle penetrates into the filling of the detonator and the friction so set up between the needle and the explosive causes the latter to detonate.

4. *By the detonation or explosion of another explosive in contact.* —This is the principle underlying the operation of the modern explosive systems for detonating H.E. shell fillings. In these the detonative impulse, commenced by a disruptive detonator, is intensified step by step, each stage being more violent than that preceding it ; thus a small amount of highly sensitive explosive can be made to detonate a large amount of insensitive explosive.

§1.10. The properties of a detonative initiating agent.

By a detonative initiating agent is meant a sensitive explosive, which can be brought to its full rate of detonation in a very short space of time, as a result of a small initial impulse. The term is confined to those explosives which are used to bring other explosives to detonation by reason of the intense localized blow which they can exert. The following are the ideal properties of a detonative initiating agent :—

(*a*) *It should be readily and rapidly brought to its full rate of detonation by flash, friction or percussion.*

(*b*) *It must be sufficiently insensitive to withstand transport, handling and shock of discharge from a gun.*

(*a*) and (*b*) are contradictory, but the necessary compromise is effected by using very small quantities of these explosives, and by compressing them sufficiently to prevent friction. Further protection is afforded by the metal sheath in which detonators are housed.

(*c*) *It must be stable in storage.*

This property is more important in tropical stations where hot, damp conditions of storage may prevail—conditions which are usually deleterious.

(*d*) *It should be unaffected by moisture and temperature.*

Many explosives become inert if damp, necessitating special precautions to prevent the ingress of moisture. Absorbed moisture may also cause chemical action with the formation of dangerously sensitive compounds.

(*e*) *It should not form unsuitable compounds with metals.*

This is important, as most explosives are exposed to the action of metals. If metallic compounds are formed, they may be very sensitive, which is dangerous, or they may be innocuous with a corresponding reduction in the efficiency of the detonant, which is undesirable.

(*f*) *It should withstand a reasonable degree of compression in filling.*

The need for compression has received attention in (*b*) above.

CLASS 1.—PROPELLANTS

§1.11. Ballistite.

Ballistite, invented by Alfred Nobel, consists of nitrocellulose gelatinized by means of nitroglycerine. The mixture is rolled several times between hot rolls, and the sheet of gelatinized propellant which results is then cut into square flakes.

Ballistite consists of :—

Nitroglycerine	39·5 per cent.
Soluble nitrocellulose	60·5 ,,

Originally, the process of gelatinization was performed entirely by the nitro-glycerine, but during the Great War a volatile solvent (acetone) was added in order to quicken the rate of production.

Ballistite was used during the Great War to augment supplies, but now it is only used for special purposes such as in the primary charges of mortars and for propelling grenades.

Ballistite in the form of flakes and lozenges is more erosive than the other propellants used in the Service, and it is inferior in stability.

§1.12. Cordite.

Cordite, evolved by Sir Frederick Abel and Professor Dewar, was introduced into the Service in 1893.

It consists of nitroglycerine, nitrocellulose and a stabilizer incorporated together and gelatinized with the aid of a suitable solvent.

By gelatinization, the fibrous nature of the nitrocellulose is destroyed, and it is converted into a colloidal form which is capable of being worked into any convenient shape.

This dense substance will burn comparatively slowly and regularly from layer to layer, even under the pressures set up during its combustion in a gun. The rate of burning can therefore be controlled. Pressure increases the rate of burning, and high initial pressures such as would strain or burst the gun are avoided by the use of a suitable size of cord.

The following natures of cordite have been introduced for the land service.

Cordite Mark I.
" M.D.
" M.C.
" R.D.B.
" W.
" S.C.

Cordite M.D. and R.D.B. are now used in nearly all equipments in the land service, though they are likely to be replaced gradually by a cordite containing a carbamite stabilizer.

Properties.—Cordite varies in colour from light to dark brown. It is poisonous and has little smell, excepting cordite W. and S.C. which have a slightly pungent odour due to their stabilizer.

Ignited in the open it burns away fiercely, though if large quantities are packed close together there is a danger of explosion.

(*a*) *Control of burning.*—Cordite burns slowly; on this account the initial pressures are comparatively low and the pressure on the projectile in the bore is well sustained.

(*b*) *Smokelessness.*—On burning, cordite is almost smokeless. The yellow colour of the smoke produced on firing with cordite charges is due to the presence of nitrogen peroxide produced on further oxidation by atmospheric oxygen of the lower oxide of nitrogen formed during explosion. A certain amount of smoke also arises from the gunpowder in the cartridge igniter. The products of explosion of cordite are entirely gaseous.

(*c*) *Freedom from flash.*—Owing to the production of a large proportion of partially burnt gases (chiefly carbon monoxide) which ignite on coming into contact with the outside air, the explosion of cordite is accompanied by a large flash. Several attempts have been made to reduce this flash, chiefly by chemical means, but no great success has been attained. With a view to obviating back-flash or the issue of poisonous gases from the breech, large guns firing from turrets are usually provided with an air blast which clears the bore before the breech is opened. With light guns back-flash is not so serious.

(*d*) *Erosion.*—Cordite Mark I was a very hot burning propellant and consequently very erosive. This led to the introduction of cordite M.D., which was a much cooler burning propellant, and the life of guns has been prolonged to a great extent by its adoption. The erosive effects of the remaining varieties of cordite are probably similar to those of cordite M.D.

(*e*) *Ease of ignition.*—Cordite is difficult to ignite; for this reason all B.L. cordite cartridges have igniters, filled with gun-powder, attached to them which reinforce the flash from the tube. The smaller the diameter of the sticks of cordite the easier it is to ignite and the quicker it will burn.

(*f*) *Stability in storage.*—Cordite is not a thoroughly stable substance.

It begins to deteriorate from the day it is made, and a slow but continuous decomposition goes on.

This decomposition results in the formation of certain acid products, and these, if they are not rendered innocuous, greatly accelerate the rate of decomposition of the explosive.

It is for the purpose of neutralizing these products of decomposition that stabilizers such as mineral jelly, carbamite, etc., are added to propellants.

The higher the temperature at which the cordite is stored, the more rapid the deterioration. For these reasons all cordite, except that in small-arm ammunition, is periodically inspected and tested, and to give the cordite a long life, it should be stored in magazines the temperature of which should not exceed 70° F. This limit does not apply to explosive stores for cordite cartridges in R.A. charge and forming part of the authorised equipment. At the same time it must not be exposed to low temperatures, as this may result in sweating on re-warming.

Contact with metals, especially iron, and also wood, affects the stability, hence the lacquering of cases, etc.

(*g*) *Effects of temperature and moisture.*—At low temperatures (below 32° F. for cordite M.D. and M.C. and below 45° F. for cordite Mark I and R.D.B.) nitroglycerine is exuded and appears as an oily film on the surface of the propellant. This film freezes at these low temperatures, and on warming again the nitroglycerine melts. This is termed "sweating." The nitroglycerine may be reabsorbed after some time at temperatures above those previously mentioned. Moderate "sweating" of this nature does not injuriously affect the cordite, though if the low temperature conditions persist for a long period the loss of nitroglycerine may be sufficiently serious to affect the ballistics and to give discordant results to some of the tests to which cordite is subjected. Felt wads or other absorbent material should not be left in contact with cordite, as after a prolonged period, nitroglycerine from the cordite may be absorbed by the wads, rendering them highly explosive. It is for this reason that wads in Q.F. ammunition have glazed board discs on the side next the cordite.

At high temperatures another form of "sweating" may occur, due to the partial liquefaction and exudation of the mineral jelly. It is harmless unless it becomes excessive, in which case the result is to deprive the cordite of its stabilizer. In tropical countries. exposure to the direct rays of the sun may cause so rapid an increase in the rate of decomposition as to render the cordite liable to spontaneous ignition.

Cordite is unaffected by moisture, and it does not absorb moisture from the air.

(*h*) *Bulk.* For a given gun the size of the charge is least with cordite Mark I, rather greater for cordite M.D. and W. and slightly greater still with cordite R.D.B. As a whole the size of charge for a given effect is less with cordite than for most other propellants.

§1.13. Cordite Mark I.

While endeavouring to avoid some of the objectionable features of ballistite, Sir Frederick Abel and Professor Dewar found that the solution of guncotton in nitroglycerine could be achieved by admixture with a liquid capable of dissolving them both and, on evaporating off the liquid (acetone), the guncotton and nitroglycerine remained behind in a gelatinous mass.

The composition of the first British smokeless powder made on this principle was :—

Nitroglycerine	58 per cent.
Guncotton	37 ,,
Mineral jelly	5 ,,

This was originally called *cordite,* on account of the cord-like form into which it was pressed.

On the introduction of cordite M.D. the name was changed to *cordite Mark I* for the purpose of distinction.

The mineral jelly was originally introduced as a lubricant to reduce the wear on the bore of the gun, but it was found subsequently by investigations on the keeping properties of cordite, that mineral jelly fulfilled the important function of acting as a stabilizer and prolonging the life of the cordite.

Properties.—Cordite Mark I is soft and pliable, owing to its high nitroglycerine content. This high content also produces a very high temperature on explosion, causing severe erosion of the bore of the gun. To minimize this, cordite M.D. was introduced and has generally replaced cordite Mark I.

§1.14. Cordite M.D.

Experience with cordite Mark I showed that, owing to the excessive erosion of the bore which it produced, shooting was rendered inaccurate after a comparatively small number of rounds had been fired. This was clearly demonstrated during the South African campaign.

Experiment led to the adoption of a modified cordite known as cordite M.D. of the following composition :—

Nitroglycerine	30 per cent.
Guncotton	65 ,,
Mineral jelly	5 ,,

The manufacture of cordite M.D. is briefly as follows :—

Small amounts of dry guncotton are placed in bags and the requisite amount of nitroglycerine is added. The contents of the bag are mixed on a lead table and passed through a sieve to give "cordite paste." This paste is placed in an incorporator with acetone and mineral jelly, and kneaded into a dough. The "cordite dough" is then pressed from a cylinder through strainers and a die, practically all the remaining foreign matter being removed by the strainers. The extruded wet cord is cut to the requisite length and dried in special stoves ; some of the acetone given off during drying being recovered. The cordite after blending is made into "lots" and despatched to a central depot for subsequent use.

Properties compared with cordite Mark I.—Cordite M.D. is harder and more brittle.

It is slower burning, therefore a smaller size is used in a given gun to obtain the same ballistics as with Mark I.

The temperatures of explosion, and therefore the pressures, are lower. Consequently the gun may be expected to wear much longer than when cordite Mark I is used. Also larger charges of M.D. are required.

M.D. is rather more difficult to ignite, and does not keep so well owing to the hardness of the material, which prevents the distribution or escape of any substances causing deterioration. These substances cause so-called "corrosions" on prolonged warm storage (see sheets 1 and 2 of the coloured plate).

§1.15. Cordite M.C.

This is similar to cordite M.D., except that the mineral jelly has been specially treated with a view to increasing its stabilizing value and so prolonging the life of the cordite.

§1.16. Cordite R.D.B.

R.D.B. is an abbreviation of Research Department "B" formula.

It was introduced into the land service during the Great War, to overcome the difficulties caused by the shortage of acetone.

The guncotton used in cordite M.D. was replaced by soluble nitrocellulose to allow of the use of ether-alcohol instead of acetone as a solvent. The percentages of nitroglycerine and of mineral jelly were adjusted so as to give ballistics similar to those of cordite M.D. It compares favourably with M.D. in stability.

Cordite R.D.B. is somewhat tougher and less brittle than cordite M.D., but not so pliable as Mark I. Its composition is as follows:

Nitroglycerine	42 per cent.
Nitrocellulose	52 ,,
Mineral jelly	6 ,,

§1.17. Cordite M.D.T. and R.D.B.T.

These are cordite M.D. and R.D.B. pressed in a tubular form to obtain increased burning surface. The former is used in small arm ammunition, and both are used with certain forms of gun and howitzer cartridges. M.D.T. is also used for certain practice charges where it is desired to arm a fuze with low velocity.

R.D.B.T. can be used wherever M.D.T. is used.

§1.18. Cordite flake.

This form of cordite was introduced for the Q.F. 4·5-inch howitzer cartridge, but is now obsolete for future manufacture. It was also used in the augmenting charge of 3-inch mortar ammunition, but has been superseded by N.C.(Y.).

§1.18a. Sizes of Cordite.

In the case of cordite M.D., R.D.B., etc., the size is given as the diameter of the holes in the die in hundredths of an inch through which the cordite is extruded. Thus, size 45 cordite has been pressed through a die containing holes having a diameter of 0·45-inch.

It must be remembered that after pressing, cordite, with the exception of S.C., undergoes prolonged stoving during which it shrinks and the finished diameter is, therefore, much less than the diameter of the die.

Tubular cordite is known by numbered sizes, which indicate the external and internal diameters of the tube as it leaves the die. Thus, cordite M.D.T. 5–2 when pressed has an external diameter of 0·05-inch and an internal diameter of 0·02-inch.

§1.19. Cordite S.C.

S.C. (solventless cordite) consists of soluble nitrocellulose, 50 per cent., nitroglycerine 41 per cent., and carbamite 9 per cent.

In this cordite there is no solvent to evaporate out, so that stove drying is eliminated, the carbamite assisting in gelatinization and forming the stabilizer for the cordite.

Regular ballistics.—S.C. cordite is uniform in dimensions, which conduces to regular ballistics.

Stability.—Carbamite is superior to mineral jelly as a stabilizer; it is therefore expected that cordites made up with this will have long life. New S.C. cordite is light in colour, but, as the carbamite is used up, it imparts a dark colour to the cordite, which is a very good indication to its " present " and " remaining life."

Carbamite gives out an emanation which softens paint and varnish.

Temperature.—At low temperatures (40° F.) S.C. cordite shows remarkable resistance to exudation of nitroglycerine.

Weight.—Charges of S.C. have to be heavier than M.D. to obtain the same ballistics in a given gun.

The plate opposite page 28 shows the colour of S.C. cordite and the appearance of corrosions.

§1.20. Cordite W.

Cordite W consists of insoluble nitrocellulose 65 per cent., nitroglycerine 29 per cent. and carbamite 6 per cent.

It is a modified form of M.D. cordite, the carbamite replacing the mineral jelly and so increasing the stability with a consequent gain in life. There is, however, a saving of acetone and there is also less time required to evaporate out the solvent in stove drying.

Sizes of cordite S.C. and W. The size is given in thousandths of an inch, *e.g.* cordite S.C. (or W.), size 057 implies that the finished cord is ·057 inches in diameter.

§1.21. Nitrocellulose powders.

Nitrocellulose powder was introduced during the Great War to assist in maintaining the large output of propellants required. It is now obsolescent as a propellant for gun and howitzer ammunition.

N.C.T. consists of soluble nitrocellulose gelatinized by means of a volatile solvent, and containing 0·5 per cent. diphenylamine as a stabilizer. It is formed by pressing and cutting into short cylinders pierced longitudinally with either one or seven holes, the number and diameter of the perforations varying with the size of the cylinders.

It is much less erosive than cordite and ballistite.

Its main disadvantage is that it is hygroscopic, *i.e.*, it absorbs moisture from the air, which rapidly affects its ballistics. It must, therefore, be retained in air-tight packages until it is required to be loaded into the gun.

It is largely used by foreign Powers.

Sizes of N.C.T.—N.C.T. is known in its various sizes by the diameter of the die used in hundredths of an inch. The sizes that have been employed in this country are 5, 11, 16 and 22.

N.C.(Y.) consists of soluble nitrocellulose 83 per cent., barium nitrate 12 per cent., and camphor 5 per cent. It is used as a propellant for the 3-inch mortar bomb in the form of small spherical grains, reddish yellow in colour.

N.C.(Z.) consists of nitrocellulose 92·4 per cent., di-nitrotoluene 6·5 per cent., diphenylamine 0·6 per cent., and graphite 0·5 per cent. It is used as a propellant in certain small arm cartridges in the form of short tubular grains.

CLASS 2.—HIGH EXPLOSIVES

A.—Shell Fillings

§1.22. Lyddite or Picric acid.

Picric acid or trinitrophenol was discovered in 1771, and was used in the dyeing industry for many years prior to its use as an explosive. It was first used as a filling for high explosive shell in 1893. It is obtained either by the action of sulphuric and nitric acids on phenol (carbolic acid), or less directly from coal-tar benzene. When cast in shell it is known as *lyddite.*

Properties.—Picric acid is a bright yellow crystalline powder of absolute density 1·76. It is slightly soluble in cold water to which it imparts an intensely bitter taste. When pure, it melts at 120·6° C. (250·9° F.) to an amber coloured liquid which on cooling sets to a mass of needle-shaped crystals.

(*a*) Picric acid is a very powerful high explosive when completely detonated. It has a high rate of detonation (about 7,200 metres/second), and its violence is evidenced by the great distortion of fragments of a lyddite filled shell after detonation.

(*b*) Picric acid is not always sufficiently insensitive to withstand impact on armour, and it is therefore unsuitable for armour-piercing projectiles.

If the detonating impulse is inadequate, a milder explosion is produced, some of the filling being unexploded. A complete detonation is evidenced by a large quantity of black smoke ; partial detonation being indicated by the smoke having a greenish yellow tinge and bitter taste due to the unexploded lyddite.

(*c*) Picric acid is sympathetic to neighbouring detonations. Cases have occurred where the whole of a dump of lyddite filled shells have detonated in sympathy with one hostile shell which has burst amongst them.

(*d*) Picric acid is very stable in storage provided it is kept dry, and is not adversely affected by temperature. It is, therefore, particularly suitable for hot climates.

(*e*) Moisture aggravates the tendency of picric acid to form dangerous compounds (picrates) with metals, and care must be taken to seal the shell against ingress of moisture.

(*f*) The great disadvantage of picric acid is its tendency to form sensitive picrates with metals. Most of these, and particularly lead picrate, are very sensitive to shock and friction, and they are capable of acting as detonators to picric acid. Great care is therefore needed to prevent their formation, by interposing a protective film of varnish between the acid and the surfaces of all metal containers, and by the use of lead-free materials. While shellac varnish provides some protection against the action of picric acid, the relatively high temperature at which lyddite is poured (about 140° C.) has the effect of softening the shellac coating, which lessens its protective character, and on this account copal varnish lead free is used. The varnish is sprayed on to the inside of projectiles and components after which they are stoved in an oven for one hour or more at a temperature of 300° F. This coating produces a somewhat elastic layer, which is much more resistant than shellac.

The material for exploder bags has been the subject of investigation, because it was found that interaction occurred between lyddite and potassium nitrate. This interaction produces nitric acid, which deteriorates cotton fabric very rapidly, but woollen vulcanized material is fairly resistant towards these conditions and has therefore been adopted in the Service for such purpose.

As H.E shell in the Service may be filled alternatively with lyddite, trotyl or amatol, all are varnished, and copal varnish is invariably used.

(*g*) Lyddite is usually filled into shell by pouring. The chief difficulty attending this operation is the high melting-point of picric acid, which necessitates the use of heated oil baths, and prohibits the use of steam or hot water for that purpose.

(*h*) As mentioned above, a complete detonation of lyddite-filled shell is accompanied by a large cloud of smoke. For ground observation this is usually good, but for air observation it is not good owing to its black colour.

Picric acid is also used as a high explosive in other countries, being known in France as melinite, in Japan as shimose, and in Germany as Granatfüllung 88.

§1.23. Shellite.

Shellite is a mixture of trinitrophenol and dinitrophenol. It is generally similar in properties to lyddite, but it has a lower melting-point, which gives it a great advantage in filling.

It is more difficult to detonate and is consequently safer in filling, transport, storage, and use.

The same precautions as with picric acid have to be taken with shellite as regards avoiding contact with metals, especially lead; and special medical precautions are taken in the case of persons employed on shellite on account of its poisonous character.

§1.24. Trotyl.

Trotyl is formed by the action of nitric and sulphuric acids on toluene, a liquid similar to benzene obtained from coal-tar or Borneo petroleum.

There are three principal grades of trotyl in the Service, the setting point, which is lowered when substances other than trinitro-toluene are present, being taken as the principal mark of purity.

Grade 1.—A pure trotyl of setting point 80·0° C. to 80·7° C.
Grade 2.—Nearly pure trotyl of setting point 79·5° C. to 80·7° C.
Grade 3.—Crude trotyl of setting point 76·0° C. to 80·7° C.

Properties.—The crude product of the nitration of toluene after washing with water is used as Grade III. This material has a yellow to reddish-brown colour, and contains from 4 to 7 per cent. of impurities, some of which may impart to it an irritating odour.

Grades I and II are prepared from Grade III by purification. They are paler in colour (Grade I being nearly white) and have no irritating odour.

Trotyl does not dissolve appreciably in water and is not hygroscopic.

In the manufacture and filling precautions are taken, as far as possible, to prevent the operator's skin from coming into contact with the explosive and the inhalation of dust. Arrangements are also made for cleansing the skin and for clean clothing.

(*a*) Trotyl is a violent and powerful high explosive, being in these respects only slightly inferior to picric acid. It has a high rate of detonation (6,950 metres/second at a density of 1·57). Complete detonation is accompanied by a black smoke of particles of carbon, but when detonation is incomplete the smoke is grey. When initiated by a gunpowder fuze and picric powder exploder, trotyl gives rise to less violent effects than picric acid.

(*b*) Trotyl is more insensitive in all ways than picric acid.

(*c*) It is not free from the tendency to detonate sympathetically, and there is considerable risk of a dump of shell filled with trotyl being fired *en masse* if one shell is exploded.

(*d*) Trotyl is a very stable explosive and, when pure, is entirely unaffected by conditions of storage. The less pure varieties are liable, however, to give rise to an oily *exudation*. Small proportions of impurities in the trotyl form mixtures which are liquid at moderately low temperatures. These impurities are allied to trotyl, and are known as isomers which, though their individual melting-points are not low, when associated together, produce an oily mixture of low melting point. This oily matter is liable to ooze out of the filling, especially under moderately warm storage, and finds its way through screw threads and other exits of the shell. The oil is explosive and is objectionable, because of the possibility of its becoming nipped between steel surfaces when certain natures of shell are fired in the gun. It may also be taken up by exploders which may be rendered unserviceable by being " deadened " ; or it may flow to the outside of packing cases, etc., and give rise to considerable fire risk. In order to lessen the risk of premature explosion from the presence of such exudation every attempt is made to have trotyl as pure as can be manufactured in the quantities required, or if it is necessary to use somewhat impure trotyl then steps are taken to prevent this oily matter from gaining access to places such as screw threads. It has been found that R.D. cement applied to the screw threads will prevent the passage of this oily exudation.

(*e*) Trotyl is entirely unaffected by damp ; the possible effects of warm temperatures have already been discussed.

(*f*) Trotyl is a relatively non-reactive substance. Being free from acid properties, when pure, it has not the tendency to form sensitive salts as in the case of picric acid. Alkalies, however, interact with trotyl and produce a body which is much more sensitive to shock than ordinary trotyl, and is more readily ignited. Contact with alkali is to be avoided in its manufacture and use.

(*g*) The low melting-point of trotyl renders it easy to melt by steam heating, and when used alone in shell it is always poured.

(*h*) The visibility of the burst of shell filled with trotyl is similar to that obtained with picric acid.

Service uses.—Trotyl is used as a filling for shell and bombs, but its principal application is as an ingredient of amatol. It is generally employed as a poured filling.

It is also used to reinforce the exploder system in amatol-filled shell. Other uses are as fillings for exploders and instantaneous fuze detonating, for which purpose Grade I is generally employed, this being the most responsive to a detonative impulse.

§1.25. Amatol.

Amatol is an intimate mixture of ammonium nitrate and trotyl. It was first adopted in 1915 to augment the supplies of the Service high explosives, lyddite and trotyl.

The amatols which have been used from time to time include 40/60, 50/50, and 80/20, these designations indicating the proportions of the ingredients, the first-named figure, or numerator, referring to the percentage of ammonium nitrate.

The ammonium nitrate, as well as the trotyl, used in amatol is graded according to its purity. The lower grades of ammonium nitrate contain saline and other reactive impurities which adversely affect the stability of the amatol, especially when lower grades of trotyl are used. The grades used are adjusted in accordance with the method of manufacture and filling of the amatol.

Amatol 80/20 is generally prepared by hot mixing ; cold mixing is also employed when the amatol is a fine-grained powder. Amatol 40/60 and other proportions up to 70/30 are prepared by mixing the ingredients in steam-heated vessels.

Properties.

(*a*) Amatol is a very powerful and violent explosive. It detonates at a rate of about 5,000 metres/second.

(*b*) It is more insensitive than trotyl and a powerful exploder system is needed in order to detonate it properly.

(*c*) Amatol possesses the great advantage over most other high explosives in that it is almost entirely free from sympathetic detonation.

(*d*) Amatol is stable in storage ; the danger from exudation is present if impure trotyl and ammonium nitrate are used, being more noticeable where the proportion of trotyl is high, as in 40/60 amatol, and being almost non-existent in 80/20 amatol.

(*e*) Ammonium nitrate is a very hygroscopic substance, and consequently amatol shares this property to a large extent. Cold-mixed amatol is more hygroscopic than the hot-mixed variety as, in the latter case, the crystals of ammonium nitrate have been more or less completely coated with a waterproofing layer of trotyl.

(*f*) Ammonium nitrate in the presence of moisture attacks metals, the corrosion products being in some cases dangerously sensitive, and in any case may finally cause the destruction of or damage to metal components.

Brass, and particularly tinned brass, is objectionable. With brass, blue and green coloured compounds are produced which are somewhat unstable and sensitive, and have comparatively low ignition points. In some cases they explode more violently than amatol. With tinned brass or bronze, ammonium nitrate produces a definitely sensitive compound (basic stannous nitrate).

On these grounds it will be found that measures are taken in the Service to prevent interaction of this kind, and where the use of copal varnish has been impracticable, the coating of the projecting parts of fuzes, etc., with R.D. cement has recently been adopted.

(*g*) The amatols, with the exception of 80/20, are easily filled by pouring. Cold-mixed 80/20 is usually filled into shell under hydraulic pressure, as it is difficult to get a sufficiently high and uniform density in the fillings by other means.

Hot-mixed 80/20 amatol has the consistency of moist brown sugar and it is suitable for screw filling. In this method a worm, working in a mass of the explosive, extrudes and packs it into the shell.

Certain types of shell have their bursting charges in the form of block fillings, which consist of paper containers suitably shaped to fit the shell cavity and filled with high explosive. The filling may be poured into the containers in the molten condition, and lyddite, trotyl, and 40/60 amatol, may be used in this way. Alternatively these containers may be filled with pellets of explosive made in moulds by means of hydraulic presses. Trotyl and 80/20 amatol are so used. Another filling consists of compressed pellets made by hydraulic presses, which are inserted direct into the shell cavity, without the medium of a separate container. This method is employed with trotyl and amatol.

(*h*) 40/60 and 50/50 amatol give a grey smoke on burst which is very suitable for observation, particularly from the air; 80/20 amatol, in which combustion is complete, yields little or no smoke. In general, amatol-filled shell contain smoke mixtures designed to give white smoke. At first these contained aluminium, but it was later found that ammonium chloride could be used.

Service uses.—During the Great War amatol ultimately became the principal high explosive used in the land service. In the naval service it was used to a smaller extent for filling mines, depth charges, etc., and it was used largely in the air service for bomb fillings.

§1.26. Baratol.

Baratol is a similar explosive to amatol, the ammonium nitrate being replaced by barium nitrate.

Properties.

(*a*) Baratol is non-hygroscopic, and not materially affected by high temperatures of storage.

(*b*) It does not attack copper to form sensitive salts, like ammonium nitrate.

(*c*) It is comparatively heavy, with a filling density for 70/30 baratol of about 2·4 (compared with 1·42 in the case of hot mixed amatol. This makes about 3 lb. difference in the weight of a 4·5-inch howitzer shell.

(*d*) It is not easily melted, but it can be incorporated for filling purposes into a thin porridge-like mass.

(*e*) It appears to detonate without any necessity for tamping.

Service uses.—It is the standard filling for grenades and bombs, in which the detonators are contained in copper sheaths.

It may be introduced as an alternative shell filling for the field army.

B.—Initiating Agents

§1.27. Fulminate of mercury.

Mercury fulminate is the mercuric salt of fulminic acid. It is prepared by mixing a solution of mercury in nitric acid with ethyl alcohol. A considerable amount of heat is absorbed in its formation from its elements, and it is probable that the state of molecular strain, of which this is an indication, is connected with its sensitiveness.

Properties.—Fulminate of mercury consists of small crystals, having in bulk the appearance of fine sand. Normally it is grey or brown in colour, but it is sometimes bleached white by the addition of copper salts during the process of manufacture.

The solubility of fulminate in water is small, and it does not absorb water from the air.

It is readily soluble in a solution of sodium thiosulphate (" hypo ") : this is the best way of destroying it if necessary.

Like many other salts of mercury, it is very poisonous.

Fulminate of mercury is very violent and, although it has not a high rate of detonation (about 4,500 metres a second) compared with some high explosives, a small initial impulse brings it to its full rate of detonation in a very short time. This almost instantaneous explosive decomposition, in a substance of such high density, is favourable to the production of the intense localized blow necessary to bring to detonation an explosive with which it is in contact. To obtain the full effect, fulminate should be well confined ; a few grains in the loose condition, when ignited, may burn without violence.

(*a*) *Sensitivity.*—Mercury fulminate possesses a considerably greater sensitivity to impact and friction than most other service explosives. It can be detonated by flash, but in order to procure greater certainty of action when ignited, it is usual to mix the fulminate with other substances which are more readily ignited.

(*b*) *Insensitivity.*—For transport in bulk, fulminate must be kept under water, only very small quantities being dried at a time as required. Drying is carried out at a temperature not greater than 140° F.

In the dry state, fulminate is only transported in the detonators in which it is used, and great care must be exercised in handling these stores.

The explosive is always used in some metal container, as, apart from the benefit attending its use in a compressed state, any leakage of so sensitive an explosive would be highly dangerous.

(*c*) *Stability in storage.*—Warm damp storage has a deleterious effect on fulminate, decomposing it, and causing interaction with its metallic envelope. In the presence of moisture, fulminate of mercury is readily decomposed by most metals (*e.g.* copper, brass and iron), with the liberation of metallic mercury, and the formation of the corresponding metallic fulminates.

Warm dry storage renders fulminate inert, especially when in small quantities, and compressed, as in the 4 and 5 grain detonators. Four months dry storage at 120° F. has been found sufficient to render these inoperative. Large detonators and loose fulminate are less rapidly affected.

Detonators should be stored, housed in their appropriate fuzes, or packed in special boxes. Great care should be exercised in handling them, and no attempt should be made to dissect them.

(*d*) *Stability under conditions of use.*—Exposure to extreme temperatures or damp *for short periods* may not affect fulminate adversely, but these conditions should be avoided if possible.

(*e*) *Formation of unsuitable compounds with metals.*—*See* (*c*) above.

(*f*) *Compression.*—Fulminate of mercury can be filled and compressed into metal containers with reasonable ease and safety. Only small quantities are handled at a time, and every precaution is necessary to avoid accident.

§1.28. Lead azide.

Lead azide is a salt of hydrazoic acid. It is prepared by mixing a solution of sodium azide with a solution of soluble salt of lead such as lead acetate. The lead azide separates out as a crystalline deposit.

Properties.—In the loose condition lead azide consists of a mass of very fine crystals like fine sand. When originally prepared it is white in colour, but it becomes greyish brown if exposed to light. It is practically insoluble in water, but it is decomposed if heated with water for a long period.

As a whole its properties are similar to those of mercury fulminate. It is somewhat less sensitive to percussion than fulminate compositions, so that detonators may require the addition of a sensitizing layer of detonating composition on top of the azide.

It is much more stable in storage than fulminate, and is quite unaffected by warm, dry storage. Moisture alone will not affect it, but carbon dioxide in the presence of moisture may attack it, liberating hydrazoic acid which will attack the metallic case of the detonators.

It has a high density and, compared with fulminate of mercury, a smaller quantity is required to initiate detonation in other explosives.

§1.29. Lead styphnate.

Lead styphnate is the normal lead salt of styphnic acid or trinitroresorcinol. It is a bright orange-yellow powder of absolute density 3·09. It is formed as a crystalline precipitate when solutions of lead acetate and magnesium styphnate are mixed.

The explosive properties of lead styphnate are generally similar to those of mercury fulminate. It is, however, somewhat less sensitive to impact and friction, the figure of insensitiveness being 18 as compared with 10 for mercury fulminate.

The chemical stability of lead styphnate is very good, and it is therefore being used to replace fulminate in cap compositions, to avoid deterioration during warm storage. It is also used in the No. 27 detonator.

§1.30. Other initiating agents.

Mixtures containing fulminate of mercury and other substances are used for the purpose of initiation. They are intended as the primary means of starting off an explosive reaction which proceeds from a process of burning, as distinct from the dynamic effects of a purely detonative impulse.

They are used in various forms of detonators or percussion caps, where compressed layers are respectively either pierced by a needle or struck on an anvil.

Nearly all these compositions consist of potassium chlorate, antimony sulphide and fulminate of mercury, and some of them contain in addition gunpowder and sulphur.

Characteristic compositions are :—

TABLE 1.30.

B. *Initiating compositions.*

Ingredients	Proportions			Cap composition	Fulminate composition
	Detonating compositions				
	"A" mixture	"B" mixture	"C" mixture		
Mercury fulminate	37·5	11·0	32·0	8·0	80·0
Potassium chlorate ...	37·5	52·5	45·0	14·0	20·0
Antimony sulphide ...	25·0	36·5	23·0	18·0	—
Gunpowder (mealed) ...	—	—	—	1·0	—
Sulphur ...	—	—	—	1·0	—
General use ...	(1) Igniferous detonators of fuzes. (2) Caps of percussion primers and tubes.			S.A. caps.	Large detonators Grenade „ Bomb „
Mode of ignition	(1) Pricking by a needle. (2) Mechanical shock.			Per-cussion.	Heat.

The reason for the inclusion of the various ingredients is as follows :—

Mercury fulminate	...	Increases the sensitivity of the composition.
Potassium chlorate	...	Increases the heat of explosion.
Antimony sulphide	...	Prolongs the flame effect.
Gunpowder (mealed) and sulphur.		Diminishes the violence of the explosion.

C.—Intermediaries

§1.31. Picric powder.

Ammonium picrate is the only salt of picric acid in use in the Service. It is an orange yellow crystalline substance manufactured by neutralizing picric acid by ammonia.

It is an explosive of considerable power, and being very insensitive it is used as a burster for armour-piercing shell in America. Its only use in the British Service is as an ingredient of picric powder, which is a mixture of finely-ground ammonium picrate and potassium nitrate in the proportion of 43 to 57.

This powder is more sensitive than picric acid, is easily ignited, and burns readily to detonation when suitably confined. Picric powder is used as an exploder for certain shell filled with lyddite, when the explosion is brought about by means of a flash from gunpowder, and not by a detonating system.

Picric powder is very stable but must be kept dry, since water gradually converts it into a mixture of potassium picrate and ammonium nitrate. The latter readily absorbs moisture, and when damp attacks metals to form compounds which are sometimes sensitive.

§1.32. Composition exploding (C.E.).

This is the Service name for the explosive trinitrophenylmethyl-nitramine. Its trade name is Tetryl. It is produced by the action of a mixture of nitric and sulphuric acids on dimethylaniline, the last named being a substance derived from benzole and methyl alcohol.

It is used in the Service in several grades, which differ slightly in their degree of purity and may be in the form of powder or " corned " grains. The latter form of the explosive is prepared from the powder by adding a small quantity of a solution of gum arabic, followed by rumbling in a mixing machine and drying.

Properties.—C.E. is a pale yellow powder. When prepared by crystallization it is in the form of small crystals. It is practically insoluble in water and is not hygroscopic. When heated it melts between 128·5° C. and 129·1° C., and at slightly higher temperatures it decomposes somewhat rapidly. On this account it is not used in the cast condition.

It is liable to produce dermatitis in certain individuals, and for this reason the corned varieties are preferred, since these give rise to less dust during the filling operations.

Chemically, C.E. is a neutral substance, and does not readily react with materials likely to come into contact with it in the Service. Alkalies, however, readily decompose it and picric acid lowers its stability.

As an explosive, C.E. contains more oxygen in the form of nitro-groups than picric acid or trotyl. It is probably on this account that it is more readily ignited than those explosives, and more sensitive to shock and responsive to initiation by fulminate. It is, however, still insensitive enough to permit of its being used in columns of short lengths in shells where it is subjected to severe shock on set-back.

Its high rate of detonation, high energy content, capability of producing very rapidly an intense pressure, and readiness of response to detonation, make C.E. very suitable for the purpose for which it is used, *i.e.* to reinforce the effect of fulminate.

Purified C.E., while less stable than picric acid or trotyl, has a sufficiently high degree of stability to free it from any suspicion of liability to spontaneous inflammation under conditions of service use. This has been established by warm climatic trials, and is confirmed by experience abroad.

Service uses.—C.E. is used largely as a filling for the magazines of detonating fuzes and gaines. In these it is employed in the form of compressed pellets to prevent it setting-back from the detonator, and to obtain the maximum output of energy from the volume available. It is also used for the exploders of shells if they are liable to be contaminated by infiltration of trotyl exudation from the filling, since it possesses an advantage over trotyl, in that its detonating properties are not seriously affected by the presence of any such exudation. C.E. is also used for exploders in aerial and mortar bombs.

CLASS 3. GUNPOWDER, PYROTECHNIC AND OTHER COMPOSITIONS, WHICH CANNOT USUALLY BE DETONATED.

§1.33. Gunpowder.

Gunpowder is the oldest known explosive. It was formerly used as a propellant, but has now been almost entirely superseded by smokeless powders.

In this country, gunpowder usually consists of 75 parts by weight of potassium nitrate (saltpetre), 15 of charcoal, and 10 of sulphur, the ingredients being ground together in a moist condition, compressed, granulated and dried.

It is manufactured in various sizes of grain, which determine the rate of burning. The most important grades are :—

	Linear dimensions of square aperture.	
	Passed.	*Not passed.*
P.3. (Pebble)	0·75-inch.	0·375-inch.
G.3 (Grain)	0·375 „	0·187 „
G.7 (R.L.G.2)	0·25 „	0·081 „
G.12 (R.F.G.2)	0·081 „	0·0395 „
G.20 (R.P.P.)	0·0395 „	0·0236 „
Mealed	0·0041 „	0·0026 „
Fuze powder	0·0336 „	0·0236 „

Properties.—Gunpowder consists of small polished grains, the grain size being determined by the dimensions of the mesh aperture through which it has been sieved. It varies in colour from black to brown according to the nature of the charcoal used. It is usually glazed, a little graphite being sometimes added for that purpose.

Sensitivity.—Gunpowder is relatively insensitive to shock, but it is very sensitive to flame. It has a fairly high ignition point (over 250° C.), but it is very susceptible to ignition by any flame or spark which may momentarily raise it to that temperature. On this account it is a dangerous explosive to handle, and special precautions must be taken that any tools used with it are not liable to give rise to sparks. To this end all tools for working with gunpowder are made of copper or bronze.

Rate of combustion.—The explosive properties of gunpowder are governed by the fact that it is a mixture of bodies none of which are themselves explosive, all interaction having to take place between particle and particle. The rate of explosion consequently depends on the size of the grains, the nature of their surface, and their density.

Thus, a fine-grained, unglazed powder, consisting of light, porous grains, would burn very much more rapidly than a large-grained, highly glazed, dense powder.

As none of the ingredients of gunpowder are explosive, and as it is only a mechanical mixture, it cannot be detonated. Its rate of combustion, however, is rapid compared with smokeless propellants. Thus, in a gun it gave a high and very rapidly attained maximum pressure, but the pressure rapidly fell off and was not well sustained during the passage of the projectile down the bore.

Stability and storage.—Gunpowder can be kept indefinitely in any climate provided it is kept perfectly dry. When wet, gunpowder cannot be ignited, and if allowed to remain damp, the potassium nitrate tends to segregate and may be absorbed from the gunpowder by paper unless this is varnished. Moreover, potassium nitrate will react readily with copper and other metals in the presence of moisture.

Service uses.—The principal use of gunpowder in the Service is for igniters of cordite charges, for which it is eminently suited on account of its inflammability, rapidity of burning, and ignition qualities. It is used as a priming, as a delay in fuzes and gaines, in the magazines of fuzes, tubes and primers, and as a bursting charge in certain types of projectiles. Varieties of gunpowder, in which the proportion of the ingredients are somewhat different from the standard, are used as the propellant in pyrotechnical stores, and in the time rings of some fuzes.

In addition to these, gunpowder is used for the following purposes :—

Blank ammunition and paper shot.

In such stores as quickmatch, safety, and instantaneous fuze.

§1.34. Pyrotechnic compositions, etc.

Pyrotechnic compositions may be divided, according to their use, into the following classes :—

(1) Illuminating compositions.
(2) Signal compositions.
(3) Smoke compositions.
(4) Incendiary compositions.

With few exceptions, pyrotechnic compositions are solid mixtures. They consist essentially of oxidizing agents, of which the principal are chlorates and nitrates, mixed with combustible materials such as sulphur, antimony sulphide, metallic powders (*e.g.* magnesium, aluminium, antimony, and zinc), and carbonaceous materials (*e.g.* charcoal, shellac, lactose, acaroid resin).

Most of these mixtures burn rapidly in the loose condition and, more or less, explosively. When used in Service stores they are usually moulded by the addition of a binding material such as shellac varnish, or compressed under high pressure. By consolidation in this manner, and also by provision of a free escape for the products of combustion, the compositions are caused to burn in a regular and controlled manner.

§1.35. Illuminating compositions.

The majority of the compositions designed for illuminating purposes, such as those used in star shell, night tracers, etc., consist of mixtures of magnesium or aluminium powder with oxidizing agents, such as barium nitrate or potassium nitrate. Other ingredients, such as strontium carbonate, are added if it is desired to modify the colour of the light, and others, such as paraffin wax or boiled linseed oil, are added to reduce the sensitiveness of the mixture and to prevent deterioration in the presence of moisture.

Illuminating compositions in general, when compressed, are not easily ignited, and it is necessary to prime them with some more easily ignited composition. Small pellets of gunpowder with quickmatch attached are embedded in a thin layer of priming composition, consisting of a mixture of gunpowder and an illuminating composition, which is pressed on top of the main filling.

§1.36. Signal compositions.

(*a*) *For day use.*—Coloured smoke compositions are used for daylight signalling in a variety of stores. The resulting smokes consist of clouds of particles of an intensely coloured material, usually a dyestuff.

In the compositions the dyestuff is intimately mixed with potassium chlorate and either cane or milk sugar. The proportions of the ingredients are adjusted so that the sugar burns in the oxygen supplied by the chlorate. The heat generated is sufficient to atomize the dyestuff, but insufficient either to decompose or to burn it.

(*b*) *For night use.*—Compositions used for signalling purposes at night are not usually required to give an intense illumination. They differ from illuminating compositions in that the light is not produced by the combustion of aluminium or magnesium powders, but by the combustion of charcoal, sulphur, orpiment, or organic materials such as shellac, lactose, etc.

By using suitable ingredients various coloured flames are produced, *e.g.* strontium salts are used in compositions which give a red flame, barium salts in those which give a green flame, copper salts give a blue and sodium salts a yellow flame.

The oxidizing agent used in most of the present service compositions is either potassium chlorate or barium chlorate. These compositions give satisfactory lights of good rich colours, but they are all very sensitive to impact and to friction, and they are liable to explode with violence instead of burning regularly.

Another disadvantage of chlorate compositions is that contact with sulphur gives rise to sensitive and unstable mixtures which, under severe conditions of storage, may inflame spontaneously. The priming compositions used with mixtures containing chlorates should therefore be free from sulphur. For instance, gunpowder should be replaced by sulphurless powder.

The signal compositions are being replaced by others in which ammonium perchlorate and potassium perchlorate are used instead of the chlorates. Such compositions are considerably safer in handling and storage, and less liable to explode.

The most distinctive colours for night signals are red, green, or white. The light given by such compositions, however, is never pure, and hence under certain atmospheric conditions white may appear yellow, yellow may appear orange, and green may appear white. The blue given by the signal compositions is feeble in intensity and is not very satisfactory as a signal.

§1.37. Smoke compositions.

Smoke-screen compositions are designed to distribute through the air a cloud of small particles which form a screen and are usually about the size of fog particles.

Until recently the standard smoke-producing substance was phosphorus, which burns in air to form the extremely hygroscopic substance phosphorus pentoxide, which is readily converted into a solution of phosphoric acid by absorption of atmospheric moisture.

Phosphorus has one advantage over other smoke compositions in that the oxygen necessary for combustion is obtained from the air, and not from an oxidizing agent mixed with it. The yield of smoke-producing substance from a given volume of filling is thus high, the air contributing more than half the weight of phosphorus pentoxide produced. Its many disadvantages, however, are causing its gradual replacement by the hexachlorethane mixtures.

Phosphorus is filled into shells which are arranged to burst on percussion at the point where the smoke screen is required. The phosphorus is sprayed into the air by an explosive burster, it then ignites and burns to form the screen. On account of the spontaneous ignition of phosphorus in air, no special igniter is necessary.

Hexachlorethane mixtures in solid form are gradually assuming more importance as fillings for smoke shell and grenades. Their dry composition eliminates the possibility of leakage, making them quite safe in storage and transport. Their superiority over white phosphorus in smoke production is also marked both in density of the cloud and period of emission.

The basic principle of these compositions is that of associating a highly chlorinated substance with zinc oxide to produce zinc chloride. The latter is volatile and very hygroscopic, picking up moisture from the air to form a dense cloud. There is little heat in the re-action and for this reason "pillaring," which was an objectionable feature of the white phosphorus types, does not occur.

Chlorsulphonic acid and mixtures of chlorsulphonic acid with sulphur trioxide, when vaporized, produce an efficient screening smoke composed of minute drops of sulphuric and hydrochloric acids in aqueous solution. A smoke screen can be readily formed by tanks, the chlorsulphonic acid being sprayed through an atomizing jet by compressed air.

Compositions, known generally as Berger compositions, depend for their effect on the formation of the hygroscopic zinc chloride. This is formed by the combustion of a composition containing zinc and carbon tetrachloride or tetrachlorethane. In this reaction the zinc combines with the chlorine, and the carbon either remains in the container or is burnt up by a small amount of an oxidizing agent, added for the purpose.

Compositions for use in shell to give smoke for observation.— Various compositions are used, which have in themselves high explosives and therefore do not in any way lower the efficiency of the shell in which they are used.

Fumyl is used as a burster and smoke composition in some smoke shell.

Mark II consists of trotyl 45 per cent., ammonium chloride 40 per cent., and ammonium nitrate 15 per cent.

Mark III consists of composition exploding (Grade I) ground 38 per cent., aluminium powder (light) 24 per cent., sodium chloride 35 per cent. and oil, linseed, boiled " lead free," 3 per cent.

Mixture, explosive and smoke No. 5, is used as a smoke composition in shell with cold-pressed amatol fillings.

It consists of : Ammonium chloride, 30 per cent. ; 80/20 amatol, 70 per cent.

Mixture, explosive and smoke No. 7, is used as a smoke composition in shell with hot-mixed amatol fillings.

It consists of : trotyl, 20 per cent. ; ammonium chloride, 40 per cent. ; and ammonium nitrate, 40 per cent.

Red phosphorus is also used with certain trotyl fillings to increase the visibility of the burst.

§1.38. Incendiary compositions.

Most illuminating compositions, on account of the high heat of combustion of magnesium and aluminium, are capable of exerting an incendiary effect, and certain of them were used for this purpose in the A.Z. type of shell used against hydrogen-filled airships.

In some small-arm bullets (incendiary) white phosphorus is used as an incendiary material.

For ground targets thermit has been used. Thermit is a mixture of iron oxide (hammer scale) and finely divided aluminium. When ignited by a suitable priming composition, interaction takes place with the evolution of great heat and the formation of a fused mass of iron and aluminium oxide at a temperature of about 3,000°C.

Owing to the fact that no gas is evolved, the whole of the heat of reaction is contained in the liquid products of the combustion and there is practically no flame. The incendiary effect is therefore local unless special means are adopted for spraying the material over a large area.

Thermit requires a high temperature for its ignition and special priming compositions have been designed for it.

METALS

§1.39. Metals.

Steel is an alloy of iron and carbon containing also manganese and silicon with small percentages of sulphur and phosphorus. Alloy steels are those which contain in addition one or more of the metals nickel, chromium, tungsten, molybdenum, etc.

H.E. shell steel contains 0·4 to 0·5 per cent. carbon. Few shell are manufactured from cast steel. H.E. shells are manufactured by punching and drawing (forging) a solid billet either cast or cut from the forged bar. With lighter natures of shell they may be bored.

Shrapnel shell steel usually contains slightly more carbon (0·6 instead of 0·5) than H.E. shell steel, and the higher mechanical properties required are obtained either by heat-treatment consisting of hardening and tempering or by the addition of manganese without heat-treatment.

Shells for attack of armour are made from alloy steels, "the composition of which is left to the manufacturer," and the points are hardened.

Cast iron used for ammunition is of the grey variety; it contains from 2 to 3·5 per cent. of carbon, part of which exists as flakes of graphite; it is consequently a brittle material having only about a quarter of the breaking strength of H.E. shell steel, with practically no ductility.

The use of cast iron for projectiles is now confined to practice projectiles and shot, small natures of proof shot, and in some cases smoke shell and bombs, and certain H.E. grenades.

Copper is a soft ductile metal universally used for driving bands. For one design of driving band 5 per cent. of nickel is added. Antimony and bismuth are impurities which in very small quantities induce brittleness in copper. Arsenic and oxygen have a hardening effect.

Zinc is brittle in the cast state but can be worked at 150° to 180° C. It has been used with additions of copper and aluminium for fuze bodies.

Tin was recently introduced as a means of decoppering guns. It is inserted with the charge and on firing liquefies and sprays on the bore, alloying with any coppering to form a brittle bronze which is readily removed by subsequent firing. An alloy of 60 tin : 40 lead is now used. This material has a lower melting-point than that of pure tin.

Brass.—The term "brass" indicates the alloys of copper and zinc. Only those brasses containing above 54 per cent. of copper are of industrial utility.

The alloy used for the manufacture of Q.F. and S.A.A. cases, tubes and various small parts contains 70 per cent. copper. It is a soft ductile material and the most suitable alloy for cold drawing. Iron, nickel, and manganese have a hardening effect and reduce its ductility. For hot rolling, extrusion, die stamping, etc., brass containing about 60 per cent. copper is used.

Class A, B, C . . . . G metals are alloys of this type, to which manganese, aluminium or iron may be added according to the mechanical tests specified. No limits for chemical composition are specified for these alloys, except that for certain purposes lead free materials are stipulated.

Class C metal represents the ordinary commercial brass.

Manganese bronze is a brass containing copper, manganese, and small percentages of iron and aluminium. Its properties are similar to those of mild steel.

Gunmetal is an alloy of 88% copper; 10% of tin with 2% of zinc. It is not easily forgeable.

Bronze.—The term bronze indicates an alloy of copper and tin. It is incorrectly applied to some other copper alloys.

Phosphor bronze is an alloy of copper and tin to which a small quantity of phosphorus has been added. The phosphorus up to 1% is added both as a deoxidizer of the molten metal and as a hardening element. The amount of phosphorus remaining in the alloy is varied according to the purpose for which it is to be used.

Forgeable alloys are 60/40 alloys of copper and zinc. In forgeable alloy No. 1 the lead content is limited to 0·03 per cent.

In alloy No. 2, 0·1 per cent. of lead is usually present. The addition of lead facilitates machining.

Aluminium bronze is an alloy of copper with about 7½ per cent. of aluminium. It is workable both in the cold and hot state, is highly resistant to corrosion, and possesses good mechanical properties.

Cupro-nickel is an alloy of copper and nickel in the proportion of 80 per cent. copper to 20 per cent. nickel. It is a hard alloy that resists corrosion. It is used for bullet envelopes.

Nickel silver (German silver) contains copper, zinc and nickel; it is resistant to corrosion.

Aluminium is not extensively used for the manufacture of ammunition; its chief advantage is its low specific gravity. By the addition of copper, nickel, magnesium or zinc, its mechanical properties can be considerably improved.

Duralumin is a light alloy containing aluminium and copper, with small quantities of manganese and magnesium. After quenching from about 550° C. this alloy exhibits the phenomenon of ageing, *i.e.* progressive hardening, with improvement of mechanical properties.

Aluminium zinc alloys are those most generally used for castings.

Lead antimony contains about 2 per cent. of antimony, which has the effect of increasing the hardness of the lead.

White metal.—The "white metal" used for cartridge lids is an alloy containing 80 per cent. tin and 20 per cent. zinc.

Iridio-platinum is an alloy made by the addition of 10 per cent. of iridium to platinum.

TEXTILE AND PAPER GOODS

§1.40. Textile goods.

Batiste is cambric, waterproofed by being coated with rubber on both sides, and is of extremely fine texture.

Uses:—Formerly used for making exploder bags.

Box cloth is new wool dyed drab.

Uses :—For washering the body and lower time ring of T. & P. and time fuzes.

Book muslin is made from long fibre cotton.

Uses :—For discs, washers, and covers for certain grenade bursters.

Cloth, all wool, vulcanized, is taffeta or similar material, proofed with rubber.

Uses :—For making exploder bags.

Cambric is a species of fine white linen.

Uses :—For the covers of propellant charges and sections of charges for 3·7-inch and 4·5-inch Q.F. howitzers.

Cotton is made from the seed fibres of the cotton plant.

Uses :—As an alternative to silk sewing for tying Q.F. cordite charges.

Cotton webbing is used as an alternative to linen tape for lifting bands for B.L. cartridges.

Cream serge is made entirely from wool.

Uses :—As an alternative to silk cloth for bags of B.L. cartridges.

Dowlas is made of flax, in the form of closely woven canvas.

Uses :—For making certain burster bags.

Fearnought is a woven woollen fabric, similar in appearance to felt, but rarely used.

Felt (white or grey) is a mixture of wool, hair and vegetable fibre. Under $\frac{1}{10}$-inch thick, it is known as " collar cloth."

Uses :—For making certain wads, washers and packing pieces. White is used in contact with explosives, grey for all other purposes.

Flax is a plant the stalks of which yield a fibre which is used for making thread and cloths, such as linen, cambric, lawn, lace, etc.

Uses :—In the form of linen it is used for many purposes.

Jean is a cotton fabric of medium texture.

Uses :—For making bandoliers for S.A.A.

Jute is the fibre of the inner bark of an Indian plant.

Uses :—For making bags, mats, yarn, and in the manufacture of safety and instantaneous fuze, etc.

Lasting cloth is made of all wool. It is closely woven and must possess a smooth even surface ; it is fireproof under certain tests laid down.

Uses :—For making bags, burster, for powder filled A.P. shell and for washers for gaines.

Linen tape and linen thread are made of half-bleached flax.

Uses :—Linen tape is used for lifting bands for large B.L. cartridges. Linen thread is used as sewing material.

Shalloon is made from all worsted. It was introduced originally as cartridge material because, being thinner than silk cloth, it formed less obstruction to the flash from the tube. It is used either dyed red or undyed.

Dyed shalloon is used for the igniters of B.L. gun and howitzer and Q.F. blank cordite cartridges.

Undyed shalloon for the smaller gun cartridges, for certain B.L. howitzer cartridges and for making bags burster for various shell, etc.

Silk cloth is made of the refuse silk from the outside of the cocoons. It is strong and of close texture, and has little tendency to hold fire or smoulder.

The cloth is free from all acids, and no substance is to be added to the silk while in the course of manufacture or after the cloth is finished.

Uses :—For making B.L. cartridges.

Silk braid is also made from the refuse from the outside of the cocoons.

Uses :—For making beckets and hooping cartridges.

Silk sewing and *silk webbing* are made from pure, long draft spun silk obtained from the interior of the cocoons.

Uses :—Silk sewing is used for sewing up cartridge bags and igniters, and for hooping and choking the smaller natures of blank cartridges ; it is also used in building up the cordite charges for B.L. and Q.F. guns and howitzers ; before such use it is first greased to prevent rotting.

Silk webbing is much stronger than *silk braid.* It is used, when existing stocks of silk braid have been used up, for securing the outer layers of cordite sticks in the building up of the heavier type of cartridges.

Silk twist is made from pure silk.

Uses :—For ties and loops, in lieu of silk sewing on exploder bags for H.E. shell.

Tracing cloth is linen highly glazed on one side.

Uses :—For placing between exploder bags and fuze.

Tape, white, is made of bleached cotton.

Uses :—For certain lifting bands, and for sealing cylinders.

Vulcanized cashmere is a woollen cloth of very fine texture, vulcanized.

Uses :—For making bags exploders filled trotyl, C.E., etc. Obsolescent.

Worsted is usually spun from wool of long staple.

§1.41. Paper goods.

Cardboard is formed by pasting together as many sheets of fine paper as will give the requisite thickness; consequently, both " middle " and " facings " are of the same quality.

Uses :—For making certain distance pieces, washers, and discs.

Glazedboard.

Uses :—Discs for tubes, fuzes, gaines and wads for small arm cartridges, etc.

Leatherboard is made from pulped waste papers and wood pulp, or from flax waste or flax straw. It is usually dyed.

Uses :—For making " cups " which are inserted after the charge in certain Q.F. cartridges. Also for washers in H.E., and smoke shell.

Millboard is produced to any required thickness by successive dips of a sheet of paper or other material into a vat of pulp; hence millboards are solid single boards.

Uses :—For making distance pieces, covers for B.L. cartridges, also washers and discs.

Paper, brown.

Uses :—For packing tubes, safety fuze, etc.

Paper, tycoon.

Uses :—For covering lighting holes in time and T. & P. fuzes, and under the C.E. channel of detonators of fuzes.

Paper, white fine.

Uses :—Discs for fuzes, tubes, primers, etc.

Paper, cartridge.

Uses :—For closing plugs of tubes.

Paper, whited brown.

Uses :—For closing plugs for tubes.

Paper, non-absorbent, is made from wrapping paper of good quality impregnated with paraffin wax and strengthened on one side with a layer of cotton webbing.

Uses :—For wrapping B.L. cartridges containing S.C. and W. cordites; also for lining packages containing cordites in bulk.

Papier-mâché is formed by glueing together sheets of strong brown paper, the upper surface being sometimes coated with hard enamel.

It is also made by running pulp into moulds and subsequently coating it.

Uses :—For making packing pieces, etc.

Pasteboard is like cardboard, but the " middles " are of an inferior quality of paper.

Uses :—For lining packages, etc.

Strawboard is a thick board made of straw-pulp, run into moulds and pressed.

Uses :—For lining packages, etc.

Vegetable paper is thin, transparent, waterproof paper; it is made mainly of vegetable substances; *i.e.* rags and wood pulp are only used in its manufacture to a strictly limited extent, the main constituent being esparto grass. Being of foreign origin it is being replaced by manifold paper.

Uses :—To waterproof the fuze composition in the time rings of all modern time and time and percussion fuzes.

Vulcanized paper is a thick transparent paper, specially treated so as to make it an excellent insulator.

Ebonite is a mixture of indiarubber, 5 parts, with 2 to 3 parts of sulphur, cured at a temperature of 167° F. under pressure.

Uses :—For electric tubes, primers, plugs, etc.

Vulcanized fibre is made either of wood-pulp or paper, and is vulcanized with the addition of sulphur and french chalk.

Chapter II

CARTRIDGES

§2.01. Definitions.

Gun and howitzer cartridges are divided into two main groups.

(*a*) *B.L. cartridges*, which are used in B.L. ordnance, *i.e.* those guns and howitzers in which the obturating devices are provided by the breech mechanism itself.

Such cartridges are simply bundles of propellant fitted into bags of suitable material, provided at one or both ends with shalloon igniters.

All cartridges of this nature are ignited by means of a tube inserted in the rear end of the vent of the B.L. mechanism. The flash from the tube ignites the powder in the igniter, which in turn ignites the propellant in the cartridge.

(*b*) *Q.F. cartridges*, in which the charge is carried inside a brass case.

These cases are used in all Q.F. guns and howitzers, *i.e.* those which carry no obturating device as part of their breech mechanisms. This is provided by the cartridge case, which expands against the walls of the chamber on firing, and is supported in rear by the front face of the breech screw.

Q.F. cartridges are fired by means of a striker in the breech mechanism firing a primer cap seated in the base of the brass cartridge case, or, in some cases, an electric or percussion tube in an adapter, the flash from the primer or tube fires the propellant contained in the case.

Q.F. cartridges are divided into two sub-groups :—

(1) *Fixed ammunition.*—The cartridge and shell are secured together by either coning the mouth of the case into a circumferential groove cut on the rear of the driving band, or on the shell just behind the driving band ; or by indenting the mouth of the case into a circumferential cannelure cut in the body of the shell below the driving band.

(2) *Separate ammunition.*—The cartridge and shell are loaded separately.

With howitzer ammunition the charge is retained in the cartridge case during transport by a leatherboard cup which must be removed before firing,

With gun ammunition the mouth of the case is permanently closed by a lid of white metal which has a low melting point and is entirely consumed on firing by the propellant gases.

§2.02. The comparative advantages of the B.L. and Q.F. cartridge systems, from the point of view of Ammunition.

Advantages of B.L. system.

1. It is economical in transport and storage space.
2. It is cheaper and easier to manufacture in quantities. Supplies of brass may be restricted in war-time.
3. There is no litter of empty cases, requiring additional labour for removal and salvage.
4. It is easier to extract and eject a tube than a large cartridge case.
5. The risk of obturation failure is probably less than the risk of a badly blown case, and the consequent repair of the gun is easier.
6. In case of a missfire, the tube can be replaced without opening the breech.

Advantages of Q.F. system.

1. The charge is better protected from flash, spray and moisture.
2. Loading is slightly quicker, owing to sponging out and riming the vent not being required, and to the tube not having to be inserted.
3. There is no risk of premature firing owing to smouldering *débris* from previous rounds.
4. *With fixed ammunition :—*
   (*a*) No ramming of the shell is required.
   (*b*) The round is more easily handled and loaded, with field equipments, and at high angle fire.
   (*c*) There is no danger of double shotting.

In some cases *separate Q.F. ammunition* is preferable to fixed, because :—

1. The cartridge can be changed without changing the shell.
2. With base fuzed shell, and with shell using tracers, the base of the shell should be accessible.
3. With howitzers, and any equipments using fractional charges, separate ammunition is at present essential.
4. Except with small calibres, the fixed type is weak at the joint of the case with the shell.
5. With certain small calibres in which there is no necessity to withdraw rounds from limbers, there is little saving in rapidity of fire with fixed ammunition, as both shell and cartridge can be loaded together on a tray.
6. With large calibres, fixed ammunition is cumbersome.

In general, it may be said that the advantages of the Q.F. system decrease with weight and size of ammunition, but, where bursts of very rapid fire may be required from a gun, and for anti-aircraft equipments, fixed Q.F. ammunition is essential.

CHARGES

§2.03. Determination of charge weight.

Generally speaking, charges are determined

(1) for a maximum muzzle velocity with a certain size of cordite without exceeding the safe working pressure of the gun ;

or

(2) in the case of reduced charges, to give a certain muzzle velocity. In this case, use is generally made of a small size of cordite in order to keep the chamber pressure sufficiently high to arm a fuze and to ensure regularity of ballistics.

The charge weight required is first calculated theoretically, and a trial carried out in a gun.

From the results of this firing, the correct charge weight is calculated for use in subsequent firings. This weight is known as the nominal weight of the charge.

§2.04. Adjusted charges.

Since lots of cordite of the same nominal size differ somewhat from each other ballistically, the actual charge used in the Service, with a few exceptions, is adjusted by weight for each lot at a standard temperature from the results of proof in a gun, with the view of minimizing the difference in ballistics between lots.

§2.05. Ballistic grouping of lots.

The adjustment of the charge will, however, give consistent ballistic results only in a new gun ; in a worn gun, different lots of cordite will still differ ballistically in spite of the adjustment of the charge.

In order to obtain the greatest uniformity in the shooting of a gun, the ideal condition would be to allocate to it a separate lot ; this is, obviously, impossible under service conditions.

Grouping is a system by which lots of similar ballistic properties can be allocated to each unit, and is considered the best method obtainable of securing consistent shooting when the lot of cordite in use has been exhausted and a new lot is required to replace it.

With recent issues of propellants for land service cartridges this system has been adopted, the classification of lots being based on the muzzle velocity given by a nominal charge fired under standard conditions. The muzzle velocities are divided into groups above and below the " mean of specification," the extent of each group being 10 f/s in the case of guns and 5 f/s in the case of howitzers.

The groups immediately above and below " mean of specification " are given the numbers 10 and 9 respectively, and the other groups higher and lower numbers, as the case may be.

For example :—In the Q.F. 18-pr. the " mean of specification " is 1,660 f/s. All lots which give at proof velocities between 1,651 and 1,660 f/s are allocated to group 9, and all lots giving between 1,661 and 1,670 f/s to group 10, and so on.

Symbols used in grouping of lots.—Each class of propellant is given a distinguishing letter, which is—

C for R.D.B. in cord or tubular form.
D for M.D. and M.C. in cord or tubular form.
B for ballistite.
N for N.C.T.
F for cordite M.D. or R.D.B. in flake form.
S for S.C.
E for cordite W.

Thus, each lot of propellant, in addition to being defined by its nature and manufacturing lot letter and number, will also be accorded a letter and number which defines its ballistic properties ; thus, C15, B12, etc.

While, therefore, the charges will continue to be adjusted as formerly, and the marking of the manufacturing lot letter and number maintained on cartridges and their packages, the above additional marking will appear. When this marking is used, the letters AC, denoting an adjusted charge, will be omitted.

It is important that different classes of propellant are not grouped together ; thus, C12 must not be grouped with N12.

§2.06. Propellants in use.

Propellants are manufactured in various shapes such as

(*a*) Cord.
(*b*) Oval.
(*c*) Tubular.
(*d*) Multi-tubular.
(*e*) Flake.
(*f*) Disc.

Oval, multi-tubular and disc are not at present used in the Service.

Each particular shape gives a different rate of burning.

Propellants in use may be suitably grouped under two headings :

(1) Propellants in stick form, such as cordite, Mark I, in cord and tubular shape.
(2) Propellants in various small sizes, such as ballistite, flake cordite, and nitrocellulose (N.C.T.), which, for want of a better word, are described as granular.

The stick form of propellant is used in all British equipments.

The other form was introduced during the Great War to assist in maintaining the large output of propellant required. It is now obsolescent except for certain S.A. cartridges.

Advantages and disadvantages of the two forms.

Stick form.—The advantage of the stick form is that it makes up into a strong and rigid cartridge. The bag is consequently but little more than a means of securing the igniter, though it also protects the propellant from contamination and to some extent from accidental ignition.

The disadvantage of the stick form is that additional labour is required to cut and bundle the sticks.

Each size of cordite is supplied in sticks to a standard length. These have in most cases to be cut to different lengths, bundled and tied to form any particular cartridge, as little cordite as possible being wasted in the process.

Granular.—The main advantage of the granular powder lies in the speed and economy of filling.

There is no waste.

In addition, a propellant of this type lends itself to automatic filling and weighing, which the long stick does not. This process is not yet fully developed, but it offers great possibilities for rapidity and cheapness of filling.

The disadvantage is the lack of rigidity in the B.L. cartridge.

The shape of the cartridge depends on the bag, which acts properly as an enclosure for the charge. A rupture of the bag would lead to loss of propellant, which is unlikely with the stick form.

§2.07. Arrangement of the propellant.

Cordite, as supplied, is cut to a number of standard lengths. In only a few instances are the sticks of the precise length required, nor is it possible to cut them to the required length without wastage.

The usual procedure is to arrange the longer sticks on the outside of the bundle; then, with the cordite left over after the ends are cut off, a number of small bundles are made, which, placed end to end, form the centre of the charge.

Bundles of very short lengths are not permitted, as they tend to increase the working pressure and upset the ballistics. The minimum length of stick permitted is one that is at least 20 times its nominal diameter. Thus the shortest length of size 16 allowed is 3·2 inches.

§2.08. Composite charges.

A point of general interest is the method of allocating the charges in the case of howitzer cartridges made up with two sizes of propellant, the so-called composite cartridges.

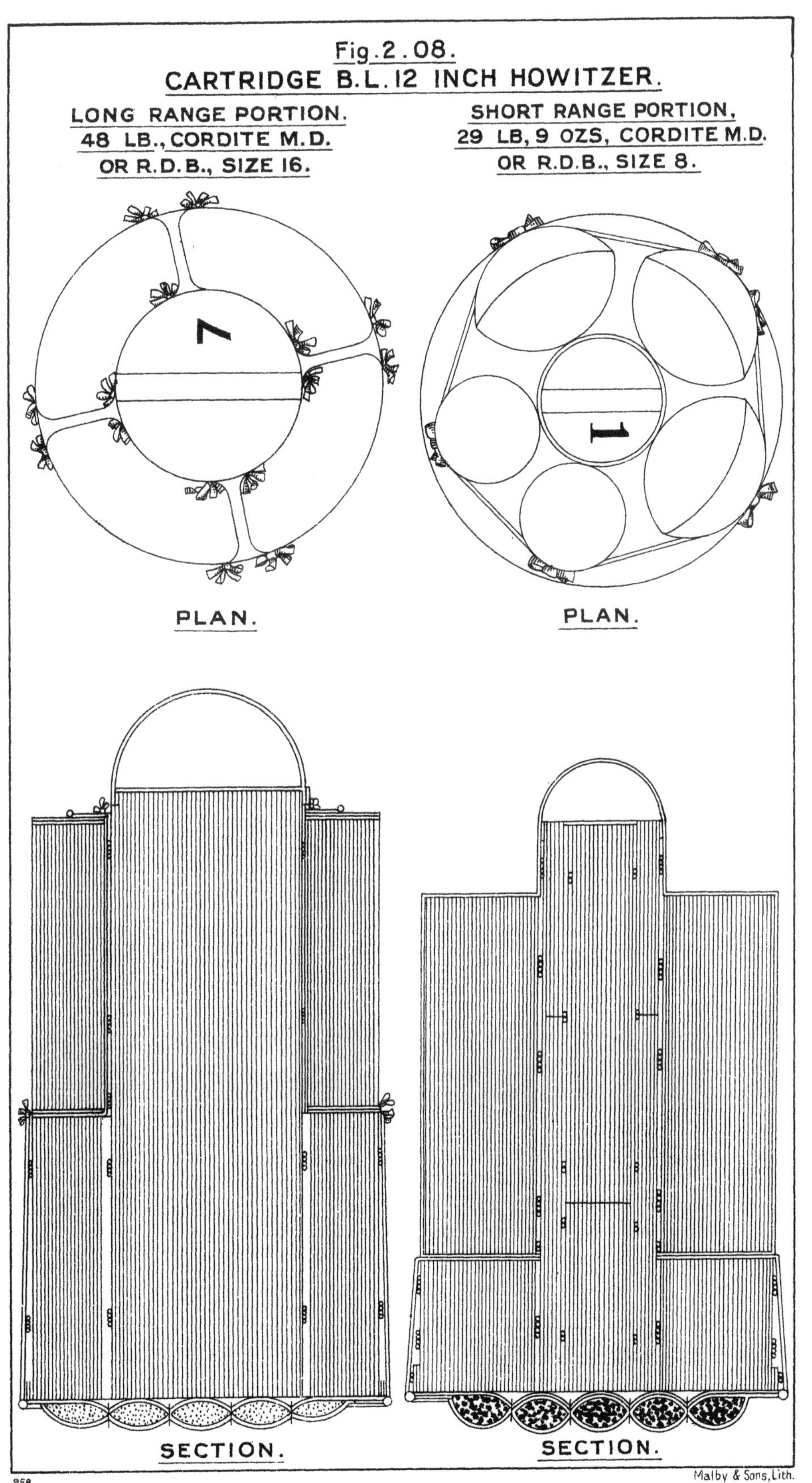

Malby & Sons, Lith.

In these cartridges, the smaller size is invariably used for the main section and also on occasions for some of the auxiliary sections for the following reasons :—

When the lower charges are used, the absolute rate of burning is decreased owing to the lower density of loading and, consequently, lower chamber pressures. Therefore, when the same size is used throughout, the point of complete combustion will move nearer the muzzle as the charge is reduced.

To secure complete combustion in the length of the bore, and preferably at a point well within the muzzle, is a matter which requires consideration when the lower charges are being selected, as it has an important bearing on the regularity of the shooting.

It is to balance this progressive movement of the point of complete combustion towards the muzzle as the charge is lowered, that with certain charges it is found advisable to use a smaller size of propellant in the section or sections used for the lower charges. Owing to the relative increase in the ratio of exposed surface to the mass of the charge, the rate of burning is thereby accelerated throughout the length of the bore.

It is also interesting to note that, with such a charge, the chamber pressures, and therefore the shock on the shell, will be relatively higher as compared with the weight of the charge.

By using composite charges, it is found that the shooting is less affected by wear in the gun and by variations in the temperature of the charge.

These advantages are probably the result of the higher chamber pressures referred to above.

Notwithstanding these advantages, the use of composite cartridges is being discontinued in many cases. There are difficulties in the use of two different lots of propellant, and in the making up of the cartridge.

By careful selection of the charges, the disadvantages of one size of propellant can be largely obviated, particularly if the lower charges are not small as compared with the full charge.

Cartridge, B.L. 12-inch howitzer.—An arrangement which secures the advantages of composite charges without imparting their drawbacks is found in the cartridges for the 12-inch howitzers.

Two cartridges are used: one, for use at short ranges—the "short-range portion"—is made up of a relatively small size of propellant; the other, for use at long ranges—the "long-range portion"—is made up of large-size cordite.

§2.09. N.C.T. charges.

N.C.T. was used to a considerable extent during the Great War as a propellant. Cartridges were introduced for the 6-inch, 8-inch,

9·2-inch and 12-inch B.L. howitzers, 4·5-inch Q.F. howitzer, and the 18-pr. Q.F. gun.

Usually N.C.T. cartridges give the same ballistics as one or other of the cordite cartridges, and may be used alternatively with these. It is necessary, however, to allow for the difference in the effect of temperature on the two propellants, the effect on N.C.T. being markedly the less.

The design of N.C.T. cartridges is influenced more or less by the loose character of the propellant. They are generally similar in design to the cordite charges in use with the howitzer, with minor modifications introduced to give rigidity and to compensate for the loose character of the propellant. These modifications are carried out by stabbing through the mushroom head, stalk, and the auxiliary sections in various places with doubled silk sewing, after the bags have been filled.

The cartridge used with the 6-inch howitzer represents a complete departure in design. In this case the charge is enclosed in a plain cylindrical bag of the full diameter of the cartridge. The sections are superimposed and suitably secured together; an igniter is provided on the outer end of the main section.

§2.10. Flash-reducing charges.

Compared with gunpowder, modern propellants produce little or no smoke, but they have the drawback of giving muzzle flash.

In the case of cordite and ballistite, this flash is most pronounced; with N.C.T. it is less so, though distinctly visible. Its intensity varies with the gun; in some cases it is so low as to be practically negligible.

The cause of muzzle flash seems to be the high temperature of the gases which escape from the muzzle. A large proportion of these gases are incompletely burnt, and on meeting the air they spontaneously inflame.

Muzzle flash is more marked with propellants containing a considerable proportion of nitroglycerine than with those composed chiefly of nitrocellulose.

To reduce this flash, the addition of various chemical substances has been tried. The idea is to use a substance, the decomposition of which requires a great deal of heat, so that the temperature of the muzzle gases may be lowered.

Flash-reducing charges were issued in the Great War for use with N.C.T. charges for the 6-inch and 8-inch howitzers.

A charge consisted of a small quantity of flash-reducing powder (equal parts of sodium oxalate and R.F.G. meal powder) enclosed in a batiste bag. It was loaded into the chamber in advance of the cartridge. No variation in ballistics was experienced when these charges were used. There was, however, a tendency to produce fouling of the bore, but this was readily removed by water.

§2.11. Super charges

were used during the Great War, when a special objective required a higher M.V. and greater range than could be obtained with the ordinary full charge. This cartridge was issued as a single charge to be loaded in place of the usual service full charge.

Augmenting charges added to the normal full charge served the same purpose as the super charge. An augmenting cartridge was a small separate portion which was added to the full charge.

§2.12. Charges for the prevention of coppering

are employed to prevent or remove coppering of the bore of B.L. and Q.F. guns. The material used is tinfoil—an alloy of tin and lead—which is placed in the gun when loading.

Action.—The action of the tinfoil appears to be as follows. On firing, the alloy melts and is sprayed on the bore of the gun. The tin combines with the copper deposited in the bore, to form a brittle alloy, which is swept away by the next projectile. The lead plays the rôle of a lubricant.

Insertion of tinfoil into cartridges :—

Various methods have been employed, and may be found in cartridges as follows :—

(*a*) In strip form, woven in among the sticks in the front of a Q.F. charge.
(*b*) Sewn in the form of flat discs to the front of the bag of a B.L. cartridge.
(*c*) Strips wound spirally round the bundle of propellant.
(*d*) Rectangular sheets wound round the propellant.

Cartridges made up in this way are distinguished by having the word " FOIL " stencilled on them in black.

NOTE.—If coppering is excessive, double the usual amount of tinfoil may be ordered ; in such cases the charge of tinfoil is designated as a " decoppering charge " and the cartridge is marked " DEC."

If it is desired to employ tinfoil in cases where it has not already been incorporated in the cartridge, it can be obtained in strip form, and the required amount crumpled into a loose ball and thrown into the front of the chamber before a B.L. cartridge is loaded.

It is important to insure that no tinfoil is placed at the rear of the cartridge of a B.L. gun or tinning of the mushroom head will occur.

IGNITERS

§2.13. General functions of igniters.

Secured to one or both ends of every B.L. cartridge is an igniter consisting of a small charge of G.12 powder enclosed between shalloon discs. The cartridge is loaded into the gun with the igniter nearest to the end of the vent.

An igniter is required as the flash from the tube is not sufficiently powerful to ignite modern propellants with certainty and regularity. To effect this the powder must be well distributed throughout the igniter, and should completely fill the stitched compartments provided, *i.e.* there must be no danger of the powder shaking down to the bottom of the pockets.

To obtain certainty of ignition, it is most important that igniters should be protected from damp. The sealed case or cylinders in which cartridges are packed should not be opened sooner than necessary and, if opened, the lids or covers should be replaced and resealed with fresh luting.

§2.14. Standardized igniters.

The old type of igniter was cross-stitched radially into two or more compartments.

It is now obsolescent and the standardized igniter is generally employed.

This type is cross-stitched into parallel compartments which are filled with R.F.G.2 or G.12.

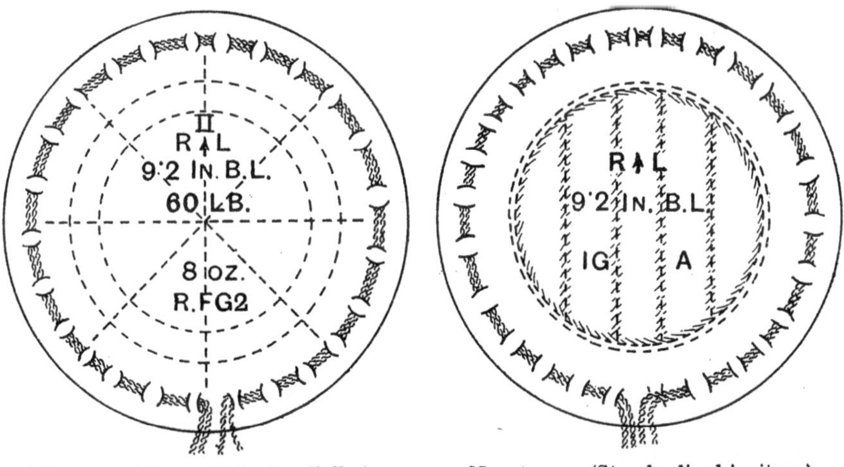

Old type. (Cross-stitched radially.) New type. (Standardized igniters.)

FIG. 2.14.

The igniter consists of two discs of shalloon between which gunpowder is enclosed. A silk cloth disc is sewn to the underside of the igniter to prevent the cordite from penetrating the shalloon. The outer disc is dyed red or is otherwise distinctively marked.

Standardized igniters for each calibre of gun may differ in weight and dimensions. The governing factor is the diameter of the charge to which they are attached. The weight of powder charge shown on the igniter (old type) is only the minimum weight which must be got in ; the igniter must be well filled irrespective of weight.

The nomenclature of standardized igniters is determined as follows :

A separate lettered series commencing with " A " is used for each calibre of gun or howitzer. A change in the size of the igniter or nominal weight of powder advances the nomenclature to the next letter, or, if the change is small, a star may be added to the letter in use.

§2.15. Auxiliary igniters

are issued in the field for use with cartridges where the igniters are suspected of damp.

At present they are only issued for the 60-pr. B.L. gun and 8-inch and 9·2-inch howitzers.

They are fitted over the end of the cartridge and are secured by a draw-string.

A special igniter is used with the 2·75-inch gun, in India only, which is filled with guncotton yarn ; these igniters, if they have become damp, can easily be dried and made serviceable.

§2.16. Igniter covers.

With the larger natures of igniters for gun cartridges 7·5-inch and above, covers are issued. During transport, the cover prevents chafing of the outer shalloon disc by contact with the cylinder; at the gun, in some cases, the cover protects the igniter from the back-flash of a previous round.

The cover consists of a millboard disc covered with silk cloth which projects over the edge. It is threaded with a draw-string by which it is secured to the igniter end of the cartridge. The ends of the draw-string are tied with a single bow and the running end of the bow is tied to a " tear-off " tab sewn to the cover.

The cover is removed before loading by sharply pulling the " tear-off " tab. This unties the bow and releases the draw-string.

A red cross is printed on the outer face of the cover to distinguish the igniter end of the cartridge.

§2.17. Q.F. igniters.

In Q.F. cartridges in which an adapter and a tube, instead of a primer, are used as the means of ignition, an igniter is also used to ensure that the charge is fired.

The earlier form of igniter consisted of a small charge of G.12 powder in a shalloon bag held centrally in the charge by a small cylinder of cordite. This form of igniter, and the cordite cylinders, are now obsolescent.

In the later form the powder is retained in a brass container (metal igniter) pierced with flash holes.

The adapter used with this metal igniter is made of bronze.

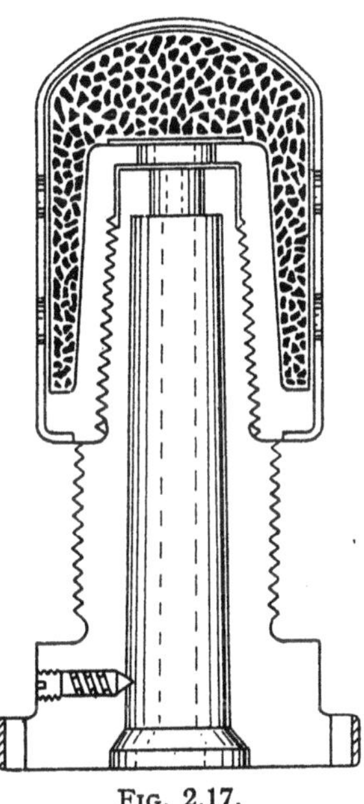

Fig. 2.17.

Q.F. Metal Igniter with Mark VI Adapter.

A small spring plunger prevents the tube from falling out of the adapter during loading.

B.L. CARTRIDGES

§2.18. Construction.

B.L. cartridges are normally of cylindrical shape.

In nearly all cases the propellant used is cordite in the form of sticks. The sticks are formed into bundles tied with silk sewing, or, in the case of the heavier cartridges, with silk-braid, in order to secure rigidity of shape and facilitate insertion into the bag.

The charge is inserted into a bag. The open end is then closed by the igniter being sewn on to it. Where two igniters are used the second igniter is sewn to the bottom end of the bag before insertion of the charge.

Lifting beckets of braid are fitted to the larger cartridges to assist handling operations.

Sub-division of cartridges.—For howitzers, cartridges are necessarily sub-divided into part charges, viz. 1st (lowest), 2nd, 3rd, etc.

For guns sub-division is only resorted to for two purposes :—

(1) Excessive weight and size of the full charge in the larger natures.

(2) The use of a full and reduced charge in one gun ; as, for example, in the 6-inch gun, where half or two-thirds charges may be required for practice or other purposes, and it is found most convenient to make the full charge up from two separate charges in the required proportion. These are tied firmly together with the ends butted by means of fairleads and lashings.

§2.19. Materials.

The requisite qualities of materials for cartridge cloth are :—

(1) They must be strong enough to stand the wear and tear of transport and handling.
(2) They should not deteriorate in store or be affected by the chemicals in the charge.
(3) They must be entirely consumed when the gun is fired and must leave no smouldering particles.
Accidents have occurred from this cause, generally when using blank cartridges or reduced charges, so that special precautions are laid down and should be carefully adhered to when using these cartridges.

The material for igniters must, in addition to possessing the above qualities—

(1) be so close in texture that the powder cannot come out ;

(2) be permeable to the flash from the tube.

Generally speaking, *silk cloth* fulfils the conditions for cartridges and *shalloon* for igniters.

Exceptions to be noted are :—

(1) *Cream serge* has been used as an alternative to *silk cloth.*

(2) *Shalloon* is used for light weights of cartridges such as the 6-inch 26-cwt. howitzer and the B.L., 2·75-inch

(3) *Cambric* is used for Q.F. howitzers.

Cream serge may be used for the protecting disc on the underside of the igniter, and for the cover of the igniter.

(4) The lifting beckets may be of *bleached cotton webbing*, or *silk braid.*

(5) In the 9·2-inch and 12-inch cartridges, the all-round lifting becket has been omitted for land service.

(6) *Silk webbing* is used in lieu of silk braid in making up the larger natures of cartridges, 7·5-inch and up.

§2.20. Diameter and length.

Diameter.—In diameter the cartridge should be large enough to ensure that some part of the powder in the igniter will encounter the flash from the tube.

Length.—The cartridge should be long enough to ensure that, under all conditions, the igniter will be close to the mushroom head, or at least that the distance between the tube and the igniter is not excessive.

In order to obtain the best possible regularity in ballistics, it is essential that the length of the cartridge should, if possible, approximate to the full length of the chamber.

For example : to meet the above two requirements, the cartridge for the 6-inch star shell is built up in dumb-bell shape.

§2.21. Cartridges for use in the field

for B.L. guns are usually made up in one charge, if possible, for convenience in handling, etc.

During the Great War a new departure was made by designing reduced charges for some of the smaller guns to give howitzer effect at short ranges and also to save the life of the gun.

§2.22. Cartridges for coast defence

are usually made up into one or more portions, for convenience in handling in the case of the larger natures, and also that a reduced charge may be used for practice purposes.

§2.23. Cartridges for star shell

are generally reduced charges. It is found that the action of these shell is improved thereby, especially at short ranges ; the remaining velocity required at burst is generally much below that which would be given by a full charge, also a more curved trajectory is required.

§2.24. Drill cartridges.

Cartridges, drill, are made to the same shape, weight, and dimensions as the Service cartridge they represent. Drill cartridges are generally stamped with the usual marking to be found on the Service cartridge, and, where necessary, fitted with lifting beckets, laced together, etc., similarly to the Service cartridge ; the higher natures are not, however, fitted with discs to tear off, but the igniter end is represented by a red cross.

§2.25. Design of B.L. howitzer cartridges.

Considerations governing design.—Cartridges for B.L. howitzers are made up of the same materials, but in a manner entirely different

Fig. 2·18.

CARTRIDGE B.L. 6 INCH GUN.

23 LB. CORDITE M.D. OR R.D.B. 16, MARK V, |L| SILK CLOTH;
FULL CHARGE CONSISTING OF 2/3 CHARGE OF 15 LB. 5. 1/3 OZ & 1/3 CHARGE OF 7 LB. 10. 2/3 OZ. LACED TOGETHER.

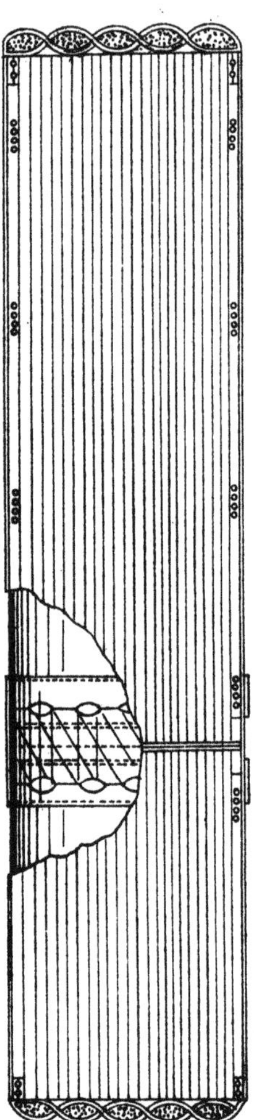

SECTION.

END VIEW.

Fig. 2·20.

CARTRIDGE, B.L. 6 INCH GUN.

12 LBS. CORDITE M.D. II MARK I/C.

SCALE 1/4.

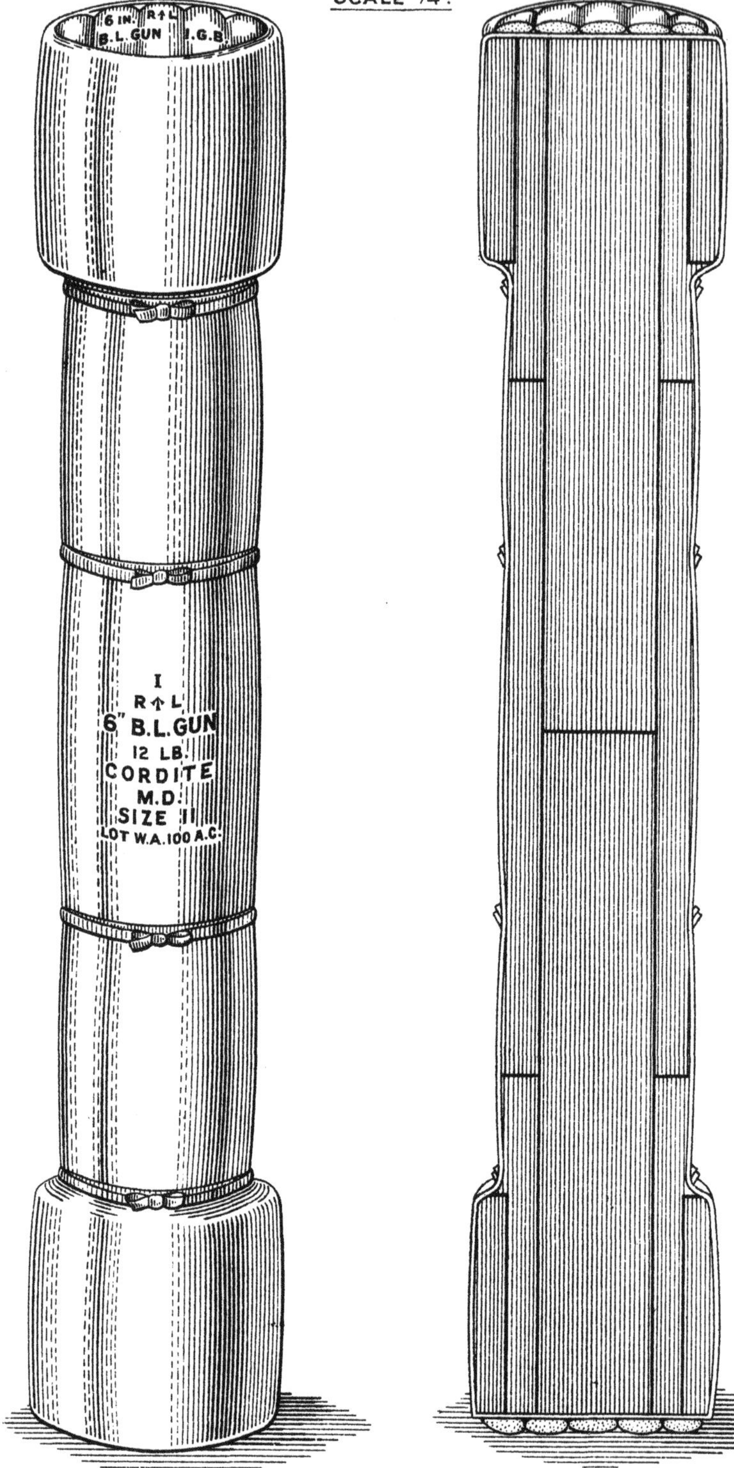

6 IN. R↑L
B.L.GUN I.G.B

I
R↑L
6" B.L.GUN
12 LB.
CORDITE
M.D.
SIZE II
LOT W.A.100 A.C.

950,

Malby & Sons, Lith.

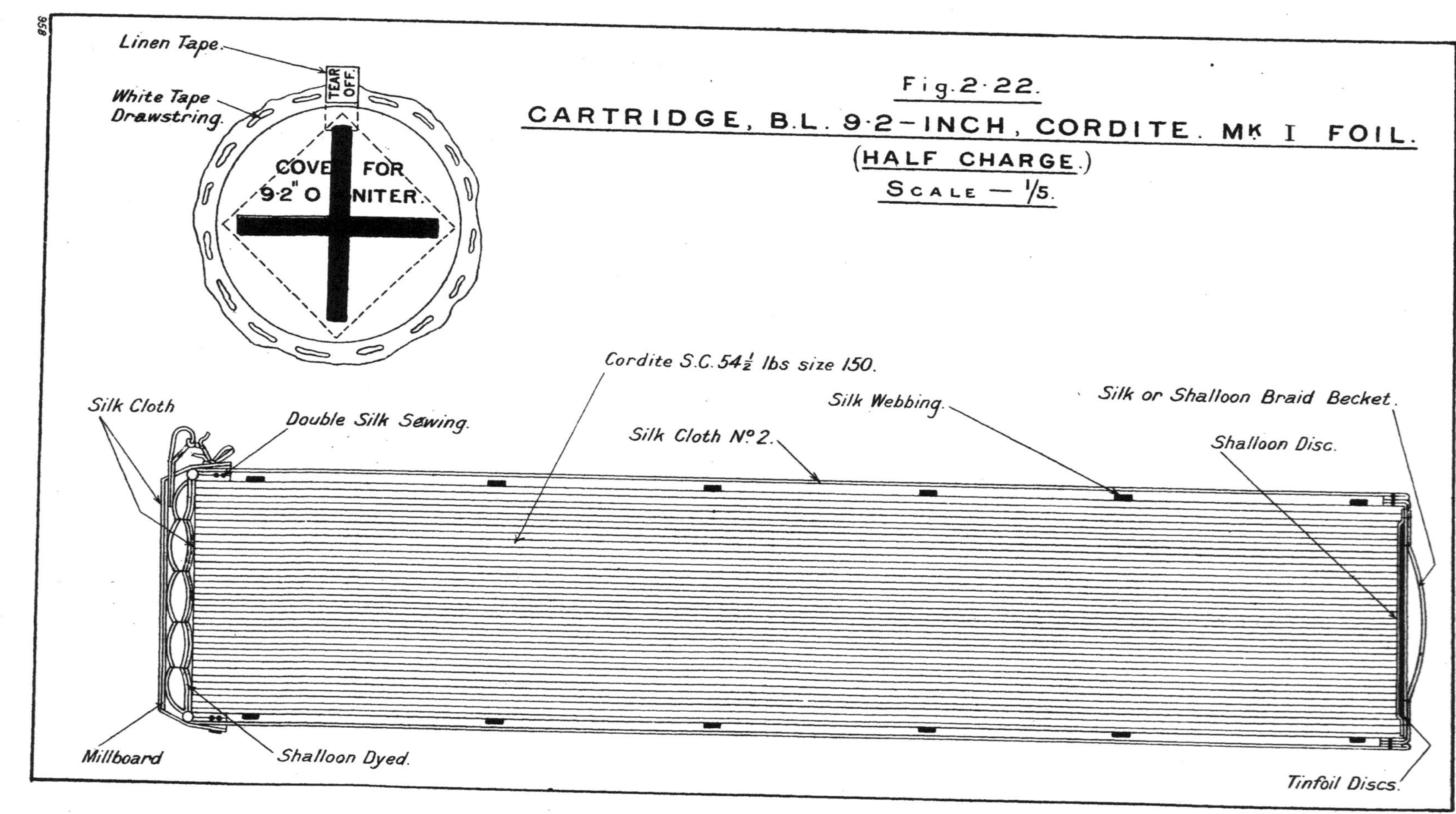
Fig. 2·22.
CARTRIDGE, B.L. 9·2-INCH, CORDITE. MK I FOIL.
(HALF CHARGE.)
SCALE — 1/5.
Linen Tape.
TEAR OFF.
White Tape Drawstring.
COVE FOR
9·2" O NITER.
Cordite S.C. 54½ lbs size 150.
Silk Cloth
Double Silk Sewing.
Silk Cloth No. 2.
Silk Webbing.
Silk or Shalloon Braid Becket.
Shalloon Disc.
Millboard
Shalloon Dyed.
Tinfoil Discs.
958

Fig. 2·25.

CARTRIDGE, B.L. 9·2 IN. HOWITZER 23 LB. 12 OZS. CORDITE M.D.T. OR, R.D.B.T. 20-10., MARK III.

WEIGHT OF CORDITE	LBS.	OZS.	DRS.
No. 1 SECTION (CORE)	8	1	0
No. 2 "	2	2	8
No. 3 "	2	11	8
No. 4 "	3	5	4
No. 5 "	3	15	12
No. 6 "	3	8	0
TOTAL	23	12	0

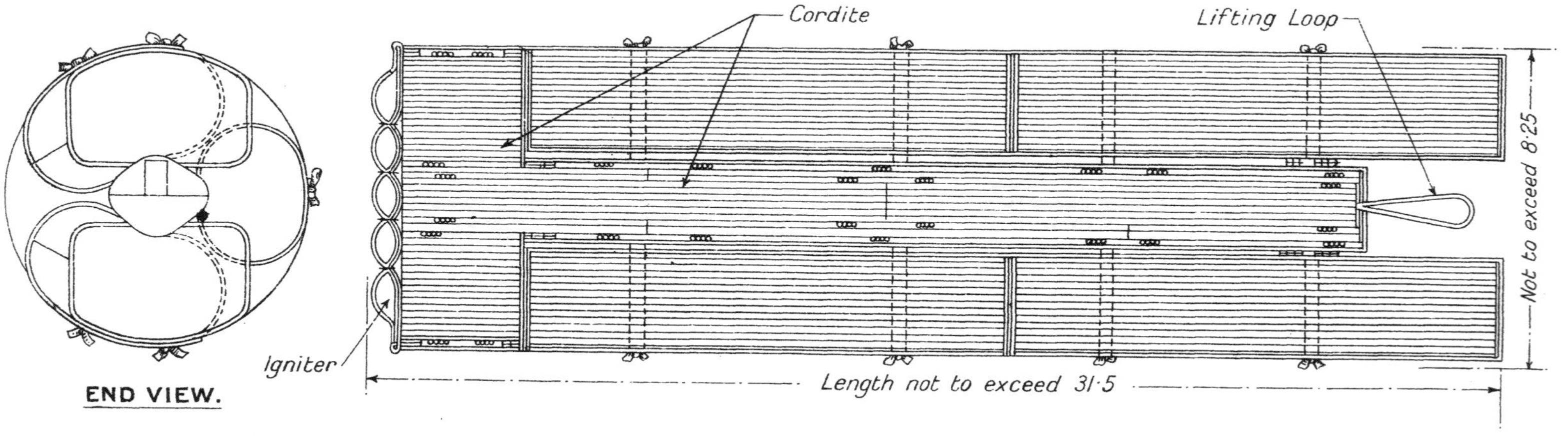

END VIEW.

958

Malby & Sons, Lith.

from that employed for B.L. gun cartridges, for the following reasons :—

(1) A howitzer is intended for high-angle fire, hence the charge is much lighter compared with that of a gun of the same calibre, viz. the B.L. 6-inch gun has a charge of 23 lb., the B.L. 6-inch howitzer has a charge of 4 lb. 11½ oz.

(2) A howitzer is a comparatively short piece of ordnance, therefore the charge must be quick-burning (small size of cord or tubular cordite) so that the whole of the charge may be consumed before the projectile leaves the muzzle. For this reason N.C.T. and ballistite have proved satisfactory as propellants with howitzers. The B.L. 9·2-inch gun takes cordite M.D., size 26. The B.L. 9·2-inch howitzer takes M.D.T., size 20–10.

(3) With a howitzer, large angles of descent may be required at short as well as long ranges. These are obtained by altering the weight of the charge; therefore, a cartridge for a howitzer is made up in sections in such a manner that its weight can be readily reduced.

Principles of design.—The main principles in design are :—

(1) The lowest charge must be stable, *i.e.* does not tend to fall over when loaded alone and so cause a missfire.

(2) The igniter, which is always fitted over the lowest charge, is made of the largest possible diameter; this is to ensure not only that the flash impinges directly on to the powder, but also that the igniter as nearly as possible covers the whole of the cartridge and causes good ignition of the charge.

For this reason the diameter of the rear end of the charge is not less than the maximum diameter of the complete cartridge, whenever possible.

In most of the cartridges a mushroom-shaped lowest charge is therefore used.

For howitzers of 6-inch calibre and above, a number of cartridges have been introduced in which, while the mushroom-shaped core is retained, the auxiliary sections are formed not as rings, but as segments to fit round the stalk.

The substitution of segments for rings is desirable because—

(1) In the segments the lengths of the sticks are arranged parallel to the core, the interstices forming a number of parallel passages proceeding from the igniter; the whole surface of the charge is therefore in the most favourable position to receive the igniting flash.

It is rendered necessary

(2) by reason of the larger sizes of cordite used in the cartridges, as it is a matter of some difficulty to bend bundles of a size larger than 4¼.

With the introduction of high-velocity howitzers with long but narrow chambers, a very long stalk is employed which is much longer than is necessary for the stability of the lowest charge used. The reason for this is, that it has been found that the firing of a small charge, seated at the end of a long chamber, is liable to cause excessive pressures due to " wave action."

So with guns and howitzers having long chambers, where small charges are used, the charge is arranged to stretch the whole length of the chamber, or as nearly so as possible.

The long stalk is very convenient for accommodating the desired number of auxiliary sections, which can be conveniently arranged round it in two tiers.

Paper collars.—In the case of 15-inch howitzers, now obsolete, it was found necessary to use paper collars in order to centre the core charge in the chamber.

This was done at the expense of the overall ignition of the complete cartridge, since the igniter only covered the cylindrical core, and the auxiliary charges lay external to it; and since the auxiliary sections formed a considerable portion of the charge, the arrangement was not conducive to regular ballistics.

This design also has the disadvantage that the whole of the paper collars may not be discharged from the howitzer on firing. Portions may lodge in the bore and cause a jam on the next round being rammed home, or a premature may be caused when the next round is fired.

With the 15-inch howitzer, in which the rate of fire was slow, precautions could be taken against the above defects.

This design was used with the first issues of cartridges for the 8-inch and 12-inch howitzers. Later, these cartridges were remodelled for the reasons given above.

§2.26. Arrangement of howitzer cartridges.

Each section of a howitzer cartridge is marked with a number to be used when the different charges are being made up.

The core or main section, which carries the igniter, is always marked No. 1; the auxiliary sections, Nos. 2, 3, 4, and so on.

Different charges are obtained by removing the auxiliary sections in the order of their numbers; thus, in the case of a cartridge comprising a core (No. 1 section) and four subsidiary sections (Nos. 2, 3, 4 and 5 sections), the charges made up from this are:—

Fifth charge	Nos. 1, 2, 3, 4, and 5 sections.
Fourth charge	Nos. 1, 2, 3, and 4 sections.
Third charge	Nos. 1, 2, and 3 sections.
Second charge	Nos. 1 and 2 sections.
First charge	No. 1 section.

The weights of the sections are selected in such a manner as to provide for all ranges. Charges made up from other combinations are therefore unnecessary.

Adequate overlap in the range is provided between consecutive charges. This suggests that, at the point of overlap, the lowest charge should, where possible, be selected, in order to conserve the life of the howitzer; it also has the advantage of giving a larger angle of descent.

§2.27 Markings on B.L. cartridges.

B.L. gun cartridges.

B.L. gun cartridges are marked on the side in black printers' ink as follows :—

(1) On the *front* :—

- (*a*) Mark of cartridge.
- (*b*) The word " FOIL " if they contain charges for the prevention of coppering.
- (*c*) Initials or monogram of the manufacturer of the empty bag.
- (*d*) Calibre of gun.
- (*e*) Weight of cartridge.
- (*f*) Nature of propellant.
- (*g*) Size of propellant.
- (*h*) Fractions $\frac{1}{4}$, $\frac{1}{2}$, etc., as applicable.

(2) On the *opposite side* of the cartridge :—

- (*a*) Lot number of the propellant.
- (*b*) Initials of the manufacturer of the propellant.
- (*c*) If the charge has been adjusted, the letters A.C. When a grouped propellant, the group letter and number will replace the letters A.C.
- (*d*) Initials of the filling firm or monogram of the filling station.
- (*e*) Date of filling (month and year).

The initials indicating the natures of propellant are as follows :—

" Cordite " ... ... for cordite Mark I.
" M.D." ... ... for cordite M.D.
" M.D.T." ... ... for cordite M.D.T.
" R.D.B." ... ... for cordite R.D.B.
" R.D.B.T." ... for cordite R.D.B.T.
" R.D.B.S.F." ... for R.D.B. square flake.
" Bal " ... ... for ballistite.
" N.C.T." ... ... for nitro-cellulose.
" S.C." ... ... for solventless cordite.
" W " ... ... for cordite W.
" W.T." ... ... for cordite W.T.

Markings on standardized igniters :—

(*a*) Maker's initials.
(*b*) Calibre of gun.
(*c*) " IG " followed by a letter denoting the pattern of the igniter.

Marking on " protecting disc " or " cover for igniter "

The " cover for igniter " on the new type of cartridge is marked with the calibre of the gun and the letter of the igniter with which used.

Example :—COVER FOR
9·2-inch " O " IGNITER.

Special marking on gun cartridges :—

Red band on B.L. 60-pr. 6-lb. 6-oz. M.D. or R.D.B. size 11, cartridges reduced charge, the stencilling being in red instead of in black.

B.L. howitzer cartridges.

On each portion :—

(*a*) Number of portion (in words or figures) on the top or side, as the case may be.
(*b*) Mark of cartridge.
(*c*) Manufacturer's initials or monogram of the manufacturer of the empty bag.
(*d*) Calibre of howitzer. If the cartridge is for a specified mark of howitzer, the mark of howitzer is given.
(*e*) Weight of total charge.
(*f*) Weight of portion.
(*g*) Nature and size of propellant.
(*h*) Lot number of propellant.
(*i*) If the cartridge has been adjusted, the letters A.C. When a grouped propellant is used, the group letter and number will replace the letters A.C.
(*j*) The words " super charge " on such charges.

On the opposite side of No. 1 section :—

(*a*) Initials of firm filling or monogram of filling station.
(*b*) Date of filling (month and year).

Igniters :—

(*a*) Contractor's initials or recognized trade mark.
(*b*) Calibre of howitzer. IG. " A," " B," or " C " on larger natures of howitzers denoting the nature of standardized igniter.

In some cases, however, the initials of the filling station or contractor may be omitted. The initials on the bag of the cartridge will then be taken to apply also to the igniter.

Special marking on howitzer cartridges.

The 12-inch How. Mk. II 48-lb. cartridge (in five parts) has two red bands across the igniter.

Marks of cartridge.—The factors which control the nomenclature of a cartridge are :—

(*a*) Weight of charge.
(*b*) Nature and size of propellant.
(*c*) Pattern of igniter.
(*d*) Arrangement of charge or alteration in dimensions.

No advance of mark is made for minor alterations, such as the use of alternative materials for the empty cartridge or the lifting band or becket, the variation of tying the cordite bundle, or the absence of an all-round lifting band.

Q.F. CARTRIDGES

§2.28. Introduction.

Cartridge cases are used in two ways, forming—

(1) Fixed ammunition.
(2) Separate ammunition.

With fixed ammunition the cartridge case is attached to the projectile : with separate ammunition the projectile is not attached to the case.

As a general rule, fixed ammunition is used for field guns and anti-aircraft ; separate ammunition is used for howitzers, and in coast defence guns except in the smaller natures, 6-prs., etc.

For field guns and howitzers, and A.A. guns, cartridge cases are fitted with a percussion primer screwed centrally into the base ; this generally contains a magazine having a sufficient charge of fine grain powder to dispense with the use of an igniter.

For coast defence guns, cartridge cases are fitted with adapters to take an electric tube. In the case of a break-down in the electric firing gear, provision is made for percussion firing, the electric tube being replaced by a percussion one.

Fixed ammunition should not be used in guns which have fired separate ammunition. In a gun which has fired separate ammunition, the chamber is usually worn away at the point where the mouth of the case is situated, and if a fixed round be then fired, the front end of the case will expand into this worn portion and will be liable, therefore, to cause a jam.

§2.29. Construction.

In the British Service the cartridge case is invariably made of standard cartridge brass (70 per cent. copper ; 30 per cent. zinc).

The cartridge is made from a brass disc by a series of cold-drawing operations. The flange is formed subsequently by a pressing or base rolling operation.

All Q.F. cases are tapered slightly towards the front, the chambers of Q.F. guns being similarly tapered. This ensures that the cartridge is well centred on insertion, and facilitates the extraction of the expanded case after firing.

The flange seats itself against the rear face of the chamber when loading, and it is against this flange that the gun extractor works when starting the case to the rear on unloading.

A screw-threaded hole is made centrally in the base of the case to take the primer or adapter and tube, which ignites the propellant in the case.

Heat treatment.—Cartridge cases are repeatedly annealed at high temperatures during the various stages of manufacture to remove stresses set up by these operations. Such stresses, if left in the cases, may in the course of time cause splits in the metal, a defect commonly but inaccurately termed " season-cracking."

§2.30. Quality of material.

The requisite properties of metal used for cartridge cases are :—

(1) Ability to stand high temperature and pressure.
(2) Sufficiently soft to expand on firing without splitting or cracking, and to act as an obturator.
(3) Should permit of being reformed to normal size repeatedly.
(4) Strong enough to resist deformation during transport, and to withstand extraction.
(5) Should not be corroded by the charge or when exposed to damp.
(6) Should not develop cracks or splits in the course of prolonged storage.

§2.31. The Manufacture of brass cartridge cases.

The first process is the manufacture of strip from a brass containing 70 per cent. copper and 30 per cent. zinc.

Ingots are cast from furnaces at about 1165° C.

After discarding the pouring head containing the dross, the ingot is passed through a series of rollings, and between certain rollings is annealed at a temperature and time of about 630° C. for 3 hours.

The resulting strip is finally annealed and cleaned.

The cartridge case is manufactured from a circular disc punched out of the strip. The process consists of a series of cold drawing operations with annealing and cleaning between the various stages. The number of draws and indents is dependent on the particular nature of the case, but in most types the various operations in general are very similar.

The following terms are used for the operations performed.

Blanking	Punching of disc from brass strip.
Cupping	First operation of making a cup from original blank.
Drawing	Extension of case by parallel cold drawing.
Annealing ...	Heat treatment between drawing operations. Temperature about 600° C. Time, about 1¾ hour.
Indenting... ...	Forming a depression in the centre of the base of the case after drawing, with the object of distributing metal in this region to form the primer hole.
Trimming ...	Removing material from the mouth of the case in a lathe after drawing.
Base rolling ...	A process used to extrude metal at the base of the case from the centre towards the outside diameter, in order to form the lip or flange on the head.
Semi-annealing ...	Heat treatment of the wall of the case. Temperature about 580° C. Time, about 2 minutes.
Tapering ...	Forming taper after the parallel drawing operation by compression into a taper die.
Necking	Reducing the diameter of the mouth of the case after tapering to the particular requirements of design in the die.
Pickling	The cleaning of cases after annealing by immersion in hot dilute sulphuric acid. Approximate strength up to 1 part concentrated acid to 40 parts water. Time, about ½-hour. Wash off in boiling-water.
Cleaning ...	Drying and burnishing case with sawdust on a running head.
Machining ...	Turning and threading of the primer hole. Finishing base of case and flange to dimensions. The interior surfaces of cases are not machined.
Special mouth annealing.	This was formerly for necked cases, but has now been extended to other types of fixed ammunition. Temperature about 550° C. Time about 1 minute.

§2.32. Testing of cases.

1. *Hardness.*—It is necessary that cases when manufactured should be of a minimum hardness. If the cases are too soft jamming is likely to occur.

Cases are tested by scleroscope on the head and on the wall below the flange after the final low temperature annealing. A minimum scleroscope figure fixed by firing trials is allotted for each type of case. Cases giving lower figures are rejected.

The part to be tested by scleroscope must not be specially burnished or prepared, or misleading results will be obtained.

2. *Freedom from internal stress.*—The final low temperature annealing removes *latent internal stresses* in the metal, which would otherwise in time lead to the development of cracks. This splitting action is known as "season-cracking"; it is accelerated by the presence of acids or alkalies.

The temperature of annealing is from 260° to 280° C. and the time is 2 hours.

One or more cases from each batch annealed is placed in a strong solution of mercurous nitrate and nitric acid. This solution will rapidly produce cracks in metal from which the stresses have not been removed. If the case does not crack within 24 hours of immersion the batch is accepted as free from internal stresses.

§2.33. Cleansing of cases.

Cases are cleaned after manufacture and before filling.

Hot caustic soda is used to remove grease, and dilute sulphuric acid to remove oxide film; the cases are then washed free of acid in hot running water and are dried in sawdust. "Congo red" paper is used as a test for acidity.

Impurities are very injurious to the stability of cordite and great care should be taken to see that all traces are removed. Lacquering of the inside of the case is a great protection in this respect.

External lacquering is also a useful protection for cases, but has the disadvantage of causing jams at high rates of fire and so cannot be used with high pressure Q.F. guns.

§2.34. Life of a Q.F. case.

Q.F. cases can be refilled after firing, but they must first be reformed and cleaned, as they expand on firing.

The operation of reforming tends to weaken the case, as a certain amount of metal is turned off the base and the lower part of the body each time.

To prevent accidents when firing with a smokeless propellant, the number of rounds a Q.F. case is allowed to fire is limited to six full rounds. Any round fired with a full gunpowder charge is counted as half a round fired with a full service charge.

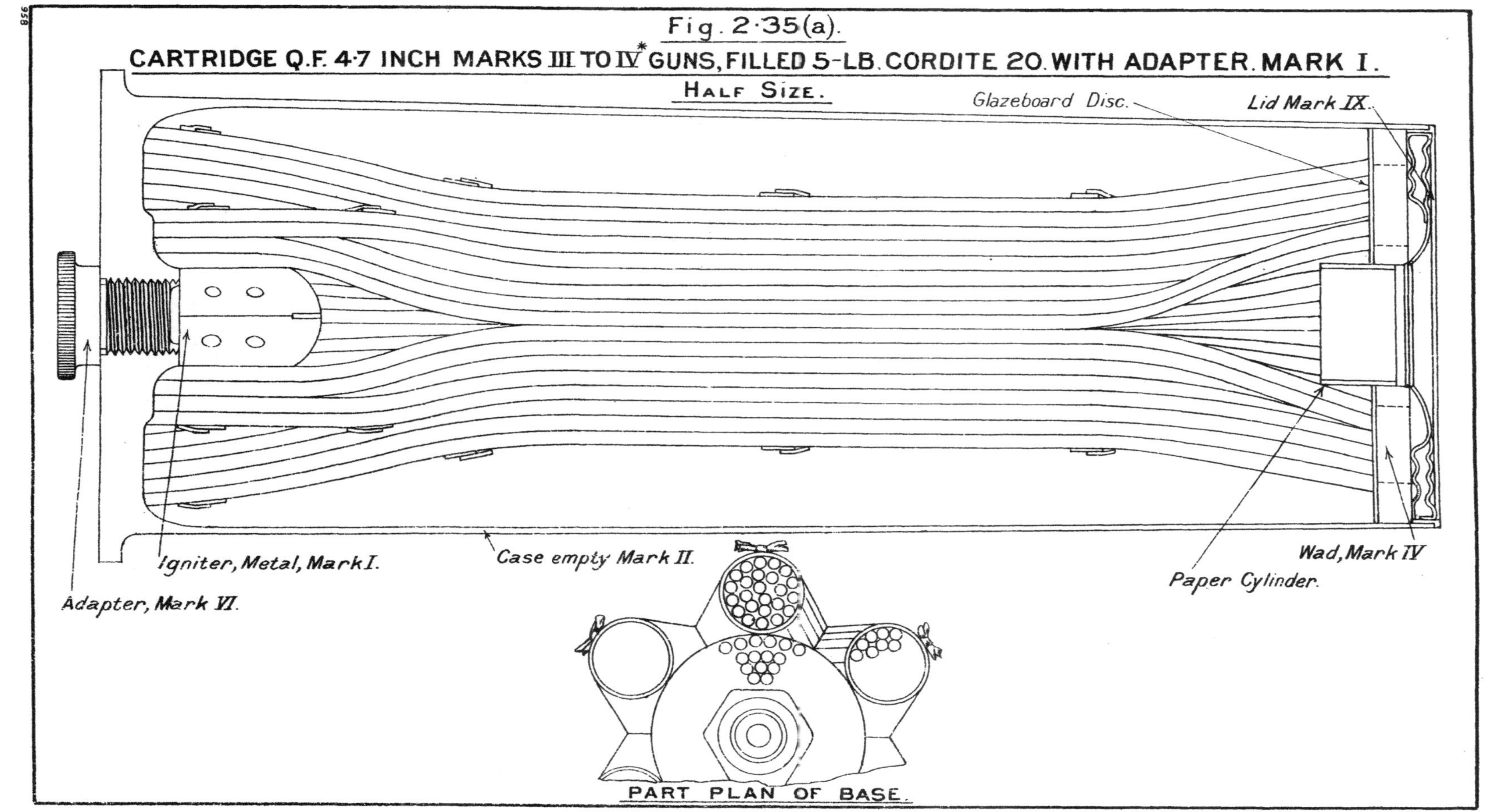

Fig. 2·35(a).

CARTRIDGE Q.F. 4·7 INCH MARKS III TO IV* GUNS, FILLED 5-LB. CORDITE 20. WITH ADAPTER. MARK I.

HALF SIZE.

958

In the case of 13-pr., 18-pr., 3·7-inch and 4·5-inch howitzer cartridge cases, the life is now determined by the thickness of the wall after repair at a fixed distance from the flange and not by the number of times they have been fired.

§2.35. Typical Q.F. cartridges.

*Cartridge Q.F. 4·7-inch, Marks III to IV**, may be taken as a typical example of separate ammunition.

The cartridge case is fitted with an adapter to take tubes to fire either electrically or by percussion.

The *case* is of the usual form of brass, and has a projecting flange at the base. A Mark VI adapter screws into the base and a metal igniter screws on to the adapter inside the case.

The *charge* is a bundle of cordite tied with silk sewing, and has a recess at the base end to fit over the igniter.

A felt wad is placed on top of the charge, and the cordite sticks are opened out in the centre to receive a paper cylinder attached to the wad. The bundle is thus kept in position in the centre of the cartridge case.

The mouth of the case is closed by a lid consisting of a cap recessed in the centre, to which a corrugated strengthening ring is soldered.

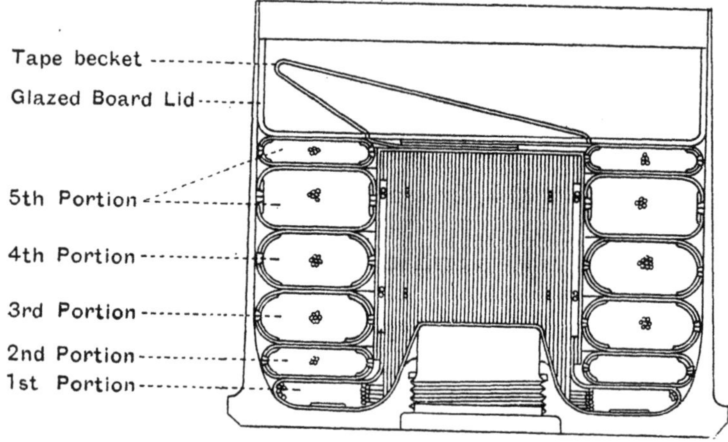

FIG. 2.35 (*c*).—Cartridge, Q.F. 3·7-inch Howitzer.

The 3·7-inch Q.F. cartridge may be taken as typical of Q.F. howitzer ammunition.

The base of the case is similar to that of the 18-pr. Q.F.

The charge consists of a core and five rings, constituting five charges. The two top rings are secured together by tape and constitute one ring for ordinary charges; if, however, star shell is to be used, the top ring (5 drs. of cordite) is discarded by untying the tape. Each ring is fitted with a loop by which it can be removed.

The mouth of the case is closed by a glazedboard cup fitted with a loop, so that it may be lifted off when it is required to remove any of the rings.

The 18-pr. cartridge may be taken as typical of fixed ammunition.

The base of the cartridge case is prepared to take a primer percussion No. 1.

The projectile is attached to the case by pressing the mouth of the brass case into a circular recess on the projectile situated just behind the driving band.

The full charge is made up in the form of a cylindrical bundle of cordite sticks tied in three places with shalloon braid. The sticks in the centre of the charge are slightly shorter than the outer ring, thus forming a recess to fit over the primer.

There is no additional igniter inside the case, as the primer gives sufficiently powerful ignition to the cordite charge.

There are various other charges, but the method of making them up does not differ greatly.

§2.36. Markings of Q.F. ammunition.

Markings for Q.F. ammunition may be considered under the following headings :—

A. Stampings all types of Q.F. ammunition.

B. Stencillings :—

(i) Symbols for adjusted charges, or when ballistically grouped.

(ii) Descriptive markings for :—

(*a*) Q.F. ammunition giving particulars of the charge.

(*b*) Q.F. fixed ammunition " batching."

(iii) Distinguishing markings for Q.F. fixed ammunition when stored in racks or limbers.

A. Stampings—all types of Q.F. ammunition.

Stampings.—These form a permanent record of certain particulars regarding the cartridge case, and are found on the base. The following are the most important :—

(1) (*a*) Calibre of gun or howitzer.
(*b*) Mark of cartridge case.
(*c*) Year of manufacture in full.
(*d*) Maker's initials or trade mark.
(*e*) Lot number of case.

(2) Particulars concerning the history of the case, including the number of charges which it has fired.

Before cartridge cases are filled with cordite, the letter " C " is stamped on the base.

According to whether the charge is " full " or " reduced," the letter " F " or " R " is also stamped after the " C."

If the propellant is removed from the case, *i.e.* not fired, the letter " F " or " R," as the case may be, is barred out.

When cartridges are filled with gunpowder (except in the case of " blank "), the case is stamped with the letter " P."

(3) *Annealing stampings.*

(*a*) The letter " A " in a circle (followed by a punch dab) indicates that the case has been subjected to *high temperature annealing* ; subsequent high temperature annealings are denoted by punch dabs after the symbol.

NOTE.—This type of annealing is not now carried out.

(*b*) *Low temperature annealing* for the removal of *internal stresses.* This is the usual method now employed, and it is denoted by letter " A " in a diamond. The manufacturer carrying out the annealing will also stamp his initials, or monogram, together with a batch letter and number. The letter indicates the year, which for 1935 is " Z ".

This type of annealing is only carried out once.

The previous high temperature symbol, initials, batch letter and number will be barred out before low temperature annealing is carried out.

(*c*) *Unannealed cases.*—Q.F. 13 and 18-prs. ; 4·5-inch and 3·7-inch howitzers, for which annealing in the past was omitted during repair, bear the letter " U " in a circle.

(*d*) *Special mouth annealing fixed ammunition.*—This is indicated by the letter " L " in a circle. This type of annealing was originally introduced for " necked " cases, and has subsequently been extended to " coned " cases.

(4) *Miscellaneous stampings, inspection marks, workmarks, etc.*

(*a*) Punch dabs near the calibre, indicate repairable defects.

(*b*) If the flange has been formed by a hammer, the letter " H " is stamped.

(*c*) The letter " S," or " S " in a diamond, indicates that the case has been tested by an instrument (scleroscope) for hardness.

(*d*) Brazing stamp. If the case is split near the mouth and it is considered that it can be repaired by brazing, a circle with a double arrow is stamped. After repair, it is barred out.

(*e*) Cases accepted with annular defects, orange peel effects, etc., are indicated by a small cross in a triangle.

(*f*) The numbers and letters " 1P," " 2P," etc., indicate cases selected for proof purposes. The letter P followed by symbols " — " or " • " indicate that a case has fired respectively one proof or one service charge.

Other small stampings such as initials of examiners, work-marks, acceptance marks, etc., may be found, and are chiefly for use in the factories.

B. Stencillings.

Symbols for adjusted charges, or when they are ballistically grouped. See Fig. 2.36 (*b*).

(*a*) Q.F. cartridges, separate loading (except Q.F. 3·7-inch and 4·5-inch howitzers); Stencilled on the side or base of the case (and formerly, also written or stamped on the label affixed to the cup or lid of the cartridge case).

(*b*) Q.F. fixed ammunition, and Q.F. 3·7-inch and 4·5-inch howitzer cartridges. Stencilled on the side or base of the case.

In the case of Q.F. 18-pr. ammunition assembled prior to January, 1921, in which a grouped propellant was used, neither the letters " A.C." nor the propellant group number and letter were shown.

The letters " A.C." may also be found painted on the side of the case, with the ballistic group letter and number stencilled (by the silver nitrate process) on the base. The latter is the current approved practice.

Descriptive markings for Q.F. ammunition, giving particulars of the charge. Fig. 2.36 (*b*).

The undermentioned " descriptive " particulars were, in the past, stencilled on all natures of filled Q.F. ammunition.

Originally these markings were placed in red paint (black for blank and paper shot) on the base of cartridge cases; at a later date, however, this was superseded by stencilling the particulars by the " silver nitrate " process on the side of cases.

This system is still current, except that for " batched " ammunition the batch particulars only are stencilled.

The particulars concern the propellant charge. They are:

(*a*) Nature of propellant charge.

(*b*) Size of propellant charge.
For cartridges 6-pr. and below, the size of the propellant is not shown, and, when cordite Mark I is used, the nature of the propellant is omitted.

(*c*) Maker's initials (propellant).

(*d*) Lot number of propellant charge.
The letter " R." may be found occasionally after the Lot number of the propellant indicating that the cordite has been reworked.

(*e*) The letters " A.C." if the charge has been adjusted. If, however, the charge has been ballistically grouped, the letters " A.C." will be omitted from the side, and the group symbols stencilled on the base.

(*f*) Mark of filled cartridge.

(*g*) Monogram of firm, or initials of filling factory.

(*h*) Date of filling (month and year).

Fig. 2·36(a).

REPAIRED CASES.

TYPICAL SKETCH OF STAMPINGS ON BASE.

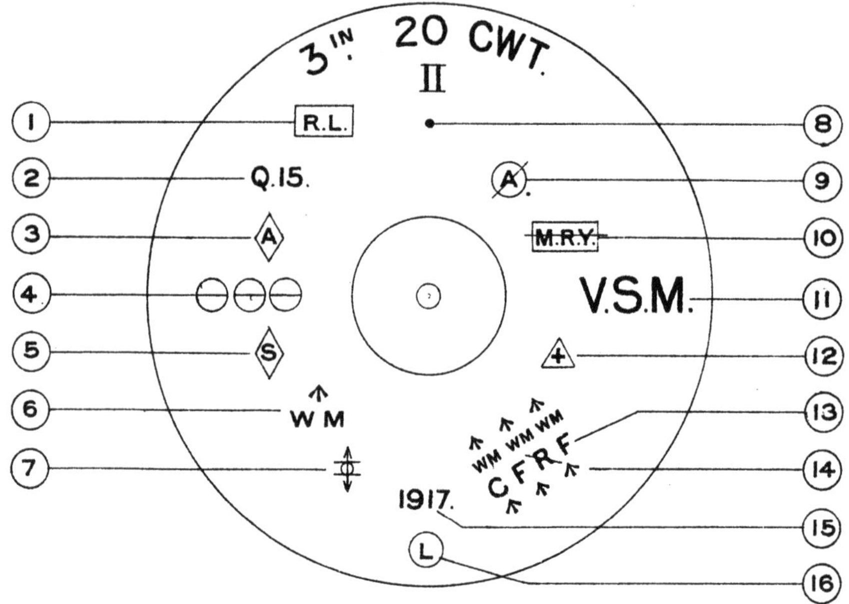

(1) *Monogram of Repair Contractor*

(2) *Low Temperature Annealing Batch Letter and Serial Nº*

(3) *Low Temperature Annealing Stamp.*

(4) *Original Lot Number.*

(5) *Scleroscope and Mercurous Nitrate Test Stamp.*

(6) *Examiner's Work Mark.*

(7) *Brazing Stamp, barred out after repair.*

(8) *Punch Dab, Rectifiable Defects.*

(9) *Previous Annealing Stamp.*

(10) *Previous Repairer's Monogram etc.*

(11) *Original Contractor.*

(12) *Accepted with Annular Grooves.*

(13) C = *Cordite Filling.*
F = *With Full Charge.* } *Each time filled*
R = *With Reduced Charge.* } *(as applicable).*
If broken down without Firing, the last indicating letter barred out.

(14) *Acceptance Stamp after Repair.*

(15) *Date of Manufacture.*

(16) *Special Mouth Annealing, fixed Ammunition.*

958

The following additional particulars for the natures enumerated :

Separate loading ammunition (except Q.F. howitzers), weight of charge.

Q.F. 13 and 18-prs. The letters " RED " denoting reduced charge.

Q.F. 3-in. 20-cwt. (*a*) Weight of charge. (*b*) The word " FOIL " if containing tinfoil for the prevention of coppering.

Descriptive markings for Q.F. fixed ammunition (" *batching* ").

Batching.—The advantages of this system are as follows :—

(i) Ammunition failures can be traced. Full particulars of components are more easily available.
(ii) It eliminates repetitions of stencilling.
(iii) To simplify ammunition supply on the line of communication and ensure continuity of performance at the guns.

Batches are *numbered* consecutively, the propellant Lot, and a complete Lot of fuzes (usually 2,000 in the case of time fuzes), governing the batch number.

The type of ammunition is distinguished by a *letter* ; the letter for each type will be found below.

Sub-batches.—If there are not sufficient fuzes in a Lot to complete a batch, then other small Lots may be added to the batch to complete it. The inclusion of these different Lots are signified by stencilling the *letters* A, B, C, etc., after the batch number.

Example :—" BATCH, C.38.B." stencilled on the side of the cartridge case would denote :—

C	Meaning	Full charge smoke.
38	,,	The 38th batch.
B	,,	That a 3rd Lot of fuzes was used.

If components are withdrawn from a batch or sub-batch, then this is distinguished by the letter " X " being stencilled on the box after the batch number, or sub-batch letter ; if the component is a fuze, then the letter " X " will also be shown on the cartridge case.

Care should be taken to keep batches and sub-batches together, and, if rounds are removed from boxes for any reason, they must be repacked in their original boxes.

Should it be necessary to ascertain particulars of any component, the label inside the lid of the box gives a full description.

The distinguishing letters adopted to indicate each type of " batched " ammunition are as follows :—

FULL CHARGES.

A	Shrapnel.
B	H.E.
C	Smoke.

FULL CHARGES—*continued.*

E ... ... A.P. and steel.
F ... ... Target.
G ... ... Case shot.
H ... ... H.E. streamline.
J ... ... Smoke, streamline.

REDUCED CHARGES.

L ... ... Shrapnel.
M ... ... H.E.

PRACTICE FULL CHARGES.

N ... ... H.E.
P ... ... A.P.
Q ... ... Common (base fuze).
R ... ... C.N.F. (nose fuze).
S ... ... Practice projectile (filled).
T ... ... Practice (weighted) and practice shot, and sub-calibre.

PRACTICE REDUCED CHARGES.

U ... ... Practice projectile.

PRACTICE BURST SHORT CHARGES.

V ... ... Shrapnel.
W ... ... Practice projectile.

Distinguishing markings for Q.F. fixed ammunition when stored in racks or limbers.

(*a*) In Q.F. fixed ammunition filled, converted or repaired before 1st December, 1929, the nature of the cartridge was indicated by *painting* the base of the cartridge case (except the cap or primer) as in the following table :—

Nature of Cartridge.	*Base of Cartridge.*
Armour piercing	Blue.
High Explosive—	
Full charge	Yellow.
Reduced charge ...	Yellow, with 1½-inch white band across base.
Burst short practice ...	Yellow, with two white quadrants.
Streamline	Yellow, with white ring ½-inch wide on outside of base.
Streamline reduced ...	White ring ½-inch wide on outside of base, remainder half yellow, half black.

Shrapnel—	
Full charge	Unpainted.
Reduced charge ...	White.
Burst short practice ...	White, two quadrants, remainder unpainted.
Smoke	Green.
,, Streamline ...	One green, three white quadrants.
,, Target	Black.
3-inch 20-cwt.—	
Practice projectile ...	Unpainted, with black ring $\frac{1}{2}$-inch wide round outside of base.
Burst short charge ...	White, two quadrants, remainder unpainted.
6-pr. cartridges for 6-cwt. guns	Bronzed.
3-pr. 2-cwt. case shot and practice projectile ...	Unpainted.

(*b*) In ammunition filled, repaired or converted *subsequent to 1st December*, 1929, " distinctive " markings are indicated by stencilling (silver nitrate process) on the base of cartridge cases according to the following system :—

Nature of Cartridge.	*Lettering.*
Armour piercing	AP
High explosive	HE
Reduced charge	R
Practice	PRAC
Practice special	PRAC. SPL.
Burst short practice	BS
Practice special, burst short charge	PRAC. SPL. BS
Streamline	SL
Shrapnel	SHP
Smoke	SMK
Target	TGT
Tracer	(T)
Tracer fuze	(TF)

In addition, BX (in conjunction with SMK) denotes the presence of a smoke box, FL the presence of a flash producer.

DEC indicates the inclusion of a decoppering charge.

§2.37. Packing of cartridges.

It is essential that B.L. cartridges should be packed in airtight cases to protect the igniters from moisture.

The primers or igniters of Q.F. cartridges are protected by the brass cases, but these also, in most cases, are packed in airtight boxes.

As a rule, cases are made airtight by sealing the joints of the lid with luting, and sometimes in addition by dermatine washers.

B.L. cartridges are normally packed in zinc, brass, tinned-plate or steel cylinders, and tinned plate or copper lined cases or boxes.

If for any reason it is necessary to use temporary wooden cases without linings the package should be lined with non-absorbent paper.

The packages are :—

(1) B.L. cartridges, 6-inch howitzers and B.L. guns below 6-inch :—

Cases, powder, metal-lined, or boxes, cartridge.

(2) 6-inch guns and howitzers above 6-inch :—

Cylinders, cartridge, which are protected by " cases wood, skeleton."

(3) Q.F. separate ammunition :—

(*a*) Boxes cartridge, made of wood and lined with tin.

(*b*) 3·7-inch and 4·5-inch howitzer cartridges are packed each in a tinned-plate cylinder and packed in wood boxes.

(*c*) Q.F. 4·5-inch ammunition may also be packed in " boxes ammunition " containing two complete rounds.

(4) Q.F. fixed ammunition :—

Boxes, ammunition, which may be either of wood or steel.

Chapter III

TUBES

§3.01. Definition.

Definition.—Tubes are the means by which the propellant charges of B.L. guns and howitzers are ignited. They are also used with certain Q.F. guns where the cartridge cases are fitted with adapters.

§3.02. Tubes general.

All tubes used in modern ordnance are of vent sealing type, *i.e.* they are designed to seal the escape of the propellant gases rearward through the vent (or adapter) in which they are fired. Such escape, if permitted, causes erosion of the vent.

Erosion is a continuous and progressive process, which creates increasing difficulty in efficient sealing and the subsequent extraction of fired tubes.

The sealing of the vent is effected as follows :—

1. The head of the tube is supported during firing by the striker, lock, or T vent, which prevents any movement of the tube to the rear.
2. The tapered body of the tube is a close but sliding fit in the tube chamber. On firing, the thin wall of the tube expands against the interior of the vent, and so prevents the escape of gas between the tube and the chamber.
3. Sealing arrangements are provided to seal the escape of gases rearward through the tube itself. These devices may consist of copper balls or cones being forced into a coned seating or the expansion of a copper cap supported internally in the tube, etc. They differ with the various types of tube.

The action of the tube in its vent, or adapter, is, therefore, analogous to that of the Q.F. cartridge in the gun chamber; in fact, the former is merely the reproduction of the latter in miniature.

§3.03. Initial ignition of the charge.

The flash from the tube should be as strong and as regular as possible to ensure the thorough ignition of the cartridge igniter.

As the distance between the rear end of the cartridge and the front end of the vent varies with the type of gun and the charge in use, the strength of the flash must be sufficient for the maximum distance that may be met with.

Tubes are, therefore, made in varying sizes.

The smaller tubes are used principally with the smaller guns, require less flash-carrying power and generally differ from the larger tubes only as regards magazine capacity.

The latest approved fillings are as follows :—

Tubes, vent, percussion, ·4 and ·5-inch.	Solid R.P. or G.20 gunpowder pellets.
Tubes, vent, electric, ·4 and ·5-inch.	R.P. or G.20 gunpowder intermixed with solid pellets.
Tubes, percussion, S.A. cartridge.	Solid R.P. or G.20 gunpowder pellets, or a mixture of grain and pellets.
Tubes, percussion " T "	R.F.G2. or G.12 gunpowder.
Tubes friction	R.P. or G.20 gunpowder.

Later marks of ·5-inch tubes, viz :—the percussion Mark V, electric Mark VIII, and " S " Mark VII, have a perforated powder disc to separate the G.20 or the priming, as the case may be, from the main filling of solid pellets and grains.

There are, however, a number of tubes still in the service with other natures of gunpowder fillings.

§3.04. Packing of tubes.

All tubes are affected to a greater or less degree by damp, which deteriorates the powder-filling and the cap composition.

In consequence, tubes are packed in tin boxes, the lids of which are hermetically sealed to the bodies by soldered tin tear-off bands.

The boxes should never be opened until actually required. The lids of the boxes containing unexpended tubes should be replaced and resealed by means of adhesive tape.

§3.05. Nature of Tubes.

Tubes may be divided into four main groups :—

1. Tubes, friction.
2. " T " tubes.
3. Tubes, percussion, S.A. cartridge.
4. Tubes, vent, percussion and electric.

§3.06. Tube, friction, Mark IV.

This tube is used for firing puffs indicating flash, for lighting safety fuze (instead of a fuzee), and for firing certain old S.B. guns.

It consists of a body made from solid drawn copper, fitted to the top of which is a copper *nib piece.* The nib piece contains a roughened copper *friction bar,* with an eyelet formed at the outer end. The roughened part of the friction bar is primed with a detonating composition made of chlorate of potassium, sulphide of antimony and ground sulphur, mixed with thin shellac varnish, a small quantity of which is smeared on the bar.

Fig 3·07.

TUBE, PERCUSSION "T" MARK I.

SCALE 1/1.

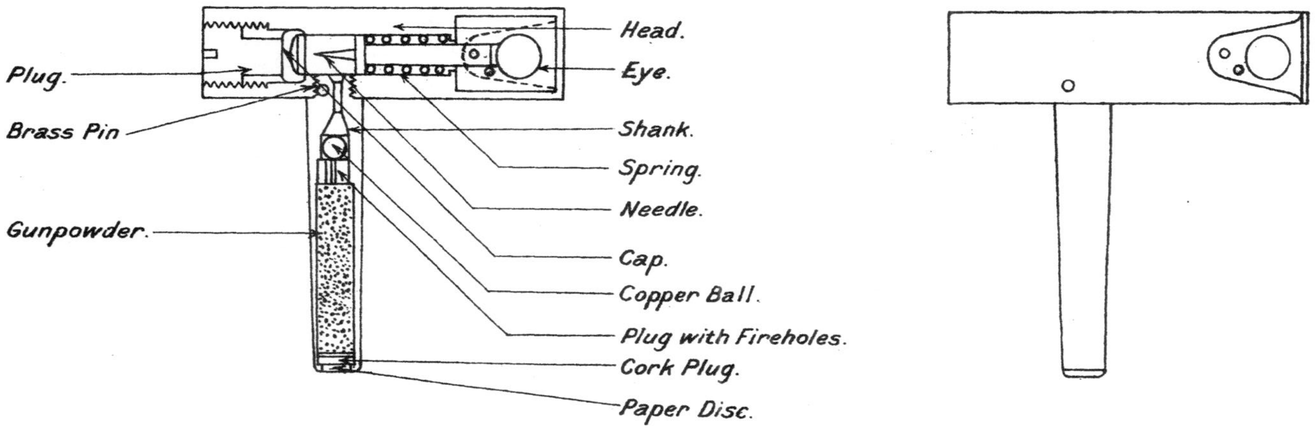

Plug.
Brass Pin
Gunpowder.
Head.
Eye.
Shank.
Spring.
Needle.
Cap.
Copper Ball.
Plug with Fireholes.
Cork Plug.
Paper Disc.

958

The body of the tube contains R.P. or G.20 gunpowder, and the bottom is closed by a cork plug, to which isa ttached a white fine paper disc. The tube is lacquered all over.

Action.—The tube is fired by pulling with a lanyard hooked to the eyelet; this draws the friction bar from the nib piece, thus igniting the detonating composition on the friction bar, the flash from which passes through a small hole in the nib piece and ignites the powder in the tube.

Packing.—The tubes are packed 24 in a tinned-plate cylinder, the outside of which is painted black.

§3.07. Tube, percussion, " T."

There is now only one pattern of " T " tube extant in the land service, viz. tube, percussion, " T." The tube, friction " T," introduced for B.L. guns and howitzers, having " T " vents, and the tube, friction, " T " push, introduced for use with B.L.C. equipments, are now obsolete.

The tube, percussion, " T " is used with heavy howitzers of calibre 9·2-inch and upwards, which are provided with a " T " vent in the breech mechanism, by which the tube is held in place.

The construction of the tube may be seen from the diagram.

Action.—The tube is inserted in the vent, and the head turned so as to engage and lock under the " T " head of the vent.

The lanyard is hooked to the eye-piece, and pulled strongly and steadily. The needle is drawn back, compressing its spring, until, at the limit of its travel, the wire connecting the needle to the eye-piece is sheared.

The needle is then released and, flying inwards, fires the percussion cap. The flash passes round the copper ball to the magazine.

The explosion forces the copper ball into its coned seating, and expands the body of the tube against the walls of the vent, so sealing the escape of gas through the head of the tube, or between the vent and tube body.

It is important to ensure a straight pull of the lanyard as, if unduly inclined to the axis of the needle, the latter may bend and jam, so causing missfires.

Re-filling of tubes.—Fired tubes, if sound, are fitted with the necessary components and re-filled; a star is then added to the numeral in the nomenclature.

Packing.—The tubes are packed ten in a square tin box painted black; there are removable lids at both ends sealed by tin strips soldered on.

The tubes are packed inside the box in such a manner that five tubes are still hermetically sealed after the removal of one lid.

During the Great War, an emergency box was issued unpainted. and with one lid only.

The ground of the exterior label on the box is orange.

Failure of " T " tubes.—During the Great War the following defects were observed in percussion " T " tubes.

(*a*) Deterioration of filling due to damp.

(*b*) Low filling in percussion cap, or defective cap.

(*c*) Roughly finished needle, or actuating spring not strong enough to force needle on to cap. Diameter of coil of actuating spring too large and pressing against the inner diameter of head of tube.

(*d*) Needle breaking or being too difficult to withdraw.

With the large output required in war, it was found impossible to ensure immunity from the above defects, and consequently the tube, percussion, S.A. cartridge was introduced for the majority of B.L. field equipments.

§3.08. Tube, percussion, S.A. cartridge. Mark II.

This tube is simply a ·303-inch, S.A. cartridge, containing a cap filled with ·6 grains of cap composition covered with a lead foil disc and varnished.

The details of the filling of the magazine can be seen from the Fig. 3.08.

The design is simple and cheap to manufacture, and gives a more powerful flash than the " T " tube.

In its present form it is unsuitable for use in equipments which give high or sustained firing pressures.

In the first instance, the tube was used in conjunction with a rifle-firing mechanism ; a special lock and slide box is now provided.

The exterior of the tube is not coloured in any way, and is distinguishable from the cartridge S.A., ·303-inch rifle grenade, which it closely resembles, in that the latter is either blackened all over, or for half its length from the mouth.

Action.—The striker of the firing mechanism fires the percussion cap, and the gunpowder in the tube explodes, giving a powerful flame to ignite the igniter of the B.L. cartridge.

The gas pressures expand the tube, and the copper percussion cap acts as a gas check to prevent the escape of gas through the base.

Packing.—The tubes are packed, heads and tails, in two brown paper packages, each containing 10 tubes ; both packages are contained in one tinned-plate box, which is painted black.

The ground colour of the label on the lid is white.

§3.09. Tubes, vent, percussion and electric.

These brass tubes are used with all B.L. ordnance provided with a suitable lock or striker mechanism, and with Q.F. equipments which use cartridge cases fitted with adapters.

The tubes are made in three sizes ; ·4-inch, ·5-inch and 1-inch.

Their shape resembles that of the Q.F. cartridge, and the body is manufactured on the same lines, *i.e.* drawn up from a disc, headed to form the flange, and tapered. The underside of the flange is bevelled to assist extraction.

Identification of tubes.

Electric tubes.—The exterior of the body of an electric tube is left plain and smooth.

Percussion tubes.—Four notches are cut out of the rim of the head, so as to enable percussion tubes to be distinguished from electric by touch.

§3.10. Tube, vent, percussion, ·4-inch, Mark VIII.

The construction of the tube may be seen from the figure.

The *body* is made of brass, drawn up from a disc. The head portion is bored out and screw-threaded to take the percussion mechanism.

Below this the solid diaphragm is machined to form an anvil for the cap, and two longitudinal holes are drilled through it to communicate the flash from the cap to the magazine.

The remainder of the body is hollow, and forms the magazine; the walls taper towards the front to permit of ready expansion on firing.

The *percussion cap* comprises a copper shell containing about ·33 grains of detonating composition "A" mixture; the surface of the composition is varnished.

The cap is placed in a recess in the cap holder, which is screwed home over the diaphragm, so that the cap is directly over the anvil.

The *striker* is seated in the striker holder, and secured to it by a copper shearing wire.

The *striker holder* is screwed into the head of the tube, being seated above the cap holder.

A brass flanged sleeve is fitted into the head of the striker stem and secured by riveting over, thus increasing the area of the striker head and allowing of a limited amount of eccentricity of the striker of the gun firing mechanism without risk of missfires.

The powder charge of G.20 gunpowder pellets is inserted into the magazine from the mouth end, which is closed by a cork plug, to each side of which is shellaced a disc of whited brown paper.

The mouth of the tube is burred over to secure the plug.

Action.—Assume the tube to be in the vent and the breech properly closed.

The needle of the lock or striker impinges on the head of the tube striker, which is driven forward, breaking the shearing wire, on to the cap.

The detonating composition thus crushed between the cap and anvil, is fired. The flash passes through the flash holes to the powder in the magazine.

The explosion expands the body outwards, sealing the escape of gas between it and the vent; the copper cap, maintained in place by the striker and striker holder, prevents the escape of gas through the head of the tube.

Mark VII differs from Mark VIII in that it has a brass washer snapped into the head of the striker in place of the flanged collar, and the magazine is filled with R.P. or G.20 gunpowder.

Tubes, vent, percussion, ·5-inch, Marks II and IV are identical except in dimensions, with the Marks VII and VIII just described, respectively.

The ·5-inch Mark V magazine is partly filled with 11 grains of G.20 which is covered by a powder disc, enclosed in fine white paper; the remainder is filled with a mixture of pellets and grains.

Packing. Ten tubes are packed, heads and tails, in a flat tin box, and the lid is secured by a tear-off band soldered on. The tear-off band should not be removed until the tubes are actually required for use.

Each box is painted black externally, and on the outside of the lid is pasted a white label with the printing in red.

§3.11. Tubes, vent, electric.

Firing electrically is the normal method employed in the Royal Navy and the coast defences.

It gives a greater rapidity of fire than the percussion method.

A tube typical of this nature is the tube, vent, electric Mark X.

§3.12. Tube, vent, electric, ·4-inch, Mark X.

The body is manufactured similarly to that of the V.S. percussion tube. The head (or rear end) being machined out internally to take the circuit and sealing devices, whilst the mouth or forward end forms the magazine.

Internal arrangements.—The solid metal in the head is machined out from the rear end to accommodate a screwed ebonite cup, and a tapered seating from the mouth end for the copper sealing plug.

These two seatings are connected by a small hole drilled centrally through the head, which forms the passage for an insulated tinned copper wire.

The wire is first passed through the copper sealing plug, which is then drawn on to it to make a gas-tight seal. The front end of the wire is bared of insulation, and forms one pole of the electric bridge.

A second pole is fitted into the front of the copper plug, and an iridio-platinum wire, giving a resistance 0·9 to 1·1 ohms, is soldered at each end to the heads of these poles, thus forming the bridge.

The forward end of the plug is shaped to form a gas-check.

The rear end of the insulated wire, after passing through the central hole in the tube, is bared of its insulation and coiled down on to a thin copper washer, which lies at the bottom of the ebonite cup in the head. A contact disc of tin and antimony is pressed into the ebonite cup to a definite depth below the head of the tube.

Filling the tube.—A small quantity of guncotton dust is pressed in round the bridge and retained in position by a perforated glazed-board cup covered with paper.

The magazine is filled with R.P. or G.20 gunpowder intermixed with powder pellets, and closed in a similar manner to percussion tubes.

Action.—Assume the tube to be in the vent, the breech closed, and the needle of the lock or striker resting on the contact disc of the tube.

On the circuit being completed through the pistol of the electric firing gear, the current passes from the battery through the striker to the contact disc, down the insulated wire, through the iridio-platinum bridge to the small copper pole and the body of the tube, back to the battery.

The bridge becomes incandescent, and ignites the priming composition and powder charge.

The body of the tube expands against the vent and prevents the gas escaping between itself and the vent. Escape of gas through the head is sealed by the copper plug. This plug is forced rearwards, the rear end fitting into the small coned seating in the body, the gas-check on its forward end expands and helps in the sealing.

Packing.—Ten tubes are packed in a flat tin box similarly to the percussion tubes, but the printing on the outside label is in black.

To prevent corrosion, tubes in open boxes may be lightly coated with mineral jelly, taking care not to coat the contact disc. Before use the tubes should be wiped clean, as mineral jelly is an insulator and might interfere with the passage of the firing current.

NOTE.—" R " following the mark of the tube denotes that it has been found electrically defective, but has been repaired.

Repaired tubes should be allocated for testing the circuits of guns.

§3.13. Tube, vent, electric, ·5-inch, Mark VIII.

This is similar, except in dimensions, to the electric ·4-inch, Mark X tube just described, but the filling differs in having a powder disc instead of a glazedboard cup.

§3.14. Drill tubes.

These are supplied for instructional purposes. They are merely empty service tubes with their mouths closed.

The bodies of these tubes are not coloured, but have four longitudinal indents ; the heads of the tubes are milled.

§3.15. Markings of tubes.

The following markings will be found on the head of a tube :—

(1) Mark of tube.
(2) Manufacturer's initials or recognized trade mark.
(3) Acceptance mark (↑).

Dummy tubes have the word "DUMMY" stamped on the head.

§3.16. Defects in tubes.

Defects in tubes may be grouped under two headings :—

(A) Lack of care in storage and handling, causing deterioration of the filling, broken bridges in electric tubes, etc.
(B) Defective manufacture, which includes bad metal, faulty machining, incorrect assembly and filling.

(A) Tubes may fail to function owing to the powder being damp and failing to give sufficient flash to ignite the charge, or the cap composition having deteriorated.

Electric tubes should be tested carefully before issue for firing in order that faulty tubes due to defective or broken bridges may be eliminated. If dropped by accident, they should not be used until retested.

These faults may be obviated by care in storage and handling.

(B) The most common causes of failure under this heading are due to faulty metal. Tubes are probably the most highly stressed war stores of their size in the Service, and it is necessary to ensure that the brass used is perfectly sound and free from impurities, and is properly treated at every stage of manufacture.

The chief failures and their causes are :—

1. "*Splits.*"—These are longitudinal slits or fissures. They are due either to unsound metal at the point of fracture, or to laps or folds in otherwise sound metal, caused in rolling down the ingot preparatory to manufacture.

An unduly hard tube may split in the wall on expansion, but such cases are extremely rare.

2. "*Bursts.*"—This defect takes the form of a hole or tear. It is nearly always due to unsound metal, but may be caused by laps or folds.

3. "*Separations.*"—The tube is broken transversely across, either partially or completely. This is now of very rare occurrence, and is usually due to metallic defects or to faulty machining, the metal having been thinned down unduly at the point of fracture.

It occasionally occurs in the tube, percussion, S.A. cartridge which is very thin in the wall, and is generally due to the lock face not fully supporting the head of the tube, which sets back on to it on firing, whilst its expanded mouth remains against the vent walls, thus creating considerable tensile stress on the body.

4. "*Blow-through.*"—This denotes the escape of gases through the head of the tube. It may vary in form from a slight escape, due to defective sealing of the cap or defective threads in the sealing mechanism (percussion tubes), to the contents of the head being completely blown out. The latter seldom, if ever, occurs in electric tubes; in percussion tubes, it is usually due to a weak cap being penetrated by the gases, the pressure of which is sufficient to strip the threads of the sealing arrangements in the head.

§3.17. Employment of tubes.

Field service.

Nature of tube	Marks of tube	Equipments
‡Tube, percussion, " T "	I† and II.	B.L. 2·75-inch gun.
Tube, percussion, S.A. cartridge.	I and II.	B.L. 60-pr. gun. B.L. 6-inch 26-cwt. howitzer. B.L. 6-inch, Mark XIX gun. B.L. 8-inch howitzer. B.L. 9·2-inch howitzer. B.L. 12-inch howitzer, Mk. IV.

Coast defence.

Nature of tube	Marks of tube	Equipments
Tube, vent, electric, ·4-inch. or alternatively Tube, vent, percussion, ·4-inch.	VI† and X. VII† and VIII.	Q.F. 12-pr. gun. Q.F. 4-inch, Marks III and III* guns. Q.F. 4·7-inch gun. Q.F. 6-inch gun. B.L. 6-inch, Marks VII and XI guns. B.L. 9·2-inch gun. B.L. 12-inch gun.
Tube, vent, percussion, ·5-inch.	II†, IV† and V.	B.L. 6-inch, Mark XXI gun. B.L. 15-inch gun.
Tube, vent, electric, "S" ·5-inch.	VIII.	B.L. 15-inch gun.

† Obsolescent.

‡ Tube, percussion, "T" may still be found for certain guns and howitzers "9·2" and above, the vents of which have not yet been modified to take the tube, percussion, S.A. cartridge.

Railway mountings.

Nature of tube	Marks of tube	Equipments
‡ Tube, percussion, " T "	I† and II	B.L. 12-inch howitzer, Marks III and V
Tube, vent, electric, ·5-inch.	IV and VIII.	B.L. 18-inch howitzer.
Tube, vent, percussion, ·5-inch.	II†, IV† and V.	B.L. 18-inch howitzer. B.L. 14-inch gun.

Miscellaneous.

Tube, friction ...	IV	Pin puff powder; also for firing radial vented R.M.L. and R.B.L. guns.

† Obsolescent.

‡ Tube, percussion, " T " may still be found for certain guns and howitzers "9·2" and above, the vents of which have not yet been modified to take the tube, percussion, S.A. cartridge.

CHAPTER IV

PRIMERS

§4.01. Definition.

A primer is the normal means of igniting the propellant charge in a Q.F. equipment.

It is screwed down flush into the rear end of the Q.F. cartridge case.

§4.02. Electric primers.

These were used in the past but their use has now been discontinued, the only remaining one being the cartridge, aiming, rifle, 1-inch electric primer. This primer will be seen illustrated with the cartridge on page 234.

§4.03. Percussion primers.

Percussion primers are used in Q.F. equipments as follows :—

Primer	Used in
Primer, percussion, Q.F. cartridge No. 1, Mark II.	Q.F. 13- and 18-pr. guns. Q.F. 3-inch, 20-cwt. guns. Q.F. 3·7-inch howitzers. Q.F. 4·5-inch howitzers.
No. 2, Marks III, IV and VII	Q.F. 3-pr. and 6-pr.
No. 5, Mark II	Q.F. 1½-pr. and 2-pr. (Royal Navy)
Nos. 11 and 15	Q.F. cartridges containing certain new type propellants.
No. 12	Q.F. 1½-pr. (Royal Air Force).
Primer, percussion, blank, Mark III	For Q.F. 6- and 3-pr. blank ammunition. (For details, *see* Fig. 4·05(*a*).)
Cap, Mark II	6- and 3-pr. ammunition with the old type of cartridge case. It has now been replaced by the No. 2 primer.

The essential features of all present types of percussion primers are :—

1. The cap, which is inserted from the mouth end, and seated so that it is practically flush with the rear face of the primer.

2. The anvil, which is screwed in after the cap, and is provided with flash holes for the flash from the cap to pass through.

3. The magazine, containing the gunpowder, which increases the flash and so ignites the propellant charge in the cartridge case.

The magazine may be either machined out of the front end of the primer itself, as in the No. 1 type, or take the form of a separate container screwed into the body of the primer as in the No. 2.

A further feature is the ball seal, which functions in a chamber formed between the anvil and a perforated plug or washer separating it from the magazine.

This is only essential for such primers as may be used with Q.F. guns in which the pressures are high or sustained. Thus, in the No. 1 Mark II primer, a ball seal is used to ensure effective sealing. Among the smaller primers, it is used in the Marks IV and VII patterns of the No. 2, as an additional but not an essential safeguard, for which room cannot be provided in the Mark III pattern, the cap itself being sufficiently strong to seal escape of gas through the body.

§4.04. Markings on primers.

Primers are stamped on the head with the following markings (*see* Fig. 4·07):—

(i) Number and Mark of primer.
(ii) Manufacturer's initials or recognized trade mark.
(iii) Acceptance mark (after filling).
(iv) Filler's initials or recognized trade mark.
(v) Date of filling (month and year).
(vi) Lot number.
(vii) " M " indicates repair and refilling.
(viii) Workmarks.

§4.05. Primer, percussion, Q.F. cartridge, No. 1, Mark II.

The *body* of the primer is of brass, Class A or B, and is formed with a flanged head, in which two slots for the key are cut.

A portion of the body is screw-threaded externally to screw into the base of the cartridge.

The percussion and igniting arrangements in the body consist of the cap, anvil and magazine.

The *copper cap* is coated with R.D. cement, prior to insertion in the body, to seal the joint in the cap seating.

The brass *anvil* is screwed home on to the cap.

The copper ball is inserted and held in position by the brass closing plug, which is screwed home after it.

A fine white paper capsule or disc is then secured over the internal boss of the primer.

The magazine is filled with G.12 powder, and is closed by a brass closing disc, to the inner face of which a paper disc is shellaced.

Six radial slits are cut in the closing disc, so that on firing the cut portions of the disc open outwards, the disc itself being held in by the mouth of the primer which is turned over on to it.

As a protection against damp, R.D. cement is applied between the body and the cap, to the outside of the closing disc, and to the joint between the closing disc and the body of the primer.

Action.—The cap is fired by a blow of the striker; the flash from the cap fires the powder in the magazine, and this in turn ignites the propellant charge.

The copper ball is blown back into its coned seating in the anvil by the pressure of the propellant gases, thus sealing the escape rearwards through the primer.

Refilling of primers.—If considered suitable, No. 1, Mark II primers may be modified and refilled. They will be distinguished by the letter " M " added to the Mark.

Packing.—Primers No. 1 are packed either 4 in a tinned plate box or 10 in a tinned plate cylinder.

Both packages are painted black and the ground colour of the label is white.

§4.06. Primers, percussion, Q.F. cartridge, No. 2.

There are three marks of this primer in use with Q.F. 3-pr. and 6-pr. ammunition, viz. Marks III, IV and VII.

Marks IV and VII are used with the latest marks of cartridge cases, which are designed for them.

The Mark III primer was designed for the older marks of cases which were originally capped, but have been converted to take primers.

It will be observed that the magazine containing a perforated pellet of R.P. or G.20 powder is cylindrical, and is screwed into the body of the primer.

A number of holes are drilled radially in the cylinder.

In the Mark III the *head* of the primer is screw-threaded, and a gas-check is formed on the front end of the body. There is no copper ball.

In the Marks IV and VII the *body* of the primer is screw-threaded and is stronger in construction than the Mark III. A copper ball is provided to seal the escape of gases through the primer.

The Mark VII is the latest type. It is similar to the Mark IV, except for small manufacturing details.

Refilling.—If considered suitable, the Mark VII primers may be modified and refilled. They will then be distinguished by having the letter " M " added to the Mark.

Packing.—Mark III, 20 in a cylinder. Marks IV and VII, 10 in a cylinder.

§4.07. Primer, percussion, Q.F. cartridge No. 11, Mark I.

This primer is similar to the primer No. 1, Mark II, except that it is longer and is provided with a screwed-in magazine.

The magazine is formed from cupped brass, threaded at the mouth to engage with the threads formed in the body. It has eight holes drilled radially to form flash holes and one through the nose to receive a white metal dome, which is secured to the magazine by riveting.

To the interior of the magazine an envelope of white fine paper is secured with shellac varnish; the charge is about 6 drams of G.12 gunpowder.

A disc of white fine paper is secured with shellac varnish over the lighting holes in the body.

Uses.—This primer is at present for use with certain propellants, but may eventually supersede the primer No. 1.

CHAPTER V

PROJECTILES

§5.01. Types of projectile.

The following types of projectiles, which are in use at the present time, will be considered :—

SHOT.

1. Practice shot.
2. Proof shot.
3. Paper shot.
4. Case shot.

POINTED SHELL.

(base fuzed)

5. Common Pointed (C.P.).
6. Common Pointed Capped (C.P.C.).
7. Common Pointed with Ballistic Cap (C.P.B.C.).
8. Armour-Piercing (A.P.).
9. Armour-Piercing Capped (A.P.C.).
   Note.—When A.P.C. shell are fitted with ballistic caps they are still known as A.P.C. and not as A.P.C.B.C.
10. Semi-Armour-Piercing (S.A.P.) and Semi-Armour-Piercing Capped (S.A.P.C.).

SHELL.

(nose fuzed)

11. High Explosive (H.E.).
12. Smoke.
13. Shrapnel.
14. Star.
15. Practice projectiles.

MORTAR BOMBS.

DESIGN OF PROJECTILES

§5.02. Introduction.

In designing a projectile it is necessary to consider :—

(1) The result to be obtained at the target end.
This will be separately discussed for each type enumerated.
(2) The capability of the projectile to withstand the forces of projection in the gun.
(3) The ballistic qualities of the projectile in flight.
(4) The suitability of the design for economical manufacture.

§5.03. Forces of projection.

(i) *Stress due to inertia of projectile.*—The chief force that a projectile is required to withstand, on firing, is the stress due to its inertia under acceleration in the bore. This stress of " set-back " has a tendency to bulge the projectile (" set up ") which must be resisted by the strength of the material of which the projectile is made.

It is readily seen that each layer of metal must be sufficiently strong to support the weight superimposed. Thus, in an H.E. shell the annulus of a section of the walls must support the weight of the walls above, the fuze-hole bush (if present) and the fuze.

The position of the most severely stressed section (known as the governing section) is dependent on the shape of the cavity and base, *i.e.*, whether streamlined or cylindrical and fixed or separate loading ammunition; it is usually situated at a distance from the base about 1 inch less than the calibre of the shell.

Experience has shown that, provided that the total calculated stress (*i.e.*, including the other stresses noted below) in this (or any other) section does not exceed the yield point in tension of the metal, the condition required is fulfilled with a sufficient margin of safety.

The maximum stress of " set-back " occurs at the governing section, and may be calculated from the formula :—

$$p_x = \frac{m}{M} \frac{R^2}{R^2 - r^2} P$$

where P = Maximum chamber pressure in tons a square inch.

M = Total weight of the shell in lb.

m = Weight of portion of the shell (plus fuze and any components carried by the shell) forward of the governing section, in lb.

R = Mean radius of bore in inches (*i.e.*, a value such that πR^2 = cross-sectional area of bore and rifling grooves).

r = Internal radius of shell in inches.

p_x = Longitudinal stress at governing section in tons per square inch.

(ii) *Stress due to inertia of the filling.*—In addition to the above longitudinal stress, the shell filling will set back on discharge, independently inside the shell, giving rise to circumferential stresses and tending to bulge the walls of the shell outwards. In the case of a solid filling such as H.E. whose weight is relatively small compared with the steel of the shell, this effect is generally neglected, but with a non-solid filling, lead bullets, or liquid, it is necessary to calculate the radial pressure p_ϵ at the governing section, due to this set-back.

If m_ϵ = Weight of the filling forward of the governing section, in lbs.

then
$$p_\epsilon = \frac{m_\epsilon R^2}{M r^2} P$$

This produces a maximum hoop tension t_ϵ on the inside of the wall, so that

$$t_\epsilon = p_\epsilon \frac{R^2 + r^2}{R^2 - r^2} *$$

$$= \frac{m_\epsilon}{M} \frac{R^2(R^2 + r^2)}{r^2(R^2 - r^2)} P$$

The value of t_ϵ thus obtained will only be true in the case of liquid filled shell (*e.g.*, chemical). In the case of a perfectly solid, homogeneous filling it will evidently be very nearly zero. For semi-solid fillings, such as shrapnel bullets, an empirical value of $\frac{1}{2}p_\epsilon$ is taken in calculating the strength of shell.

Equivalent simple stress in the wall of the shell.—We therefore have three principal stresses, p_x, p_ϵ, and t_ϵ in the wall of the shell. Arranged in algebraical order, the value of p_ϵ always lies between the other two and therefore,† on the simple shear theory, the equivalent simple stress in the wall

$$S_r = t_\epsilon + p_x$$

This stress should not exceed the specified yield of the steel employed. It is probable, however, that with this limit, the use of the simple shear theory may give an unnecessarily strong wall to the shell, thereby decreasing its capacity and rendering it not as effective as it might be. It has been customary in the past to calculate the equivalent simple stress by the formula

$$Sr = \sqrt{t_\epsilon^2 + p_x^2}$$

which has no logical basis, but appears to give adequate results.

(iii) *Stresses due to rotational forces.*—A projectile must also be sufficiently strong to resist the circumferential stress due to its angular velocity (centrifugal force) and also the torsional couple caused by its angular acceleration on being rotated by the driving band. Both of these stresses are very small, and in calculating the strength of a shell they may in general be safely neglected.

(iv) *Shear stress.*—(*a*) There must be sufficient metal behind the driving band to support it during engraving and rotation. For the purpose of calculating this amount, the engraving pressure is assumed to be 3 tons to the square inch and the material to resist this pressure is taken as cast iron having a shear strength of 5 tons to the square inch. The result obtained cannot, however, always be strictly adhered to for reasons which will be discussed later in para. **5.07**.

* *See* Textbook of Service Ordnance 1923, bottom of page 161.

† *See* Textbook of Ballistics and Gunnery, Part II, Sect. III, 1933.

(*b*) *Stress on the central portion of the base due to gas pressure.*—The force of the explosion of the charge tends to punch the base into the shell.

Let w = weight of filling,
p_w = internal pressure on base due to inertia (set-back) of filling.

We have

$$p_w = \frac{wR^2}{Mr^2}P$$

Internal force on the base $= \pi r^2 p_w$
External ,, ,, $= \pi r^2 P$

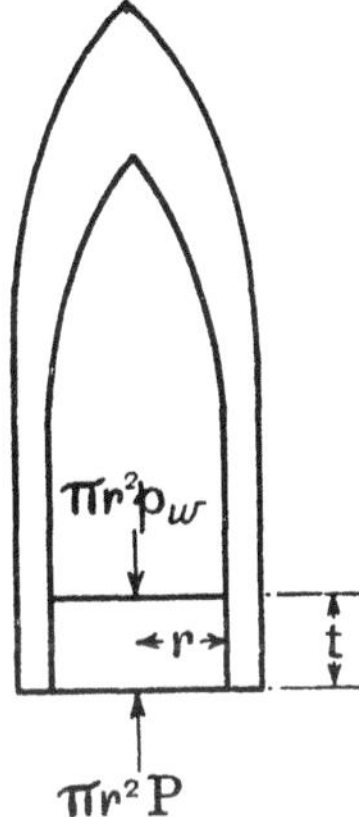

Resultant force on the junction between the base and the walls of the shell $= \pi r^2(P - p_w)$

If t = thickness of base
S_B = shear stress at junction

$$2S_B t\pi r = \pi r^2(P - p_w)$$

$$\therefore S_B = \frac{rP\left(1 - \frac{wR^2}{Mr^2}\right)}{2t}$$

A factor of safety of 2 is employed, and the weight of the filling is in practice generally neglected in comparison with the total weight of the shell. We therefore have for practical purposes the simple formula

$$t > \frac{Pr}{S}$$

where S = maximum shear stress allowed in the material in tons a square inch.

The yield point in shear is taken as three-quarters of the yield point in tension. Hence if the yield point in tension is 19 tons a square inch, the yield in shear (S) equals three-quarters of 19.

There will also be a certain amount of bending action due to the varying stress on the base, but shear is the major effect.

The tapering or gradual decrease in thickness towards the head of the walls of an H.E. shell enables the walls to be lengthened, and hence the capacity of the shell to be increased, without an increase in the weight of the walls and consequently in the stress of the set-back in the governing (or other) section as compared with the stress which would be present in a parallel walled shell.

In practice, however, parallel walls may be required to simplify manufacture or filling, if solid blocks of explosive are to be used. In any case there is a limit in practice to the amount of taper that can be imparted, as, if over-emphasized, the weakness of the head compared with the body may cause unequal and unsatisfactory fragmentation when the shell is detonated.*

NOTE.—The weight of a cubic inch of the following materials is:—

	Lb.
Lead	0·4
Brass	0·31
Steel	0·282
Lyddite	0·056 to 0·0585
Trotyl	0·056
Amatol 80/20 (hand stemmed)	0·044
Amatol 80/20 (machine filled)	0·051
White phosphorus	0·066
Water	0·036

To obtain the quality or strength required in the metal, forged steel is generally employed for projectiles, but semi-steel, cast steel, and cast iron have been used in certain cases.

Steel is tested to see that it is up to the required standard, the specification laying down the minimum yield and the breaking strain in tons a square inch, and the minimum final elongation. For example, the figures for H.E. shell steel are—

Yield	19 tons a square inch
Breaking strain ..	40 to 55 tons a square inch
Elongation	15 per cent.

These figures are obtained, using a test-piece 2 inches in length and ¼ square inch cross sectional area.

§5.04 Ballistic qualities in flight.

The ballistic qualities of a shell in flight are expressed by the "ballistic coefficient," which is a figure representing the *efficiency* with which the projectile is able to overcome the air's resistance. It is given in the formula

$$C_o = \frac{M}{\kappa \sigma d^2} \dagger$$

where C_o = Standard ballistic coefficient
M = Weight of projectile in pounds
κ = Coefficient of shape
σ = Coefficient of steadiness
d = Diameter of projectile in inches.

The factors in this expression will be considered in turn.

---

* NOTE.—Experiment has shown that the high explosives used as shell fillings generally require a certain degree of confinement (or tamping) to develop their full power on detonation.

† *See* Textbook of Ballistics and Gunnery, Part I (1927), para. 3.14.

§5.05. Weight.

If two projectiles of similar shape but of different weight be fired with equal muzzle velocities, then the heavier will range the further and have the flatter trajectory.

Apart, however, from considerations of stress on gun and mounting there is a limit to the weight which can usefully be adopted in a projectile of any given calibre, as undue increase of weight entails undue lengthening, which in turn introduces ballistic disadvantages, and experience has shown that a suitable weight is given by the formula

$$\frac{W}{d^3} = \text{a constant}$$

where d = calibre in inches, and W = weight of projectile in lbs· This constant varies slightly, but may be taken as being equal to 0·5 for most projectiles in use in the land service.

The various types of projectile used in any individual gun are, as far as possible, made of similar weight, in order that the muzzle velocity may remain constant ; this necessitates variation in length. Variations in length and in shape (*e.g.*, a nose-fuzed shell as against a pointed shell) would cause them to range differently, and, in order to simplify the range tables, efforts are made to modify the various projectiles of individual guns so as to range the same at any rate at the " critical range," that is, the range at which it is expected that fighting will mostly be carried out. This is generally done by slight alterations to the shape of head of the different projectiles, the weight remaining constant.

§5.06. Shape of projectile—external.

For a given velocity " κ " may be taken as a function only of the external contour of the head and base of the shell, and is a measure of the resistance which such contour offers to a steady stream of air in the direction of the longer axis of the shell. At the present time the effects of the shapes of both head and tail are included in " κ " ; when knowledge of these effects is more complete it may be possible to distinguish between them, and it will then be desirable to have two factors, one for the shape of the head and the other for the shape of the tail.

Shape of head and tail.—The governing factor determining the shape of the head, and to a certain extent that of the tail also, is the velocity of the projectile through the air. A ship is " streamlined " both at bow and stern to give maximum speed through the water, *i.e.*, to reduce resistance. The shape of the projectile is governed by similar considerations, but in the case of the air there is found to be a critical velocity, that of sound, at which conditions alter.

Shape of head.—At velocities below that of sound (about 1,100 f/s) the pressure waves caused by the nose of the projectile travel away faster than the projectile itself in ever increasing circles, whereas at higher velocities the nose is continually piercing the spherical waves and forming a new wave front, the effect being a conical envelope of high pressure with an apex angle varying with the velocity. As the velocity increases, the apex angle becomes more acute, and if the wave envelope presses against the side of the head of the projectile, the resistance to its motion is increased. The shape of the head is therefore of great importance and the higher the velocity the more pointed it must be.

The exterior of the head (including the fuze, if any) must also be smooth. Any irregularities produce further waves and absorb energy. The best shape has previously been considered to be an ogive, in which the ends of the curve are tangential to the parallel sides. This shape has the advantage that it can be simply made by fitting a radial arm to carry the cutting tool on the lathe that is used to fashion the head. It is evident, however, that to make an ogive more pointed, it must be made longer. This means either increasing the length of the projectile or decreasing its cylindrical portion. Either may cause instability (see later under " Length "), so in some cases it is necessary not to have a true and continuous ogive. As a matter of fact, it has been found experimentally that, within fairly wide limits, the exact shape of the curve appears to be of less importance than (*a*) the angle of the apex, and (*b*) the length of the head from the nose to the shoulder. It is therefore important that the latter length should be emphasized in the nomenclature.

Nomenclature for describing the shape of head.—The shape of the head is described as of " n " c.r.h. or " n/m " c.r.h.

In the first case, the head is a true ogive, struck with a radius of " n " calibres with the centre in the plane containing the shoulder or the forward end of the parallel portion of the shell.

In the second case, the length from the shoulder to tip is the same as the length of a true " n " c.r.h., but the head is struck with a radius of " m " calibres with the centre not necessarily in the plane containing the shoulder.

Reference Fig. 5.06(*a*), the modern shell is designed so that the height CD (often referred to as the " ballistic length ") is the governing factor in the shape of the head, the radius of the curve AED being of secondary importance.

Thus, true 4 c.r.h. would mean that circles of 4 calibres radius are struck with their centres on the line AB produced.

4/10 c.r.h. would mean that the length CD was due to 4 calibres radius and the curve AFD was of 10 calibres radius.

4/∞ c.r.h. would mean that the length CD was due to 4 calibres radius and the head was conical.

This system of nomenclature was adopted in order to enable an exact definition to be given of any shaped head for design purposes.

Previously, shell were referred to by the calibre length of the main striking radius. This was of use only when the head was a true ogive. If the main radius, however, happened to be struck from a point below the line of shoulder AB, it might be anything up to infinity for any height CD. Thus quoting the main radius gave no indication of the true shape of the head.

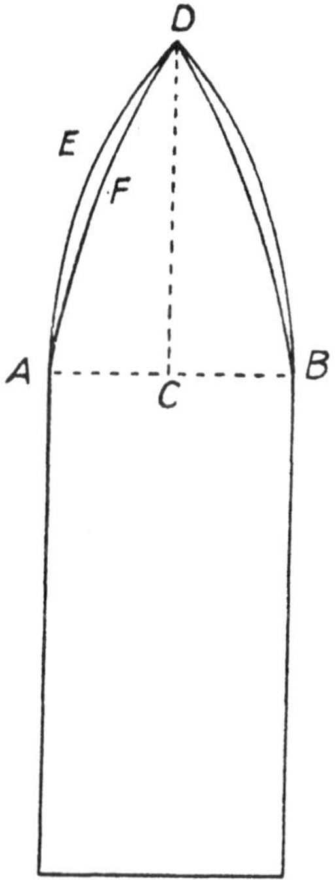

FIG. 5.06 (*a*).—Shape of Head.

The system gives latitude in design to enable the contour of the shell and fuze to be made to coincide, since, having the c.r.h. required the length CD can be determined, also the length and breadth of the fuze being known, the curve AFD can be easily adjusted.

The length CD for any calibre may be determined from the formula :—

$$\frac{l}{d} = \frac{1}{2}\sqrt{4n - 1}$$

where l = ballistic length C.D.
d = calibre of shell, and
n = c.r.h. of the true ogive.

Thus for a 6-inch calibre

The length CD for 2 c.r.h. = 7.94 inches
,, ,, 4 ,, = 11.63 ,,
,, ,, 6 ,, = 14.4 ,,

Service nomenclature.—Shell having heads of more than 2 c.r.h. are identified by the letters A, B, C, or D after the Mark of the shell.

For example :—

Mark III	denotes a shell of 2 c.r.h. and under.			
,, IIIA	,, ,,	over 2 c.r.h. and up to 4 (incl.)		
,, IIIB	,, ,,	,, 4	,, ,,	6 ,,
,, IIIC	,, ,,	,, 6	,, ,,	8 ,,
,, IIID	,, ,,	,, 8	,, ,,	10 ,,

These code letters representing the approximate c.r.h. are intended for use in the field when sorting ammunition, as shell with the same letter will range similarly at the fighting range of the equipment and nearly the same at other ranges.

As regards shell with compound shape of head the procedure adopted for nomenclature is as follows :—

New designs of projectiles, if designed to range the same as an existing type, are given the same letter as that type. If, however, a new type is not designed to range with any existing Mark, its ballistic qualities are assessed with a view to giving it an equivalent or Service c.r.h. Thus the designation of the 60-pr. 5/10 c.r.h. 56-lb. shell, which was given a service nomenclature of 10 c.r.h., is Mark ID, while the 9·2-inch A.P.C. 5/10 c.r.h. was allotted a service nomenclature of 6 c.r.h. and is known as the Mark XIB.

It will be seen, therefore, that the nominal c.r.h. sometimes differs materially from the actual c.r.h. and it is dangerous to use the formula on page 102 without ascertaining the true conditions.

§5.07. Steadiness during flight.

The stability of a projectile during flight is fully dealt with in the Text Book of Anti-Aircraft Gunnery, Vol. I. Briefly it involves two conflicting requirements, viz : (i) stability at the beginning of flight, and (ii) stability during the maximum curvature of the trajectory.

Instability at the beginning of flight is caused principally by (*a*) faulty centering of the projectile in the bore, and (*b*) muzzle vibrations. The effect is mainly a loss of range due to increased air resistance.

Instability due to the curvature of the trajectory is caused by a faulty gyroscopic balance between the transverse moment of inertia of the projectile and the couple exerted on the shell by the displacement of the centre of pressure of the atmosphere from the centre of gravity, relative to the direction of motion. The effect may be a loss both of range and of accuracy, but is very much smaller than the effect of initial instability, and up to a spin of 1 turn in 20 calibres, can scarcely be detected.

To counteract instability, spin is imparted to the projectile by the rifling engaging in the driving band. The greater the spin, the sooner will initial instability be damped out. On the other hand, if the spin is too great, yaw (the angle between the axis of the shell and the tangent to the trajectory at the point considered) increases with increase in the curvature of the trajectory. The best rate of spin to be given to the projectile is therefore a compromise and is determined largely by trial. Once it is fixed, the shell must be designed to give the greatest steadiness conformable to that rate of spin.

The factors affecting steadiness, therefore, are :—

(i) the centering of the projectile, that is to say, the manner in which it is supported in the bore ;

(ii) the transverse moment of inertia ; involving primarily the length of the projectile, secondarily the distribution of its mass ;

(iii) the position of the centre of gravity, in order to obtain the maximum effect of the couple caused by the resistance of the air.

As regards centering, the forward shoulder of the projectile cannot be made an accurate fit in the bore owing to manufacturing tolerances and the effect of wear. But it is evident that the greater the distance between the shoulder and the driving band, the better supported it will be and the less chance there will be of the shell axis lying transverse to the axis of the piece. This distance can be increased (*a*) by increasing the length of the cylindrical portion, and (*b*) by placing the driving band as close as possible to the base of the shell. For a given length of shell, the length of the cylindrical portion is determined by the shape of the head, already discussed. The position of the driving band is limited (*a*) by consideration of shear stress mentioned previously, and (*b*) by the fact that if it is put too close to the base of the shell it is found to have an adverse effect on shooting, owing to the formation of additional turbulent eddies.

§5.08. Length of projectile.

The lowest admissible length for accurate shooting is found to be 2 calibres. Most projectiles are from 3½ to 4 calibres in length, but 4½ calibres is in use in certain cases, and this length is exceeded in shell fitted with ballistic caps, though the actual shell is no longer than the ordinary shell, so that, though ranging further, there is no gain in bursting power.

Ballistic Caps.—As an alternative to lengthening the actual head of a projectile, it may be fitted with a ballistic cap, in order to increase the calibre radius head (c.r.h.) and so decrease the air resistance.

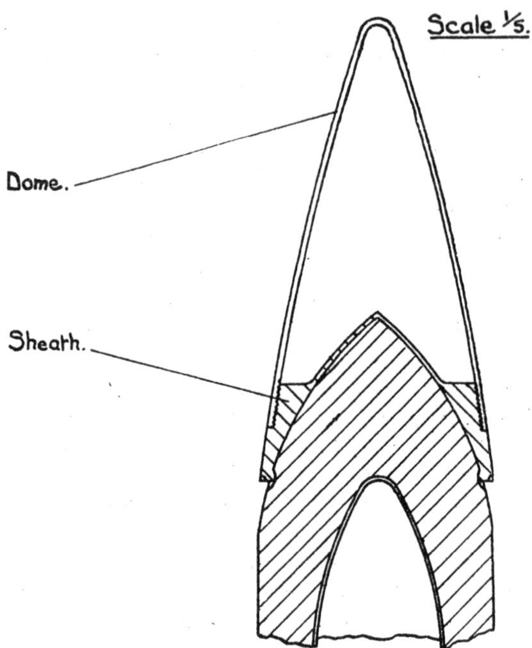

FIG. 5.08 (*a*) C.P.B.C.

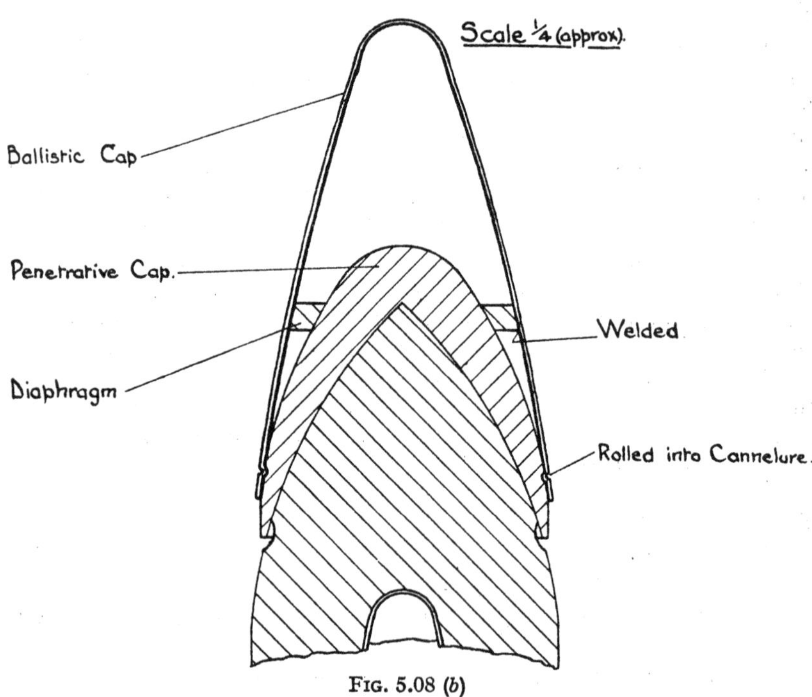

FIG. 5.08 (*b*)

The cap, which may be either ogival or conical in shape, is usually made of thin steel, but aluminium alloy has been used also.

Methods of securing ballistic caps :—

(*a*) To the head of a C.P. shell. The sheath portion is sweated to the head, and then notched into indents formed in the head. The dome portion is screwed and shrunk on to the sheath and then soldered at the junction.

(*b*) To the head of an A.P.C. shell. Fig. 5.08(*b*) shows a method of converting a 6 in. A.P.C. Mark VII shell to Mark VIIB, by fixing a ballistic cap to the penetrative cap.

The ballistic cap may also be secured by being screwed on to the penetrative cap.

The ballistic cap, being light, enables a long head to be used without adversely affecting the balance of the projectile, and the disadvantage of giving a large c.r.h. to the shell itself is minimized, but difficulties arise in ensuring functioning of the fuze in the case of nose-fuzed shell. Its use is confined in land service to base-fuzed shell.

To ensure concentricity of the tip, each cap has to be assembled and adjusted individually, and care must be taken in transport and storage that it is not injured by rough usage.

Shape of Tip.—It is a general rule that the tips of caps should be rounded off with a radius of 1/15th of the calibre.

§5.09. Diameter of projectile.

The diameter of the body of the projectile is in all cases slightly less than that of the bore across the lands, thus providing a clearance or " windage " that ensures the free passage of the projectile along the bore. The clearance varies for the different calibres, but is kept as low as possible. The " windage " has been reduced in order to obtain more accurate shooting, but projectiles of the former " windage " are still supplied to guns liable to steel choke.

Bands.—Some projectiles are made larger in diameter at the shoulder and lower portion of the body than at the centre. These parts of greater diameter are known as the front and rear bands, respectively, and must not be confused with the driving band.

Bands facilitate manufacture, as the central part need not be brought so accurately to gauge, and the weight of the empty shell can be adjusted, within limits, by varying the amount of metal removed between the bands.

§5.10. Driving bands.

The functions of a driving band are :—

(i) To rotate the projectile ;
(ii) To centre the projectile as well as possible in the bore, and
(iii) To prevent the escape of gas past the projectile.

In the earlier types of driving band a considerable escape of gas, leading to erosion, occurred, and this was believed to take place before the band was properly seated. The gas-check band was therefore introduced, which included an undercut lip or second slope near the rear of the band, and usually of a larger diameter than the rest of the band. The intention was that gas should enter under this lip and force it outwards on to the shot seating. Actually as the band travels forward and is compressed by the shot seating the lip is forced hard down on to the metal in rear of it, and as will be seen later, for the band to be successful, no gas should be able to get underneath it. The increased diameter of the lip is useful in reducing overram in worn guns, but apart from this it therefore appears to have no particular value. "Hump bands," similar in general shape, but without the gas-check "grave," have since been introduced and are quite successful.

General Charbonnier, of the French Army, has suggested that wear of guns is due, not so much to escape of gas past the driving band, as to turbulent eddies of hot gas in the bore due largely to sudden changes in dimensions at the front end of the chamber. If this is so, reduction in wear will be brought about more by redesigning guns than by insisting on gas-sealing driving bands. At the same time it seems likely that though *wear* may be due largely to eddies, *scoring* probably owes a good deal to escape of gas, and there is no doubt that the gas-sealing qualities of a driving band are of importance.

Material of the Band.

The metal used must be soft, to give the least wear possible in the gun. But it must not be so soft as to strip under the rotational and engraving stresses in the bore. Copper has hitherto been found best for most purposes, and it should be as free as possible from impurities to ensure uniform hardness. Variations in the hardness of bands, due either to the presence of impurities such as oxygen or arsenic, or to cold work done in pressing on to the shell, cause appreciable variations in ballistics, particularly in the case of smaller calibres. A little manganese is included to assist in getting rid of oxygen and to make the copper less brittle.

Cupro-nickel bands are used in the Navy for some high-velocity guns, but it is doubtful if much advantage is gained therefrom.

Gilding metal (Cu 90 per cent, Zn 10 per cent.) has also been tried. The Americans have used it in 3-inch shells with apparent success. The advantages of gilding metal over copper are :—

(*a*) Coppering of the bore is reduced ;
(*b*) Zinc is a de-oxidizer, and would eliminate an element that affects the performance of copper bands ;
(*c*) Economy in copper, which may be important in war time ;
(*d*) Gilding metal is slightly cheaper.

Position on the shell.—To give the maximum steadiness in the bore, the band should be placed as near the base as possible. In practice this is limited by (*a*) the tendency for a band placed too near the base to cause eddy waves and so reduce ballistics, (*b*) the tendency for a pressed on band to tear off the metal behind it owing to the force of engraving. Sufficient metal must be left in rear of the band to give the necessary shear strength. (*c*) Streamlined shell must have the band in front of the streamlining, and (*d*) in fixed ammunition, the driving band must be far enough forward to allow the projectile to be firmly secured to the cartridge case.

Attachment of the driving band.—The driving band is firmly attached by being forced into a groove round the shell by a press.

In order to prevent it slipping round the shell, there are a

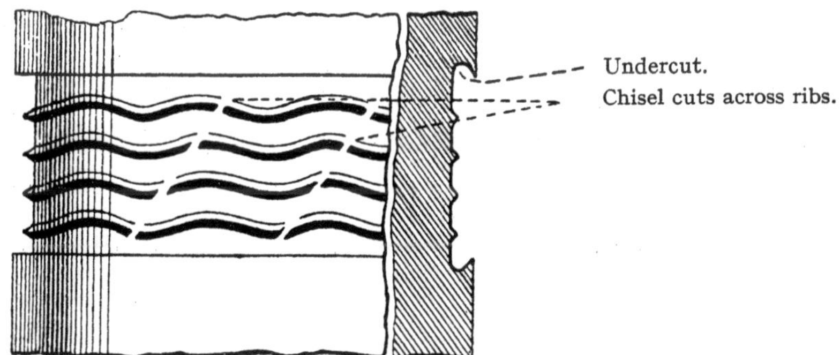

FIG. 5.10 (*a*).—Ribs triangular in section.

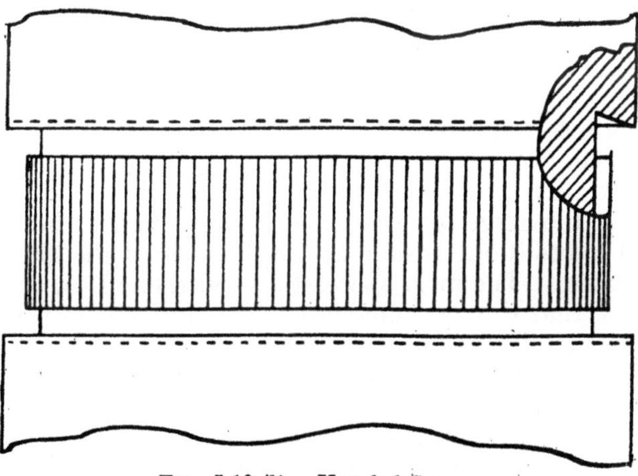

FIG. 5.10 (*b*).—Knurled Groove.

number of waved ribs at the bottom of the groove ; chisel cuts are made across them to allow the air in the channels between the ribs to escape when the driving band is being pressed on.

As an alternative, the bottom of the groove for the driving band of steel projectiles may be knurled with single groove knurling.

Originally the ribs were straight and ran circumferentially round the groove ; they were cut away at intervals to prevent the band slipping. This method is now employed for those shell the grooves of which are too narrow to permit of waving.

Dimensions and shape of driving bands.

A well-designed band is one that gives good accuracy with both new and worn guns. The principal causes of inaccuracy due to the band are—

(i) Fanning, or the setting out of the extruded copper due to centrifugal force, which causes increased air resistance, and

(ii) Shearing, or wearing smooth of the band, owing to the rotational inertia of the shell.

Subsidiary causes are—

(iii) Bad centering, with consequent effects of groove marking on the body of the shell itself, and eccentric fanning, and

(iv) Impurities in the copper, already mentioned.

(i) *Fanning* may be of various kinds, viz :—

(*a*) *Skirt fanning* occurs when the working copper of the band is forced to the rear and extends in a solid skirt beyond the original rear end of the band. It is an indication that there is too much working copper in the band for its width or in the rear section, and it is then usually produced when firing is carried out from a new or nearly new gun. In a worn gun, the surplus copper is lost before full engraving is attained.

To assist in overcoming skirt fanning, *cannelures* are provided in many bands into which the surplus copper should be swaged. A suitable capacity for the cannelure is one that will accommodate about 80 per cent. of the displaced copper.

In " gas-check " bands the *grave* in rear of the gas-check acts in the same way as a cannelure, but it is important that the grave shall not be too large, or gas finds its way under the displaced copper and forces it out as soon as the shell leaves the muzzle, thus causing fanning. The most successful service bands of this nature have a grave that can accommodate approximately 74 per cent. of the copper on the gas-check above groove diameter. Another feature which appears to assist in counteracting fanning in bands of this nature is to provide a small ridge of non-working copper in rear of the grave. This ridge affords a good bed on to which the swaged copper can be forced, and there is then less likelihood of gas getting underneath it.

(*b*) *Ribbon fanning* may occur either in front or rear of the band. A ribbon of copper is extruded under the land, and stands out in flight under the influence of centrifugal force. This type of fanning is not easily detected in recovered shells, owing to the ribbon having broken off on impact and the line of fracture being straight and close up to the band. It may be detected by firing through jump cards.

Front ribbon fanning is due to the rapidity with which swaging begins, accompanied by a sharp local rise of temperature, when a small amount of copper will be " splashed " forwards. To prevent it, the front slope of the band must in general not be much steeper than that at the beginning of the rifling, thus giving a surface for this forwardly extruded copper to bed down on.

Rear ribbon fanning probably occurs when a column of copper is cut or driven out of the band by the thrust of the land of the rifling. To avoid this, the metal must be swaged out sideways into the cannelures and grave, or on to the rear slope of the band. The radial pressure on the band must be high enough for the copper to behave plastically.

The principal factors influencing this effect appear to be (i) the angle of slope of the beginning of the rifling, which must be a gradual one (with most guns it does not exceed 12°), (ii) the dimensions of the rifling lands, and (iii) the speed of engraving. Guns that show the greatest tendency towards this trouble are small calibre high velocity guns such as Q.F. 3-inch 20 cwt. The larger the gun (and the greater the ratio depth/width of land) the less likelihood does there appear to be for rear ribbon fanning to develop. It has also been found by trial to be slightly worse with sharper twists of rifling.

Fanning of all kinds becomes much less troublesome with a worn gun than with a new one.

(ii) *Smoothing of the band.*—The area of copper against which the driving edge of the rifling presses to give the necessary rotational thrust to the shell is known as the " working copper," and apart from the cross-sectional shape of the band, this area is the most important feature that has to be determined when designing a new band.

The thrust intensity on the band can be calculated from the formula

$$R=\frac{\pi k^2}{CG}P\sin\theta \quad *$$

where R = Thrust intensity on the band in tons a square inch,
k = Radius of gyration of projectile,
C = Area of working copper in the band,
G = Number of grooves,
P = Powder pressure in chamber of gun,
θ = Angle of rifling, *i.e.*, $\tan\theta=\pi/n$, where the rifling has a twist of one turn in n calibres.

* This formula is a simplification, omitting the effect of friction. It gives results that are close enough for practical purposes.

The thrust R sets up a shear stress S in the base of the copper lands, such that

$$S = \frac{d}{w} R$$

where d = Depth of rifling groove, and
w = Width of rifling groove.

The value of S must never exceed the shear strength of the copper, usually taken as 12 tons on the square inch. In a worn gun, heavy loss of copper takes place owing to friction and gas wash. This results in a reduction of the area of the band in contact with the lands and so increases the value of R. The more worn the gun, the greater the loss of the copper, and there may come a time when the thrust intensity rises to a height at which shear takes place and the band is left smooth.

Another factor is also present. As the gun becomes worn, the lands near the beginning of the rifling tend to become smoothed out. Their driving edges are then no longer normal to the surface of the bore but laid over at an angle.

New rifling.
Thrust tangential to shell.

Worn rifling.
Thrust no longer tangential.

The effect is as if the copper of the band was being nipped in a sharp corner between the rifling of the gun and the wall of the shell. A much higher intensity of stress is set up in this sharp corner, and if it goes on too long, as happens in a badly worn gun, the whole surface of the band is smoothed away and the shell is no longer properly rotated.

The lower the thrust intensity R, the further can the driving edge be laid over by wear before the band fails. The connection between the two quantities has not been clearly established, but it is probably of the form

$$W = \frac{A}{R + B}$$

where W = Wear of the gun, measured as a percentage of the calibre, at which band failure takes place.
A and B = Constants, depending on the material of the band and the dimensions of the band and rifling.

At present the dimensions of driving bands are based on past experience, but are largely a matter of trial and error when any new design is being introduced.

Shape.—The avoidance of fanning has already been discussed, and from a consideration of the points brought out it is clear that a wide shallow band will in general give less fanning than a deep narrow one. If the band is too shallow, however, it may not grip the beginning of the rifling in a worn gun, so bringing the life of the piece to a premature end before the ordinary condemning limit has been reached. Also it is more difficult to secure a wide band than a narrow one firmly to the shell. The ideal design should therefore have the greatest depth possible consistent with avoidance of fanning.

Types of Driving Bands.

G.1.—This is a band typical of those used in 6-inch, 12-inch, 13·5-inch and 15-inch guns. It has given satisfaction in the heavier natures. It is generally known as the " gas-check " driving band, from the formation of the central part, where a second slope is formed in rear of the cannelures projecting beyond the remainder of the band; it is cut away at the rear to form a lip or gas-check. The pressure of the gases on the underside of the lip tends to force it against the surface of the bore and seal the escape of the propellant gases.

Although this type of band prevents erosion to a certain extent, it has certain drawbacks. The lip is very liable to damage during transport and loading, hence these bands must be carefully protected in store and transport by grummets or other means.

G.2.—This design is typical for 9·2-inch and 7·5-inch guns. It is a departure from the gas-check type; the raised portion at the rear gives the name of " hump " band. It is much more resistant to damage in handling than the gas-check type, and probably the " hump " serves the purpose of checking the gas flow quite efficiently.

H.1.—This is a special band for the 18-inch howitzer. It is similar to G.2, but has cannelures.

H.2.—This type is typical for howitzer projectiles, 12-inch, 9·2-inch, and 8-inch, of which the last only are without cannelures. The driving band has a comparatively long slope, which is serrated to give a good grip to prevent the shell slipping back at high angles of elevation.

G.3 was introduced for use with projectiles of naval high velocity 6-inch guns, as mentioned above. It is made of cupro-nickel. It has no cannelures and the metal in rear of the second slope is not grooved out to form a gas-check lip. Projectiles with these bands are distinguished by a white band painted round the body immediately above the driving band.

H.3 is a similar plain band introduced for projectiles for 6-inch howitzers and below. The slopes are gradual, permitting flow of metal; and no cannelures are necessary.

G.4.—This is a combination of the " hump " type and gas-check type with the " grave," *i.e.*, the space behind the gas-check lip, much reduced. It was designed by the Elswick Ordnance

Fig 5·10(c).

TYPES OF DRIVING BANDS.

GUN.

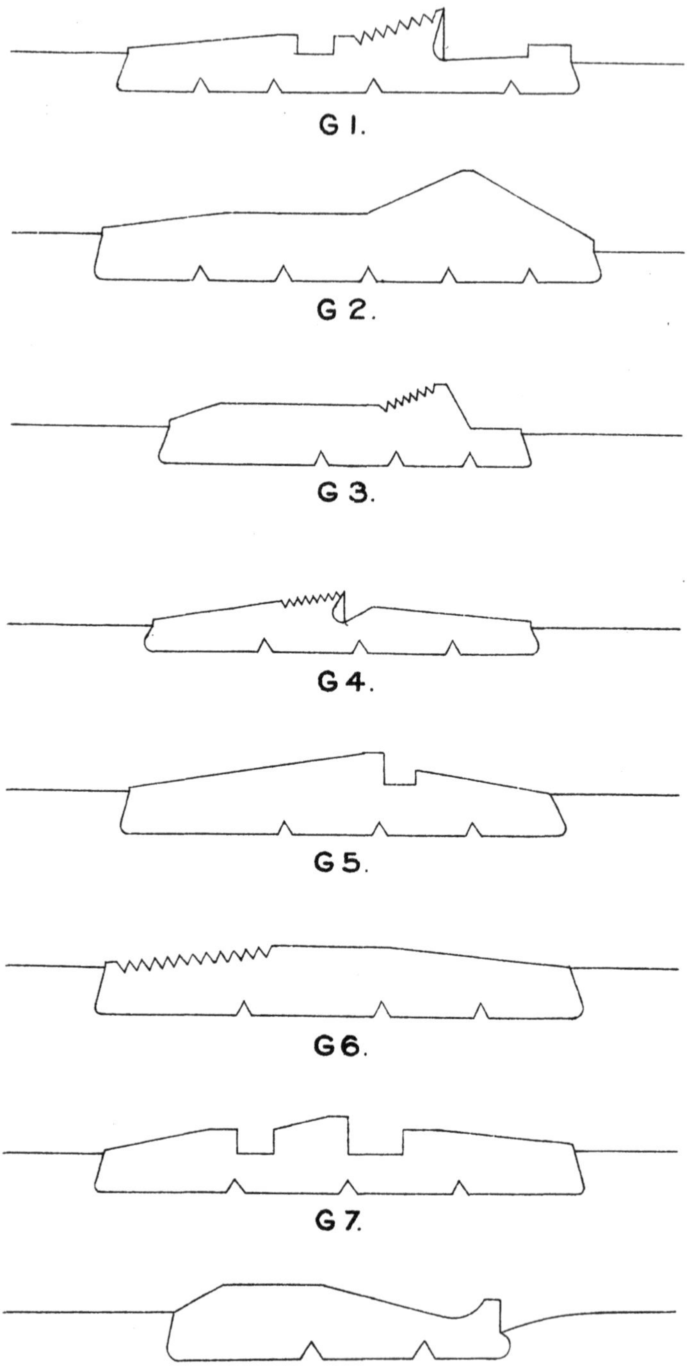

G 1.

G 2.

G 3.

G 4.

G 5.

G 6.

G 7.

G8.

958.

Fig. 5·10 (d).

TYPES OF DRIVING BANDS.

HOW^{R}.

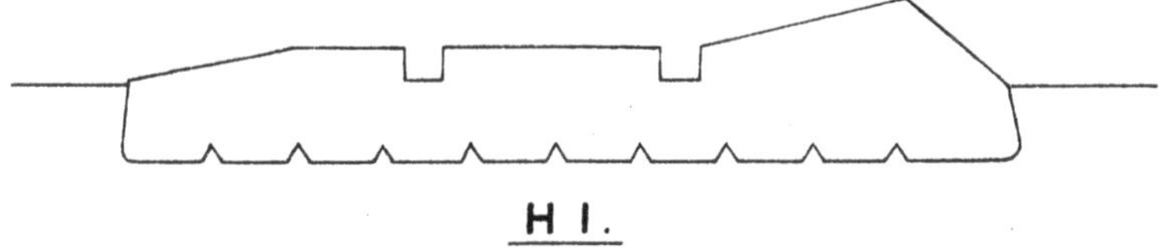

H 1.

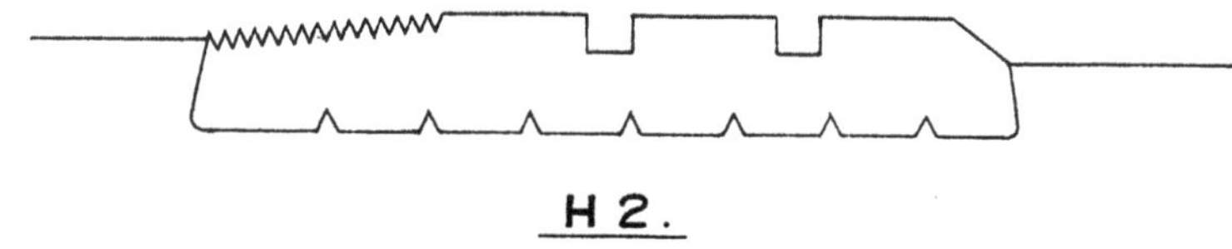

H 2.

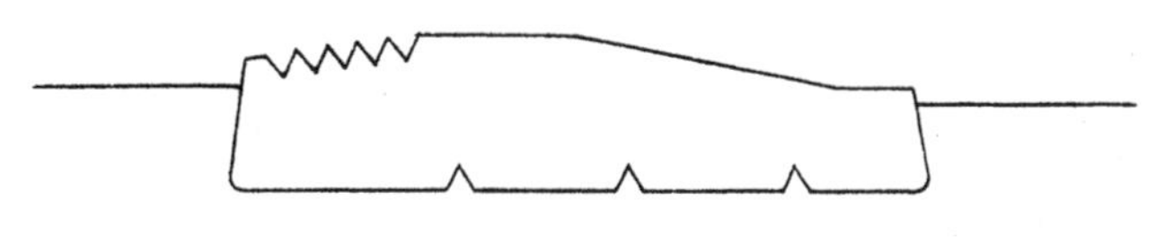

H 3.

Company and is generally known as the E.O.C. band. It has been adopted for use with high velocity B.L. 5·5-inch and 4-inch guns.

G.5.—This is a design of band at present applied to naval service 4·7-inch and 5·2-inch guns. It is a hump band of simplified contour, easy to manufacture, and sturdy.

G.6 was introduced for land service B.L. 60-pr. and Q.F. 4·7-inch guns. It is a simple strong design, suitable for conditions of field service.

G.7 is a plain cannelured band, not unlike the original Vavasseur. It is used for certain naval fixed ammunition guns. It has a tendency to rear fanning, which will be eradicated when certain contemplated alterations are applied.

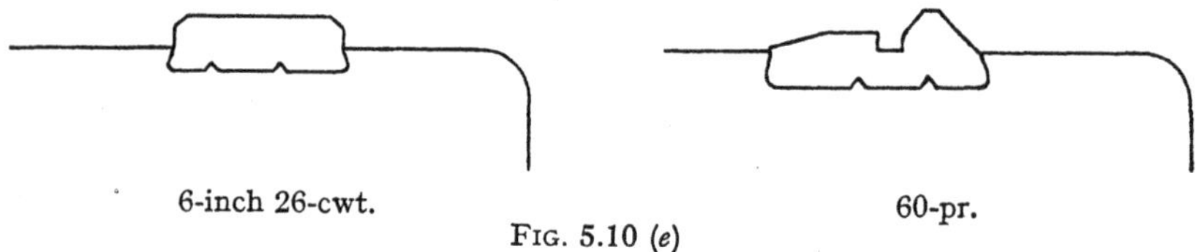

Fig. 5.10 (e)

G.8.—This is the Q.F. 18-pr. band. It is provided with a raised cartridge stop in rear. The profile is plain, and similar bands, some with and some without the case stop, are used in all lower calibres.

Economy Bands (*see* Fig. 5.10(*e*)).—Owing to the shortage of copper during the Great War, so-called " economy " driving bands were fitted on some shell, chiefly those used with B.L. 60-pr. guns and the B.L. 6-inch and Q.F. 4·5-inch howitzers. These bands were made considerably narrower than the service band in order to economize copper, but the diameter was slightly increased to prolong the life of the gun by giving better sealing effect and to decrease the loss of velocity due to over-ramming. There was

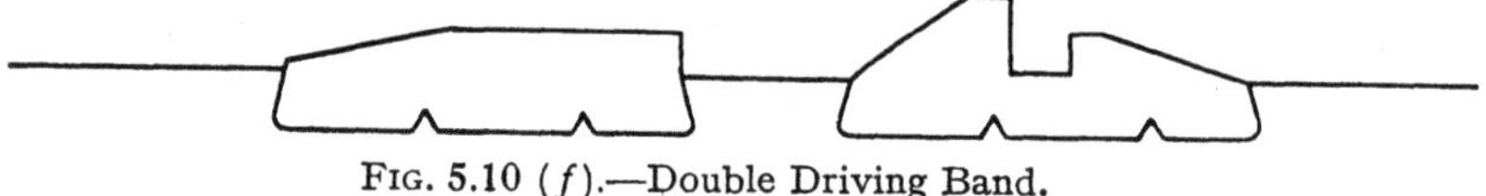

Fig. 5.10 (*f*).—Double Driving Band.

considerable saving of copper and of the life of the gun, but accuracy was affected in some cases by unsteadiness in flight due to the narrow band causing the shell to be badly centred, and to fanning.

Since the war a return has been made to the former type of band in the case of the B.L. 60-pr. and 6-inch howitzer projectiles.

Double driving bands.—These are now being introduced on 8-inch and 6-inch natures of shell. Fig. 5.10(*f*) is a typical design. The 6-inch band for naval service is of cupro-nickel, but for land service it is of copper.

This type of band was introduced to meet the mechanical difficulties of pressing a very wide single band on shell.

Double driving bands, in which the forward band is situated near the shoulder of the shell, have been proposed and tried for specially long-range projectiles, the object being to give greater steadiness to projectiles during the passage of the bore.

Augmenting strips and rings.—The augmenting strip was originally introduced in the days of powder charges to prevent over-ram in worn guns. On the introduction of cordite charges its use was extended to act as a gas-check. It consisted of a strip of pure copper of even section and grooved on one side.

The strip was placed in the cannelure grooved side downwards and hammered round the shell until the two ends met, the can-

Scale $\frac{1}{1}$.

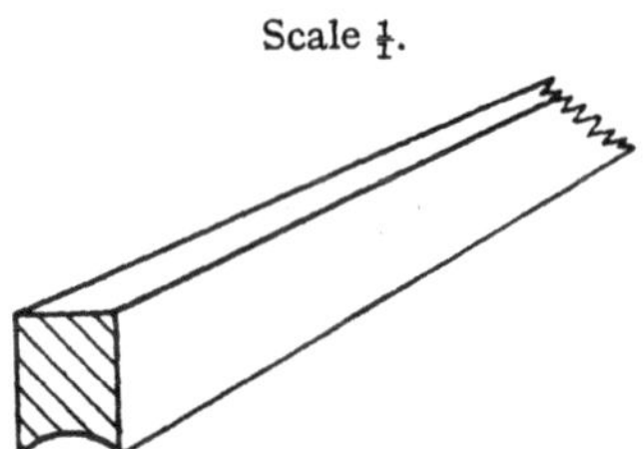

Fig. 5.10 (*g*).—Augmenting Strip.

nelures on the Vavasseur bands being undercut to take the strip if necessary.

During the Great War a different type of strip was introduced for use with 12-inch howitzer H.E. projectiles. As modern cannelures are not undercut, the strip was retained in the cannelure by means of eight inclined grooves formed alternately at equal distances on the sides of the strip when fitted in the groove. For future manufacture the strip will be formed in the band, *i.e.*, there will be no forward cannelure, but a slight hump.

The augmenting ring is intended to provide more copper for use in a worn gun. It used to be slipped over the base of the shell and seated in the recess under the gas-check.

§5.11. Shape of projectile—internal.

Although the external contour of a shell is more or less determined by the requirements of external ballistics, the strength of a shell is governed by internal design and the quality of the material specified. Besides the obvious requirement that a shell should arrive at its objective intact, strength is of the greatest importance with shell to provide against the danger of premature explosion in the gun, especially when filled with H.E.

Armour-piercing shell have to resist the severe shock of impact with a steel armour plate, and are therefore made so as to offer the maximum possible resistance to break-up. It is clear that the steel of the projectile should be so distributed that it is employed with the greatest efficiency as regards strength.

Internal design of head.—The usual practice is that the wall in the head follows the external contour, keeping approximately the same thickness of metal as at the shoulder. The metal is sometimes slightly thickened towards the fuze-hole in order to allow sufficient depth to take the fuze, etc. The head thus forms a naturally strong structure which is unlikely to fail on penetration of earthworks, etc.

As already mentioned, A.P. and semi-A.P. shell have specially strengthened heads which are hardened and in the former case are practically solid. The projectile will, on arrival at the target, be submitted to crushing forces acting in the opposite direction to those set up on firing, and a shell should be of such strength that it does not break up on impact, but bursts as a result of explosion initiated by the fuze.

Base Adapters.—Base adapters are fitted to the majority of base-fuzed shell. They were introduced in order to give an aperture of increased diameter to facilitate manufacture. They also facilitate the insertion of the container and the filling of the shell. Their sizes generally approximate to the full bore of the cavity. In some cases they may obviate distortion of the fuze and exploder system at plate perforation (*see* Fig. 5.11). Base adapters have been used with H.E. nose-fuzed shell to facilitate manufacture, but the manufacture of such shell has been discontinued in peace time; a disadvantage of these shell is the difficulty of sealing the joint.

Base adapters are turned with a flange and are screwed externally to fit the shell, the plain part in front of the thread having a reduced diameter to accommodate the container, and they are bored and screwed internally to take the fuze. The length of the external thread on the adapter is determined by similar conditions to those governing the strength of the base of the shell, but the thread and flange must also be of sufficient strength to withstand the impact of the projectile on the plate without allowing the adapter to set forward under stress of its weight and that of the fuze.

(*See* Appendices for the formulæ that have been used in connection with the attack of armour.)

Fuze Sockets.—Shrapnel shell invariably have sockets, which are made either of metal or steel.

Fuze-hole bushes.—In certain cases these are permitted as an alternative to the solid head in H.E. shell. They are made of steel; metal was permissible at one time, but its use is now discontinued.

§5.12. Capped shell.

Caps of steel (*see* Figs. 5.20(*b*) and 5.08(*b*)), fixed over and attached firmly to the head of the projectile, have been introduced for armour-piercing (A.P.) and common pointed (C.P.) shell for use against armour.

To perforate an armour plate, a shell has to punch its way through the thin hard face and the thick and tough supporting back.

The object of the cap is to hold the head together at the moment it meets the hard face, and so cushion the blow which tends to shatter the point of the projectile, and there is consequently less tendency for the point itself to be broken or crushed. In a certain type of cap the forward portion of the cap itself is hardened. This penetrates the hard face of the armour and thus opens the way for the real point, so further increasing the penetrating power of the shell. With a high striking velocity, at normal impact, the cap is found to add materially to the penetrating power of the shell against a cemented plate, but the relative assistance given by the cap to the shell decreases as the striking velocity drops, and at low velocities, say under 1,000 f/s, becomes inappreciable.

There are various ways of attaching penetrative caps to shell :—

(*a*) By cotter pins. Tapered holes are formed in the head of the shell and cap, into which are driven cotter pins.

(*b*) By notching the cap into indents made in the head of the shell, usually about six.

(*c*) By pressing or rolling the cap whilst hot into an annular groove.

(*d*) By interrupted raised ribs on shell and in cap.

(*e*) By tinning the cap and the head of the shell at a temperature of about 450° F., sweating them together, finally notching the skirt of the cap into indents of the shell.

Methods (*a*), (*c*) and (*d*) are not now employed.

§5.13. Shell for use in guns and howitzers.

Though guns and howitzers of the same calibre do not as a rule fire shell of the same weight, projectiles can be designed for use in both types of ordnance, provided the difference in chamber pressure and in the twist of the rifling is taken into account.

Where both a gun and howitzer of the same calibre exist, the words " GUN," " HOW.," or " GUN and HOW." are marked clearly on the shell.

Heavy and Light Shell.

Certain guns and howitzers are issued with " heavy " and " light " projectiles. They are distinguished by the letters " H " and " L " stamped on the base and stencilled on the shoulder after the numeral of the shell.

§5.14. Manufacture of shell.

The manufacture of shell does not differ in general principle in the various types. The normal method is by forging from steel ingot or bar, though in certain cases such as practice and smoke shell the shell are cast direct to the required shape.

The steel is manufactured by the acid open-hearth or electric furnace processes, though for gun shell below 6-inch calibre and all howitzer shell, the basic open-hearth process may be used.

The steel is top-poured into ingot moulds, after which the upper end of the ingot (about 25 per cent. by weight) is removed to get rid of piping. The fracture is then examined.

All casts and ingots are numbered. Each cast is analysed. The ingots are then issued. If more than one shell is to be made from the same ingot, the ingots are rolled to their approximate section, and are fractured into their required lengths. The following procedure is then carried out on the ingot or billet :—

(1) Forged hollow by punching or drawing, limiting temperature 1,150° C. The base of the ingot must form the base of the shell, except with piercing shell where it forms the head.
(2) Forgings gauged and examined for flaws. One per cent. of forgings are tested for mechanical properties.
(3) Surplus metal at the open end cut off, leaving the shell a definite length.
(4) Base centred concentrically with interior cavity.
(5) Turned on the exterior to plan dimensions.
(6) Cavity bored out and the length of the shell corrected if necessary. Shrapnel shell are only machined inside where the tin-cup and disc fit into the cavity.
(7) Base faced and open end tapered.
(8) Bottling. The open end is heated and closed in under a press.
(9) Mouth faced and fuze-hole bored.
(10) Radius turning of head.
(11) Thread of fuze-hole cut.
(12) Weight adjusted by removing metal from the base.
(13) Groove for driving band formed.
(14) With H.E. shell, the base is recessed for base-plate.
(15) Cavity cleaned by sand blast.
(16) Preliminary examination.
(17) Driving band pressed on.
(18) Machining of driving band to plan dimensions.
(19) With H.E. shell, base-plate fitted, caulked, or riveted.

NOTE.—Base plates of H.E. shell are made of material in which the grain runs parallel to the face of the plate, that is, at right angles to the axis of the shell.

(20) Base plate faced flush.
(21) Cavity again sand blasted and varnished with copal varnish.
(22) Shell stoved at 300° F. to dry the varnish thoroughly.
(23) Shell stamped and greased outside.
(24) Examination and proof.
(25) Shell painted.

TYPES OF PROJECTILES

§5.15. Practice shot.

Used for practice over sea ranges. These projectiles are usually solid cast-iron of the same weight as the service projectile. A steel flat-headed practice shot with tracer has been introduced for tank practice from the 3-pr. 2-cwt. gun over land ranges.

§5.16. Proof shot.

For the proof of guns, howitzers, and charges. They are made of forged steel of the same weight as the corresponding service projectile, and are cylindrical in shape and flat-headed, so that they shall not penetrate too far into the butt.

§5.17. Paper shot.

These are used to test the mountings of guns, which cannot, owing to their position, fire service projectiles in time of peace. They are designed to cause the same amount of recoil as a service projectile and to break up in the bore. The body is made of wood pulp or rolled brown paper, and is filled to the correct weight with small shot and sawdust.

§5.18. Case shot.

These generally consist of three or more long steel segments held in position inside a thin tinned-plate canister, the whole being filled with bullets. The top and bottom are formed of steel plates, over which the serrated edge of the canister is turned. In the larger natures the shot is strengthened by a central bolt. The object of this construction is to facilitate the shot breaking up on leaving the muzzle, allowing the bullets to scatter.

§5.19. Common pointed (C.P.) and common pointed capped (C.P.C.) shell.

General design.—These shell are designed for the attack of lightly armoured vessels, concrete emplacements, dug-outs, etc., and are not intended to penetrate thick armour. They are made of forged steel, and are usually about 3·5 to 4 calibres in length. The walls are thicker than those of nose-fuzed H.E. shell, in order to enable the shell to hold together on the shock of impact and to penetrate before bursting. The thickness at the point is not much greater than that of the walls, as the main consideration in design is to give as large a bursting charge as possible, and therefore a large capacity, though latterly the tendency has been to make the head slightly more solid.

C.P.C. shell were used in the larger calibres, 6-inch and above, and these shell have hardened points which increase their penetrating power.

Shell with hardened points are liable to spontaneous cracks. They are therefore subjected to a keeping trial in the open to allow latent cracks to develop after hardening and before final acceptance.

Filling of C.P. and C.P.C. shell.—C.P. and C.P.C. shell may be filled either with gunpowder or H.E.

In the case of gunpowder fillings, with C.P. above 12-pr., the

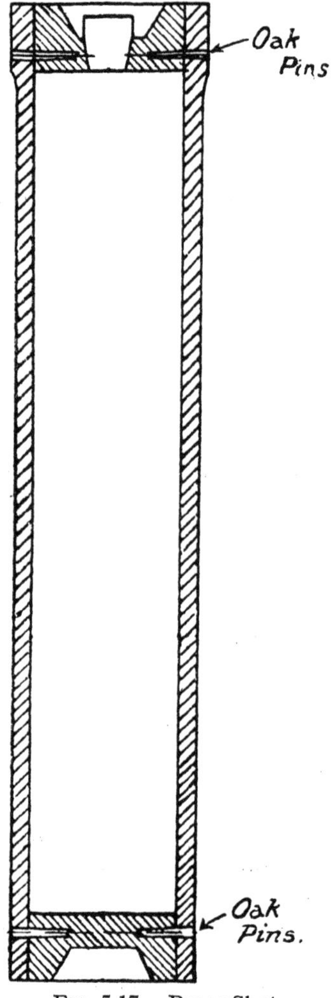

FIG. 5.17.—Paper Shot.

powder bursting charge is enclosed in a dowlas bag, the neck being made of shalloon to facilitate ignition, as shalloon is more permeable to flash. The base of the shell is packed with primers filled powder to make a compact filling and facilitate ignition by the fuze. C.P.C. shell are lined with copper containers to prevent accidental prematures through spontaneous cracking of the head. C.P. shell 12-pr.

and below differ from the remainder in that there is no dowlas bag, the powder being poured straight into the shell and there are no primers.

C.P. shell are coated internally with a special paint, " velvril," in order to prevent friction between any loose powder and the walls, causing a premature. For a similar reason the inside of the copper container in a C.P.C. shell is varnished.

The nature of the powder used for filling C.P. and C.P.C. shell is :—

Q.F. up to 6-pr.	G.12.
Q.F. 12-pr. and above	P.3 and G.12.
Primers are filled with	G.12.

In the case of H.E. fillings except C.P., the shell are fitted with an aluminium container, the interior of which is coated with copal varnish.

The 6-in. C.P.B.C. is filled with a mixture of trotyl and beeswax in the proportion of 93 to 7, a block of R.D. composition 1006 being first inserted into the apex of the container.

A cavity lined with a paper tube is formed in the filling to receive the exploder system consisting of two $2\frac{3}{4}$-oz. pellets of trotyl. Fig. 5.20 (*c*), showing filling of 9·2-in. A.P.C., is almost identical with 6-in. C.P.B.C., except that the trotyl surround is omitted.

Closing the bases of C.P. and C.P.C. shell.

(*a*) C.P. shell are screwed internally to take the fuze or plug, and in some cases a steel bush is fitted.

(*b*) C.P.C. shell are fitted with a base adapter, large enough for the copper or aluminium container to be inserted.

NOTE.—C.P. shell filled powder or H.E. do not have containers, as the points are not hard.

The closing of the base of a C.P. or C.P.B.C. filled H.E. is similar to that of an A.P. shell filled H.E.

§5.20. Armour-piercing (A.P.) and armour-piercing capped (A.P.C.) shell.

General design.—The definite object of these shell is to perforate armour and then burst effectively ; every other consideration is subordinated to this end. At proof a successful shell is one that perforates the plate and emerges at the other side with the cavity for bursting charge intact.

To attain this purpose an A.P. shell is made of forged or cast steel, and has a pointed and solid head which is specially hardened. The walls are considerably thicker than those of C.P. shell to enable them to withstand the shock of impact, and are of toughened steel, in order to afford greater tenacity to ensure the shell holding together when striking hard-faced armour, especially at oblique impact. Its capacity is small in comparison with C.P. shell, as this is sacrificed in order to provide strength for perforation.

The c.r.h. of an A.P. shell is small, about 1·6 calibres, as the hole through the armour has to be punched as quickly as possible in order that the striking velocity may not be entirely lost before the shell gets through.

Owing to the hardening process to which the heads are subjected, they are liable to spontaneous splits. A.P. shell are stored, therefore, in the open for three months before final acceptance, to allow any latent crack to develop.

A.P. capped shell (*see* page 115) are used with the larger calibres, 6-inch and above.*

Filling of A.P. and A.P.C. shell.—A.P. and A.P.C. shell were formerly filled powder, and stocks of powder-filled shell are still

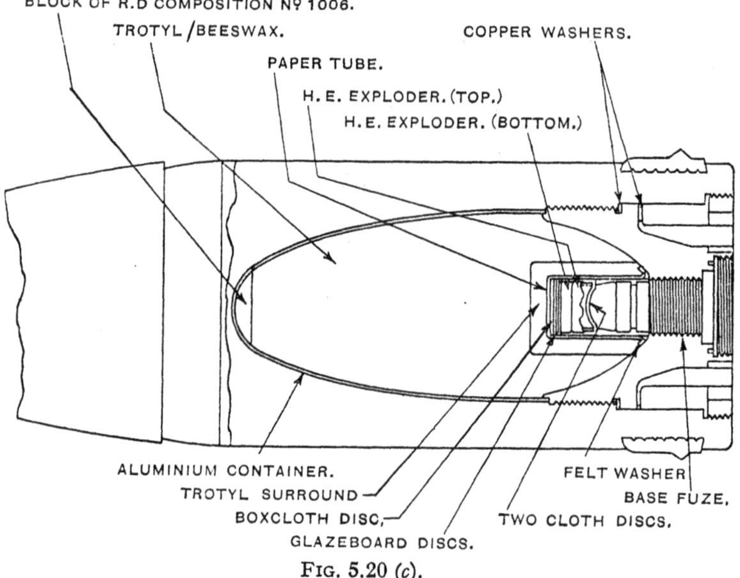

FIG. 5.20 (*c*).

held in the coast defences. In future, however, the filling will be H.E. There are also a number of shell in existence filled lyddite. The present approved filling for this type of shell is, however, trotyl or a mixture of trotyl and beeswax.

The filling of A.P. shell with powder is very similar to that of C.P. shell. The differences are :—

(*a*) Lasting cloth is used instead of dowlas, as it is closer in texture. There is no shalloon neck.

(*b*) G.12 is used instead of P.3 mixture.

A.P.C. shell filled powder are similar to A.P. shell, but have less capacity. There are no copper containers.

---

* For further information on the attack of armour, *see* Appendices.

A.P.C. shell filled H.E. are lined with an aluminium container to guard against premature detonation due to spontaneous cracking of the hardened head. A cavity is left in the base of the filling, lined with paper, into which is placed an exploder containing picric powder for lyddite or shellite fillings, and trotyl for trotyl fillings. The space left by the shrinkage of the filling is filled with beeswax composition; 3-pr. A.P. shell are filled lyddite without a container.

Armour-piercing shell with cap, filled H.E., Trotyl and Beeswax. (Fig. 5.20 (*c*).)

Fig. 5.20 (*c*) shows the latest type of A.P.C. shell; it is also typical of the 6-in. C.P.B.C.

There exist, however, several earlier marks which will be found to differ in details regarding adapter fittings.

Closing the base of A.P. and A.P.C. shell (gunpowder filling).—A.P. and A.P.C. shell filled powder have a steel bush screwed into the base, which is screwed internally to take the fuze or plug.

Closing the base of all pointed H.E. shell (except 3-pr.).—Pointed shell are fitted with a base adapter of approximately the same diameter as the cavity.

The base of the fuze, or plug, is covered by a copper gas-check plate to prevent the propellant gases penetrating through or over the fuze or plug. The gas-check plate is held in position by a base cover plate consisting of a perforated steel plate and screwed ring, which engages with a thread cut in the adapter. Gas-check plates are essential for these shell.

§5.21. Semi-armour-piercing (S.A.P. and S.A.P.C.) shell H.E.

These shell are intended for use against lightly armoured vessels, such as submarines and light cruisers. They are similar in design to A.P. shell, but there is less material in the head, and the capacity is greater. Their penetrative power is superior to that of C.P. They are filled lyddite or trotyl, and the base is closed in a similar manner to other pointed H.E. shell.*

§5.22. High explosive (H.E.) shell.

Design.—H.E. shell are designed to cause damage to material by the force of their burst, or to personnel and aircraft by fragments.

Against earthworks, etc., the main consideration in design is that the bursting charge should be as large as possible. For example, in a large mortar bomb, the weight of the charge forms a

---

* NOTE.—A table showing the fuze for each type of shell will be found on page 201.

A list of service plugs will be found in R.A.O.S., Part II, Pamphlet No. 1, Appendix VIII.

high proportion of the total weight, as the walls can be made very thin, owing to the shock on discharge being small.

Against personnel and aircraft, the maximum effect is obtained when the shell breaks up into a large number of pieces, each of which should retain sufficient velocity and be large enough to kill or disable a man at a reasonable distance from point of burst.

Against personnel a quick-acting fuze is required, whilst against material a delay or non-delay fuze, according to the depth of destruction required.

Capacity and strength of walls.—To obtain a maximum capacity, the walls may be tapered, and made as thin as practicable consistent with adequate strength. It must be remembered, however, that a shell which gives excellent mining effect by virtue of its large capacity, may, owing to the lack of metal in its walls, give inferior fragmentation. For other reasons also thicker walls than those theoretically possible may be adopted, as, for example, in the case of parallel walled shell (*see also* page 99).

Avoidance of prematures.—It is of the utmost importance that the risk of prematures should be minimized. H.E. shell are therefore made of forged steel, with walls and base (usually) in one piece ; and, in order to guard against the possibility of flaws in the centre of the base due to piping in the ingot, a shallow recess is bored out, and, after examination, a base plate is screwed or caulked in.

When using amatol 40/60, in which Grade III trotyl was used as a filling for shell fitted with screwed-in bases, an exudation of trotyl oil sometimes took place through the threads of the base adapter, which was thought to have been the cause of prematures. In order to prevent this exudation, the base adapter was caulked internally and externally, and the bottom of the shell cavity filled with a pad of cement. The use of Grade III trotyl has ceased.

The interior of the shell is copal varnished in order to give a smooth surface and to prevent possible corrosion of the steel by the filling.

With lyddite and shellite fillings, it is essential that all materials used in the manufacture of fuze-hole bushes, fuzes, plugs, etc., and also paint, varnish, luting, etc., should be lead-free.

§5.23. Filling of H.E. shell.

The explosives used are :—

(1) Lyddite.
(2) Trotyl.
(3) Amatol.
(4) Shellite.

Before the Great War, all common (H.E.) shell in the land service were filled lyddite. Just before the War, trotyl was introduced for H.E. shell for field equipments, but owing to the enormous increase of H.E. shell required during the War and the difficulty of producing enough lyddite and trotyl to meet the demands, amatol, in which trotyl is combined with the cheap and easily produced ammonium nitrate, opened up a large supply of high explosive, and will be the filling chiefly employed for field equipments in any future war, though for peace time, amatol is not normally employed owing to its liability to cause corrosion to shell and components.

A table showing the future fillings for land service shell will be found on page 129.

General remarks on filling.—The principal requirements for a satisfactory filling are :—

(*a*) That the explosive can be filled into the shell in a sufficiently dense form and in such a manner that cavitation does not take place, and that the filling does not set back on the shock due to the acceleration of the shell.

A fairly high density of the filling is essential. Firstly, because the detonation is more violent, and, secondly, because, if the filling is not dense enough to prevent setting back on acceleration, with a sensitive filling such as lyddite prematures may occur ; with less sensitive fillings, such as amatol, the air gap left between the exploder system and the filling may be sufficient to decrease the intensity of the detonating wave and cause loss of power in the detonation. It must be realized, however, that an explosive is more difficult to bring to a state of detonation if it is compressed. The ideal condition would be a density progressively increasing with distance from the exploder system.

The effect of confinement on a high explosive shell is of interest. In the case of high explosives possessing high velocities of detonation, for instance, exceeding 6,000 metres per second, the effect of tamping, as determined by the fittings used for closing the shell, such as fuze, etc., is not of great moment, although even in such cases the use of, say, aluminium in place of brass fuzes has an adverse effect. In the case of explosives of the amatol type, however, where the velocity of detonation is little over 5,000 metres per second, the effect of tamping is distinct.

(*b*) That the exploder system is suitable and will amplify the detonating wave initiated by the fuze sufficiently to detonate completely the mass of explosive in the shell.

The exploder system is employed to amplify the impulse given by the fuze. Such an impulse may be one of two kinds : (i) detonative, when provided by a fuze or gaine filled with C.E., and (ii) ignitory, when a flash-giving fuze is used. It is convenient to consider the exploder systems used with these two types of impulse separately.

(i) *Exploders for detonative impulses.*—In all detonating systems, an impulse of sufficient intensity must be imparted to the main charge to ensure its complete detonation. A comparatively insensitive filling, such as cast trotyl, may be brought occasionally to complete detonation by the fuze alone, but not invariably, and some additional impulse is normally necessary.

The simple exploder consists of a bag or pellet of explosive which augments the impulse given by the fuze or gaine. Exploders of this type are used when the main filling is such that it is unlikely that it will retreat when the shell is fired.

As is mentioned later, the steel exploder container was introduced with amatol fillings in order to exclude damp. This container is in the form of a steel pocket attached to the fuze-hole, the pocket being surrounded by a sleeve of trotyl. It was found later that its use permitted the employment of an amatol filling of a density somewhat lower than that considered necessary to obviate cavitation on firing. This was of great value to the land service as it enabled supplies to be augmented. Fillings of this character may possibly retreat to some extent under the stress of set-back on firing, but the continuity of the explosive system is maintained by the sleeve of trotyl surrounding the exploder container. This sleeve of trotyl can follow any retreat of the main filling, and so preserve an effective train of explosive between the exploder and the main filling.

Just as it is necessary to have the main filling in such a condition that firing in the gun does not bring about an excessive amount of movement or retreat, so it is necessary for the exploder to remain in contact with the fuze. It has been found that the wave of detonation loses power rapidly in the air, and, therefore, any air gaps in the system of fuze, exploder, and main filling are detrimental to the efficiency of the shell as a whole. In order to maintain such contact between the fuze and its exploder, the practice of compressing the exploder was adopted.

(ii) *Exploders for igniferous impulses.*—The base fuzes in the Service are at present mainly of the igniferous types. With these, the exploder system is ignited and burns with an increasing degree of violence until detonation occurs. The degree of this detonation depends upon the nature of the filling and the size of the projectile. The time required to bring about explosion is such that the burst does not take place until the target has been penetrated. With igniferous base fuzes in shell filled H.E. the exploder is normally picric powder.

§5.24. Development of the modern exploder system.

The methods used for filling H.E. shell with lyddite before the Great War were as follows :—

(1) *Long central cavity filling.*—A long central cavity was left in the lyddite for an exploder of picric powder. This was the

method of filling for all calibres up to B.L. 10-inch. The substitution of compressed pellets of trotyl for the P.P. exploder was ordered for use with a detonating fuze, but has not been carried out to any great extent owing to the lack of suitable fuzes. The cavity at the nose of the shell would be filled in this case with exploders of trotyl in bags. This type of filling will now only be found in small Coast Defence shell.

(2) *Solid filling.*—There was no central cavity, but a small space was left on top of the lyddite in the nose of the shell, which was filled with exploders either of P.P. or trotyl.

(3) *Short central cavity filling.*—This method was adopted for the smaller natures of Q.F. guns. A short cavity was left in the top of the filling to take an exploder sufficient to make contact between the fuze and the bursting charge. (*See* Fig. 5.24(*a*).)

These three methods of filling will still be found in H.E. shell of pre-war manufacture of which there are stocks in coast defences at home; for future fillings, however, only (3) will be retained as the method of filling for H.E. shell for coast defence, except in a few special cases. (*See* Fig. 5.24(*b*).)

With the introduction of amatol filled shell different methods of filling had to be adopted. In the first place the explosive requires a stronger send-off, and secondly the adoption of sensitive fuzes with gaines necessitated alterations in the size of the cavity.

The following factors led up to the present methods of filling H.E. shell:—

In the first instance, the shell were prepared with a cavity exactly to accommodate the No. 1 gaine, or, as an alternative, a 1½-oz. or 1¼-oz. exploder of trotyl crystals according to the type of fuze used. As this gaine proved unsatisfactory, the No. 2 gaine was introduced, which was shorter. The space thus formed in the cavity was filled with a 14-dram trotyl exploder in a batiste bag. In order that it should not be necessary to remove the 14-dram exploder from the shell when fuzes without gaines were used, the 1½-oz. exploder was discarded and a 10-dram exploder issued to fill the space above the 14-dram exploder. The bag of the 10-dram exploder was provided with a silk loop for its removal when replaced by a No. 2 gaine.

The necessity of inserting these bags choke downwards in the cavity was proved by the fact that detonations were seldom obtained with the choke upwards. It was also found that, owing to insufficient compression, the exploder bags set back on the acceleration of the shell from the bottom of the gaine or fuze, so creating an air space which led to explosions only. As a remedy, a trotyl pellet ·75-inches* long was inserted before the bag, and the

* In some designs of fillings without exploder containers, to obviate the use of this pellet a cavity of 5·55 inches was introduced.

screwing-in of the gaine or plug compressed the exploder bag tightly below the gaine, and eliminated the result of set-back. The top exploder of 10-drams used in place of the No. 2 gaine was also replaced by one of 14-drams. Two batiste discs were placed between the exploder and gaine to prevent the bag being torn whilst the latter was being inserted.

These changes were still found to be unsatisfactory with amatol 80/20 fillings, and it became clear that the root of the trouble was the want of power in the exploder system. The system, however, worked satisfactorily with lyddite, trotyl, 40/60 and 50/50 amatol fillings. (*See* Fig. 5.24(*c*).) It was decided, therefore, after lengthy experiment, to introduce the above arrangement as the standard exploder system for all fillings, but to give additional power by surrounding the exploder with trotyl for 3·7-inch and above filled amatol 80/20. (*See* Fig. 5.24(*d*).)

In order to simplify the system, a single exploder was substituted for the bottom 14-dram exploder bag and ·75-inch trotyl pellet, the latter having been really in the nature of a makeshift. This exploder was distinguished by the letter " A."

Following on this, all exploders used for shell fillings in the field service were lettered. These exploders may be filled with trotyl crystals or C.E., these explosives being suitable for use with detonating fuzes, the material of the bag being batiste, cashmere, paper, or in the latest types vulcanised wool, the latter being found to suffer least from deterioration in storage. The following are the principal exploders in use, with their nominal weights :—

	Trotyl Crystals.	C.E.
	Drams.	Drams.
" A "	18·5	17.5
" B "	16·5	15.5
" C "	14.0	13·0
" D "	11.0	10·25
" F "	5·0	5·0
" S "	8·0	8·0
" Y "		9·0

The difficulty of detonating amatol satisfactorily, if the filling has become damp owing to the hygroscopicity of the ammonium nitrate, led to the introduction of exploder containers. These are steel containers and are screwed into the fuze-hole of the shell, the threads being carefully sealed with cement. This container serves many useful purposes. Besides excluding damp from the filling, it prevents the possible contamination of the exploder by the exudation of trotyl oil from the filling ; it also ensures contact being maintained between the exploder and fuze or gaine, should the filling set back during the acceleration of the shell. Its use has therefore been extended to trotyl fillings.

Whenever possible in modern fillings, a thick or thin card tube is now used to line the exploder cavity. Those used in lyddite or trotyl fillings are copal varnished. For amatol the tube is impregnated with paraffin wax. To complete the sealing of the cavity, the top of the filling is sealed with a waxed felt washer inserted whilst hot between the closing washers. Where practicable, the tube is left sufficiently long and provided with serrations cut on the edge. These are turned back and sandwiched between the felt and closing washers. This method affords a very effective protection against contamination to exploder containers and explodering and fuzing components.

Owing to the fact that amatol produces little or no smoke on detonation, with 3·7-inch and above, smoke composition may be embodied in the bursting charge (with amatol 40/60 the smoke composition is contained in a small box at the base of the filling). Smoke composition may also be employed in trotyl filled shell if conditions of observation necessitate.

For 13-pr., 18-pr., and anti-aircraft H.E. shell (*see* Fig. 5.24(*e*) and Fig. 5.24(*f*)) an exploder container is considered unnecessary both because of the small size of the filling and because there is less chance of the ingress of damp. For convenience of filling block charges are sometimes used in the case of 13-pr. and 18-pr. shell. They may consist of three blocks, the top and bottom being of amatol and the centre of trotyl, or the complete charge may be of lyddite contained in a paper carton.

An exception to the above system is the 6-inch howitzer streamline shell. This shell is fitted with a burster container made of steel and filled with trotyl, which takes the place of the exploder container with trotyl surround. A small cavity is made in the trotyl filling of the burster suitable for a " B " exploder (*see* Fig. 5.24(*g*)).

NOTE.—All bottom exploders are placed in the shell in the filling factory, where they are suitably compressed.

The shell are then either plugged or fitted with a top exploder, if applicable, and fuzed. The top exploder is used with certain fuzes and is issued with the fuze when required; it is compressed by screwing in the fuze. The following shell are normally issued fuzed: 4·5-inch, 3·7-inch, 18 pr., 13-pr. and anti-aircraft.

Table 5.24 shows the methods of filling and exploders to be used in H.E. shell, but it must be realized that there are still large stocks of H.E. shell, held over from the War, which may embody any of the different steps that have been described in the building up of the present systems.

In the case of H.E. shell filled lyddite (coast defences) the exploder should normally be of picric powder for use with igniferous fuzes (*see* page 131) or trotyl crystals for use with detonating fuzes.

TABLE 5.24

A.—FIELD SERVICE

Nature	Figure No.	Filling	Depth of cavity	Exploder system	Exploder*
3·7-inch–12-inch except 6-inch how. S.L.					
3·7-inch, 4·5-inch, 60-pr., 6-inch and 12-inch	Figure 5.24(*c*)	Trotyl	6·2-inch	Exploder contr.	A.
3·7-inch, 4·5-inch, 4·7-inch to 9·2-inch	Figure 5.24(*c*)	Amatol poured	6·2-inch	Exploder contr.	A.
3·7-inch to 9·2-inch	Figure 5.24(*d*)	Amatol 80/20 trotyl surround.	6·2-inch	Exploder contr.	A.
3·7-inch to 12-inch how.	—	Lyddite.	6·2-inch.	Non-expr. contr.	A. Trotyl.
6-inch how. S.L.	Figure 5.24(*g*).	Amatol. Trotyl. Lyddite.	3.9-inch.	Burster contr. filled trotyl.	A.
13-pr. and 18-pr.	Figure 5.24(*e*)	Amatol 80/20 pressed centre trotyl.	6·2-inch.	Non-expr. contr.	A.
18-pr.	—	Trotyl poured.	6·2-inch.	Non-expr. contr.	A.
13-pr. and 18-pr.	Figure 5.24(*e*).	Trotyl	6·65-inch.	Non-expr. contr. Smk. Bx.	B.

B.—COAST DEFENCE

Nature	Figure No.	Filling	Depth of cavity	Exploder system	Exploder*
12-pr. and above.	Figure 5.24(*a*)	Lyddite.	3·65-inch.	Non-expr. contr.	5 drs. P.P.
4·7-inch and above.	Figure 5.24(*c*)	Trotyl.	6·2-inch.	Exploder contr.	A.
6-inch gun, Mark XXVI/XXB.	Figure 5.24(*b*).	Lyddite. Trotyl.	4·62-inch. 4·62-inch.	(No. 9 gaine over expr.)	26 drs. TNT 26 drs. C.E.

C.—ANTI-AIRCRAFT

Nature	Figure No.	Filling	Depth of cavity	Exploder system	Exploder*
3-inch, 20-cwt. 16-lb. shell.	—	Trotyl	5·13-inch.	No. 8 gaine over expr.	C.
Do.	—	Trotyl.	7·94-inch.	No. 8 gaine over expr. and Smk. Bx.	D.
Do.	Figure 5.24(*f*).	Trotyl.	7·27-inch.	No. 9 gaine over expr. and Smk. Bx.	Y.

* The top exploder, if any, will depend on the type of fuze used.

§5.25. Efficiency of H.E. shell.

The maximum effect of an H.E. shell is obtained by complete detonation of its filling, though this is not always attained; and before being passed into the Service a percentage of shell are proved to determine their detonative qualities. The degree of detonation or of explosion obtained is judged by the appearance of the burst or "plume," and the colour and amount of smoke produced. Complete detonation of lyddite and trotyl gives a black smoke, and the "plume" should be regular about a central vertical line and thrown well up into the air. Yellow streaks, particularly noticeable with lyddite, mean undetonated material. The black smoke is due to free carbon in the products of detonation.

With amatol the ammonium nitrate supplies extra oxygen, and in the case of amatol 80/20 the proportions are worked out to give complete combustion of the products, consequently there is very little smoke at all. This has led in some cases to the addition of special smoke-producing material to assist observation. The smoke of amatol forms a bluish-grey haze, which becomes darker as the proportion of trotyl rises.

It has been pointed out that the exploder system for H.E. shell has been arranged to produce certain effects, and mention has been made of the igniferous and detonative impulses used in the Service. It should be realized that high-explosive shell, to be efficient in producing damage, must not only be fragmented into pieces of sufficient size, but that these fragments must have additional velocity imparted to them beyond that given them by the translational velocity of the shell. It is possible that a shell may be fragmented, but that the fragments may have little damaging power because they possess little velocity. In the case of a high-explosive filling which is completely detonated, a large amount of energy is imparted to the fragments, and it is this which makes the fragments of an H.E. shell, with suitable size of filling, so effective.

In the case of lesser degrees of violence, which are termed "partial detonation (P.D.)" or "explosion only (E.O.)"—resulting, for instance, from the use of the igniferous system of impulse in small-calibre H.E. shell—a considerable proportion of the filling is used up in the ignitory phase of the explosion. Thus, it is found that 2 lb. of H.E. brought to complete detonation produces as much damage to the surroundings as 10 lb. of similar H.E. which is merely ignited.

In all cases of detonation the container of the high explosive concerned is broken up, irrespective of its mechanical strength, and the fragments produced are distorted or stressed according to the degree of violence produced. In the case of lyddite the disruptive power is great, and this is shown by the effect on material in the neighbourhood of its detonation. Thus, if the fragments of a high-explosive shell filled lyddite and 80/20 amatol respectively be compared, it will be found that, while the fragments do not differ greatly in actual size, the indications of mutilation and distortion on the lyddite fragments is much greater than in the case of the amatol, and this reflects the velocities of detonation, which are of the order of 7,250 metres a second for lyddite and 4,620 metres a second for amatol.

With lyddite fillings and igniferous fuzes, a picric powder exploder is usually employed, which when ignited burns to detonation. With a detonating fuze trotyl exploders are used. Picric powder is not suitable as an exploder for trotyl and amatol. Its use as an exploder is therefore restricted to lyddite fillings where an igniferous fuze is employed.

With amatol 80/20 it was found necessary, except in very small calibres, to provide a " trotyl surround " in addition to the exploder, owing to the difficulty of detonating this mixture satisfactorily.

It should be noted that C.E. exploders are not used with lyddite or shellite filled shell except that with 6-in. howitzer streamline shell, filled lyddite, which has a burster container filled trotyl, a C.E. exploder is inserted when the fuze used is not fitted with a gaine

Owing to the use of amatol, and the introduction of graze and time fuzes requiring gaines, it was found necessary to increase the size of the exploder cavities used in pre-war designs. In order, however, to standardize filling operations as much as possible, lyddite and trotyl cavities were increased to agree with those required for amatol. By standardizing the cavities and exploders used, the possibility of failures due to an air space being left between the fuze and the exploders was minimized, the exploders being made of such a size that they would be well compressed by screwing in the fuze.

The material of the shell walls has a bearing on the efficiency of the shell. The stronger the material used, the greater the size of the filling in relation to the total size of the shell and the more violent will be the detonation. But if the walls are made too thin, the fragments formed on detonation will be less effective than they might be. The walls of a H.E. shell are therefore generally made thicker, particularly at the front end, than they need to be to withstand the stresses of discharge. Steel with very high tensile strength is not therefore generally employed; the normal steel for H.E. shell has a yield point of 19 tons per square inch. Generally speaking, cast iron is unsuitable for H.E. effects, because the material is incapable of withstanding tensile stresses of any magnitude; the effect of a high explosive on cast iron is to produce an inordinate amount of what is practically dust, and the metal is therefore broken up into useless particles. Where shell have only to be broken open, such as is the case with smoke shell, then cast iron may be used.

§5.26. Smoke shell.

These are used for the production of smoke screens. Various smoke producing mixtures have been tried, the commonest being until recently white phosphorus.

Design.—The shell are made of steel, semi-steel or cast iron. They may be divided into three types :—

(*a*) Double diaphragm type. (white phosphorus)
(*b*) Burster container type. (,, ,,)
(*c*) Base ejection type. (smoke composition)

In the first two cases the chargings are introduced through a small aperture in the side of the body, which is closed by a tapered steel plug driven in tightly. The bursting charge is small and just sufficient to open the shell, as the smoke charge must not be scattered too far. A quick-acting fuze is used to prevent the shell burying itself before the burst takes place.

(*a*) *Double diaphragm type.*—This shell has a separate head screwed on to the body. Both body and head are formed as closed containers, the head containing the bursting charge, and the body the smoke charge. The filling in the head is of fumyl or trotyl which acts as a bursting charge.

This type is obsolete for future manufacture. It is complicated in manufacturing details and difficult to seal owing to the large joints.

(*b*) *Burster container type.*—In this design of smoke shell the smoke charge is contained in the body of the shell, but the nose is closed by a burster container screwed into the body and screw-threaded to take a 2-inch gauge fuze or adapter. The bursting charge is placed in the container, which is closed by the fuze or plug.

This type offers facilities for standardization of manufacture and filling, and is easier to seal effectively. The best method of sealing is by rolling or caulking the container to the body. The screw threads are cemented in addition.

NOTE.—There was another type of smoke shell "Tinned Plate Container Type." The white phosphorus was charged into the container apart from the shell, it had a pocket in the centre for the burster. It is now obsolescent.

White phosphorus smoke shell are provided for the following calibres as under :—

Calibre	Mark of Shell	Type
18-pr.	IC, IIC, III, IV, V and VI.	Burster container.
3·7-inch howitzer. ...	III, IV and V.	Burster container.
3·7 inch mortar.	I.	Burster container.
4·5 inch howitzer. ...	III.	Double diaphragm.
Do. Do.	IV, VI, VIII, IX and X.	Burster container.

(*c*) *Base ejection type.*—An improved type of smoke shell similar in principle to star shell has recently been approved for use in the field.

The shell is fitted with a separate base retained in position by steel pins or a brass screwed ring.

The filling consists of a number of canisters containing smoke composition of the hexachlorethane type together with a small bursting charge.

The principle governing their employment is that of bursting the projectile in the air by means of a time fuze and thus igniting the composition in the canisters which fall to the ground, burning in a manner similar to the Generator Smoke Ground No. 5 or the No. 63 Grenade.

The base ejection projectiles have several advantages over the old type, notably, improved safety in storage, since white phosphorus is no longer employed as a filling.

Further, they are more efficient in smoke production and have less tendency to become buried in soft ground, as happened sometimes in the case of shell fitted with percussion fuzes.

Target shell are special smoke shell used to produce smoke clouds forming targets for anti-aircraft practice purposes. The filling was originally gunpowder but is now white phosphorus. For the 3-inch 20 cwt. the design is as shown at Fig. 5.26(*c*).

Storage.—Shell charged with phosphorus must be stored apart from other ammunition, and when transported by sea must travel as deck cargo. If leakage occurs, the ammunition should be copiously drenched with water, and measures taken to destroy it.

Since phosphorus melts at 111° F., it must be kept at a cool temperature, and should not be exposed to the direct rays of the sun. A round should not be kept longer in a gun than necessary.

§5.27. Shrapnel shell.

Shrapnel shell are intended primarily for use against personnel. They were also used in the Great War for wire cutting, but were superseded for this purpose by H.E. as soon as a quick-acting fuze (No. 106) was available.

Design.—The main object in design is to carry the greatest possible weight of bullets, each bullet being large enough to be effective, *i.e.*, to put a man out of action. A striking energy of some 60 ft./lb. is considered suitable.

For field guns and howitzers there are 41 bullets to the lb., but old designs of shrapnel for medium field guns and howitzers took bullets of 35 to the lb., and heavier natures of shell 14 to 27. An increase in c.r.h. decreases the space available for bullets.

The kinetic energy of a bullet is given by $\frac{1}{2}mv^2$, where m is the mass of a bullet, and v its velocity; hence the necessity for the use of heavy metal in the bullet. In the case of the bullet for field guns and howitzers

$$\tfrac{1}{2} \times \frac{1}{41} \times v^2 = 60 \times 32 \text{ foot poundals.}$$

$$\text{so } v = 397 \text{ ft./secs. approximately.}$$

To be effective, the velocity for the bullet should not be below about 400 f/s. In this respect it should be noted that very few howitzers are provided with shrapnel for their equipment, on account of the large angle of arrival and the low remaining velocity at burst.

Cone of dispersion.—Another point that requires consideration in the design of shrapnel shell is the " cone of dispersion." This applies more particularly to the smaller type of shell, where the number of bullets is comparatively small.

When the shell is opened in the air the bullets are ejected in the form of a cone due to the forward velocity of the shell and its spin. With the smaller shell, if the opening is too big, a large proportion of the bullets will be wasted owing to the cone of dispersion being too great, so that a considerable portion of the bullets will not go in the direction of the target, but be thrown out on either side of it.

The angle of opening is the angle between the lines of flight of the outer bullets. This angle increases with the range, because the forward velocity of the projectile falls off faster than the spin. With modern shell of 2-inch fuze-hole, the opening left by the blowing out of the fuze socket gives a convenient size of opening for small natures.

Q.F. 13-pr. and 18-pr. shrapnel were at one time filled with pressed powder pellets in the central tube for the purpose of increasing the cone of dispersion and facilitating observation, but owing to the difficulty of keeping up the supply of these pellets during the war they were dispensed with and they will probably not be reintroduced, as the advantage gained is not sufficient to warrant the additional cost of manufacture.

Construction.—The *body* is made of forged steel as thin as possible consistent with standing the pressure set up on discharge, so that it may contain as many bullets as possible. The opening charge is small, as it is only required to blow off the head, which is lightly attached to the body, and eject the bullets which receive their velocity from the remaining velocity of the shell.

The bursting charge is of gunpowder contained in a tin cup fitted in a recess in the case of the shell; the tin cup is intended to confine the powder, which might explode on shock of discharge if it were loose in the base and liable to get nipped between the disc and the shoulder.

Above the tin cup is fitted a steel disc which rests on an annular shoulder formed on the interior of the shell. This protects the tin cup from damage when the bullets set back on the shock of discharge.

A *central tube* screws into the disc, and one end fits tightly into the mouth of the tin cup; the other end is soldered or expanded into the *fuze socket.* The socket is secured to the body by three or four screw-threads in 18-pr. shell and under; in larger natures it is screwed into the head or is made in one piece with the head. All modern shell have a socket of 2-inch gauge; shrapnel of the old type with the narrow G.S. gauge are now obsolete.

Filling.—The body of the shell is filled with bullets which are kept in place by resin poured in in a molten state and allowed to set. The resin prevents the movement of the bullets and also provides more smoke for observation purposes. In the larger natures a lining of brown paper is sometimes placed between the bullets and the body of the shell to prevent the bullets and resin adhering to it.

Composition of bullets.—The bullets usually used for shrapnel are made of an alloy called " mixed metal," *i.e.*, seven parts lead and one part antimony, the latter being the hardening agent. In some of the heavier shell 12-oz. steel balls or 2-oz. bullets made of cast iron, called " sand shot," have been used.

Method of closing the head, etc.—The larger natures of shrapnel are made with separate heads, held on by twisting pins and by shearing screws or rivets, which are sheared on the shell functioning. On top of the bullets is a felt washer, and the head is filled up with a wood block. In the smaller shell the fuze socket alone is used to close the head of the shell.

In order to prevent the powder working up into the fuze socket, the end of the central tube is closed by means of a shalloon disc. A fixing screw is provided in the fuze socket for retaining the fuze.

Action.—The fuzes used with shrapnel are designed to burst the shell either in the air by a time arrangement or by percussion on impact. In either case the action is the same. The flash from the magazine of the fuze passes down the central tube and fires the opening charge, which blows off the head or socket and ejects the bullets.

Anti-aircraft shrapnel.—Shrapnel was used during the Great War against aircraft, fitted with time fuzes only. It has been found, however, that H.E. shell with time fuzes are far more effective at large angles of sight owing to the high velocity of the fragments and the all-round direction of the burst. Anti-aircraft shrapnel is now only used against low flying and ground targets.

§5.28. Star shell.

General.—The function of the star shell is either to illuminate a particular area of limited extent or a definite target. It is used with 6-inch howitzers and below.

The star shell now in the service is of the single star and parachute type.

The rate at which the star will fall is determined by the size of the parachute; therefore, provided this is of sufficient area, the shell can be burst well up to the target, and so give efficient illumination at long ranges. At short ranges, if the parachute is ejected at too high a remaining velocity, it is apt to break away; it has been found necessary, therefore, to fire these shell with reduced charges at the shorter ranges in order to ensure correct functioning.

The remedy for this disability is a matter of design, and a shell has been introduced which is suitable for use with full charges. The body of this shell is made of high tensile steel, to secure capacity without loss of strength, and the parachute has a stronger attachment to the star.

Design.—The body is made of forged steel, and the shell is fitted with a burster in the head, supported on a baffle plate, which is prevented from rotating in the shell by means of screws. Below the baffle plate is the star, consisting of a steel case filled with star composition, with priming composition at the top. To the bottom of the star case is fitted a swivel which forms the means of attachment for the parachute below. A ball bearing is fitted which reduces friction and allows the parachute to open more readily. The parachute is made of fabric, strengthened by means of wire rope, and a series of wires are connected to a grummet, which is attached to the star case swivel. The parachute is folded up and pressed into the shell, the rear end of which is closed by a steel base, fixed with mild steel shearing pins and prevented from turning by a pin. A lead washer makes the joint between the base and the shell. The set-back of the star is taken on the base by means of two semi-circular steel supports.

Action.—On the time fuze functioning, the magazine of the fuze ignites the powder burster below, which sets up sufficient pressure to shear the mild steel shearing pins in the base, and which at the same time ignites the priming composition of the star. The star and parachute are ejected from the base of the shell, and the parachute then opens out and, righting itself, allows the flaring star to fall gradually to the ground, open end downwards.

§5.29. Practice projectiles.

Projectiles are specially made for anti-aircraft practice. Hitherto, these have been made of cast iron, and completely filled with gunpowder and a pellet of magnesium mixture to give a flash. This type has been superseded by one designed to give an indication of the burst without breaking up the shell body. The new type has a steel body, the threads in the fuze-hole being reduced in number to ensure blowing out of the fuze readily. The filling consists of a wood block with a small charge of gunpowder and the flash producing pellet.

For practice other than anti-aircraft.—Various types of service shell are used which for some reason have become unfit for use as service shell but which are suitable for use at practice. Such shell are called practice projectiles and may be fired, fuzed or plugged. In some cases, service shell are filled with inert material and fired plugged.

§5.30. Tracer shell.

Projectiles to which tracers are to be fitted are made with a socket in the base to receive the tracer. In some cases the socket is entirely within the base, which is thick enough to enclose the tracer. In others the tracer and socket are external to the base which is of normal thickness. In the former internal tracers are used, and in the latter external tracers, but their construction is similar, viz. a metal body, steel cap and brass closing disc. The tracer is filled with tracer composition and, at the rear, a thick layer of gunpowder S.F.G_2.

§5.31. Drill shell.

These have varied uses in " Drill " and different types are employed according to the requirement. Some of these, up to 4·7-inch calibre, are specially made in hard wood, and in one case—the 3·7-inch howitzer—a special canvas covered shell is used. In recent years quite a number of service shell, either empty or weighted, have been utilized for such purposes as " Fuze Drill " or " Loading Teacher."

§5.32. Mortar ammunition.

Mortars are designed to throw bombs at short range. The design of these bombs is governed by similar considerations, as regards set-back, as are applicable in the case of shell fired from guns, but the chamber pressure and muzzle velocity are far smaller. They are never fired at low angles of elevation, as an almost vertical angle of descent is required.

During the Great War rifled muzzle-loading mortars were tried ; they were not so successful, however, as the smooth-bore mortar, which was found to be quicker and more convenient. All modern mortars (except the 3·7-inch) are smooth bore.

Modern mortar bombs are made of forged or cast steel. In shape they are streamlined and are fitted with vanes on the base end, or on a rearwardly projecting tail. The air pressure on the vanes during flight prevents the bomb " toppling " and ensures a reasonably stable head-on flight.

During the Great War three main types of mortar were evolved :—

(1) Light, 3-inch and 4-inch.
(2) Medium, 2-inch and 6-inch.
(3) Heavy, 9·45-inch.

Of these, only the 3-inch is likely to be employed in future.

Three-inch bomb.—The 3-inch, or Stokes mortar, bomb was originally cylindrical. It contained a bursting charge of 80/20 amatol, which was fired by a detonator in the head.

A container for the propelling charge, consisting of a 12-bore ballistite cartridge, screwed into the base or, alternatively, was made in one piece. It was pierced with a number of holes to allow the gases from the cartridge to escape, in order to reduce the pressure inside the container.

Augmenting charges, consisting of rings of flaked cordite or of N.C.(Y.) powder enclosed in a cambric bag, were used with the cylindrical bomb to increase the range. These were placed as required round the container.

The cylindrical 3-inch bomb has now been replaced by a new vaned and streamlined bomb.

§5.33. 3-inch mortar bomb, 10-lb.

This bomb has a streamlined body fitted with a perforated container for the propelling charge, to this are attached six vanes to stabilize the bomb in flight.

For the propelling charge, a 95-grain ballistite cartridge is used. This can be augmented by secondary charges of N.C.(Y.), which are made up in celluloid cartons and are placed between the vanes. They are held in position by a coiled wire spring hooked around and through the holes of the vanes.

There are H.E., Smoke and Practice types of this bomb, on similar lines to shell. For H.E. and Smoke a steel exploder container is fitted, which constitutes the burster in the case of the smoke bomb.

The fuzes at present allocated are : Fuze D.A. No. 138 for H.E., Fuze D.A. No. 139 for Smoke and Fuze D.A. No. 139P for Practice.

§5.34. Insertion of plugs and fuzes.

Shell should not be fuzed until required for use, though this practice is departed from with certain calibres and natures of shell for reasons either of storage or supply. A plugged shell is safer in transport and storage than a fuzed one, and fuzes are best preserved in a serviceable condition when packed in their cylinders. Shell should never be left without a plug or fuze.

Projectiles taking nose fuzes.—If the shell is to be fired immediately, the fuze or plug may be inserted without lubrication, but if not, the threads of the fuze or plug should be lubricated with thin, lead-free luting and a fillet of thick, lead-free luting should be placed under the flange of the fuze or plug if it is not fitted with a washer.

Time and time and percussion fuzes are generally retained by a fixing screw, the hole for which should be filled with luting and the surplus wiped off after insertion of the screw.

Some fuzes for H.E. shell have a groove round the base of the body; the lip of metal below the groove is in this case punched into shallow slots cut in the nose of the shell.

Projectiles taking base fuzes.—The plugs or fuzes must always be coated with lead-free luting, thin on the threads and thick under the flange immediately before screwing home into the shell. The luting acts as a lubricating and waterproofing medium, and also prevents the propellant gases entering the shell.

§5.35. Causes of blinds and prematures.

Blinds.—The fact of a fuzed shell failing to burst may be attributable to faults in either the fuze or the filling. Those due to the fuze are enumerated on page 160, Chapter VI.

The faults in the filling are :—

(1) Exploders not detonating or exploding due to damp, especially those made from picric powder.

(2) During the Great War exploders often failed to detonate, owing to their being affected by the exudation of trotyl oil from low-grade trotyl in the bursting charge.

(3) With powder-filled shell a blind may occur owing to the filling having become damp.

(4) Shell may be incorrectly filled, with a consequent gap in the exploder system.

Prematures.—These may occur in the gun or outside it, the former being the most serious, especially with H.E. shell, and every endeavour is made to avoid them by means of safety shutters, which prevent fuzes from being able to function until the shell has left the muzzle. The causes of prematures in shell may be due to the shell, to the filling, or to the fuze. Those due to the fuze are enumerated on page 161, Chapter VI.

Prematures due to the shell.

(1) By a faulty shell breaking up in the gun.

(2) By the penetration of propellant gas into the shell through a flaw.

(3) Distortion of the shell due to—

(*a*) Oil in the bore or an obstruction. Oil, if too liberally applied, may collect on the driving band as the projectile travels up the bore, and as it cannot escape, forces in the walls of the shell, causing "*set-down.*"

(*b*) Weakness of the shell walls. The walls of the shell, being unable to withstand the pressure due to set-back, bulge outwards, causing "*set-up.*"

(4) A shell used in ordnance for which it is not intended, *i.e.*, a 6-inch howitzer projectile being fired from a 6-inch gun.

(5) Friction due to—

(*a*) Unvarnished walls.
(*b*) Absence of burster bags in powder filled shell.
(*c*) A rusty tin cup in shrapnel shell.

(6) Cavitation due to faulty filling, permitting set-back on firing. This is most likely to occur with lyddite filled shell.

(7) Exudation of trotyl oil from the filling in shell fitted with base adapters and filled with low-grade trotyl, or amatol made from low-grade trotyl, becoming ignited.

§5.36. Markings on projectiles.

Markings described are chiefly the latest, all those markings on existing stocks of projectiles which refer to out-of-date methods of filling, etc., will be found in R.A.O.S. Part II.

Markings on projectiles may be divided into two main classes :—

1. **Stampings.**
2. **Painting and stencilling.**

Stamping generally indicates facts concerning the empty shell and its inspection ; the type of shell, or filling, is sometimes shown by stamping, so that in the event of the paintings on the shell becoming obliterated the shell can be identified.

The position of the stampings is as follows :—

(*a*) On projectiles used for separate loading ammunition on the base.
(*b*) On projectiles used for fixed ammunition on the body above the driving band.

The following stampings will be found :—

1. Calibre and Mark.
2. Letter indicating approximate c.r.h.
3. H or L where a heavy and a light shell are used in the same gun.
4. GUN & HOW or G & HR on projectiles suitable for both gun and howitzer of the same calibre.
5. Manufacturers' initials (or recognized trade mark) and date of manufacture.
6. Lot No. of empty projectile.
7. C.S., F.S., or B.S. denoting that the projectile is made of cast steel, or forged steel, or bored from the billet of steel.
8. $\overline{\text{A}}$. on 6-pr. and 3-pr. steel shell that have been annealed.
9. $\overline{\text{C}}$.I. denoting that the projectile is made from cast iron.
10. S.S. denoting that the projectile is made from semi-steel.

11. M.C.I. stamped close to the top of the head of shrapnel shell fitted with malleable cast iron heads. (The use of this stamping has been discontinued.)
12. SMK., or C.P.B.C., denoting the filling or type of shell as applicable.

Painting and stencilling.—Projectiles are painted and stencilled for the following reasons :—

(*a*) To enable the nature of the projectile to be easily identified in the field.

(*b*) So that the sorting of projectiles can be done by unskilled labour.

(*c*) Provides a protection against rust or corrosion.

(*d*) Give details of filling ; date and place of filling ; Lot No., exploders, etc. ; and facilitate identification in case of failures and enable sorting and withdrawal to be readily affected.

Painting.—The bodies of projectiles are painted as follows :—

Yellow ...	... Shell with H.E. fillings.
Black ...	... Practice projectile, filled inert or practice shot.
Black ...	... Shell containing a powder filling, or a bursting charge of powder, *e.g.*, shrapnel and star, etc.
Green ...	... Smoke and smoke target shell.
Unpainted...	... Case shot.

Special cases :—

18-pr. shrapnel for use with reduced charges.	White from driving band to shoulder remainder black.
18-pr. H.E. for use with reduced charges.	White from driving band to green band, remainder yellow.
3-inch 20-cwt. shrapnel burst short charge.	White from driving band to shoulder, remainder black.
3-inch 20-cwt. H.E. burst short charge.	White from driving band to green band, remainder yellow.
3-inch 20-cwt. practice burst short charge.	White from driving band to yellow band, remainder black.
A.P., A.P.C., and C.P. shell filled shellite.	Green from tip to shoulder for uncapped shell, and on cap for capped shell.

Painting.—**Tips** of projectiles.

Red ...	... Shrapnel shell.
White ...	... A.P. and practice shot.
Black ...	... Absence of smoke-producing compound in amatol shell.

Painting.—**Rings** round head of projectile.

Red	Filled wholly or partly with an explosive.
Red crosses ...	H.E. shell filled amatol or trotyl and are suitable for use in hot climates.
Black ring above red ring.	Powder filled fuze should be used.
Two black rings below red ring.	H.E. shell fitted with P.P. exploder and powder primer, for practice with powder filled fuzes.
Green ring above red ring.	Trotyl exploders (will not be indicated in future).
Light brown ring above red ring.	Smoke shell made of cast iron or semi-steel.

Painting.—**Rings** round heads of piercing shell.

No white ring ...	Indicating C.P., C.P.C. or C.P.B.C. shell.
White ring above red ring.	,, S.A.P. shell.
Two white rings one on each side of red ring.	,, A.P. or A.P.C. shell.

Painting.—**Bands** round the body of shell.

NOTE.—Shell, filled lyddite.	No band to indicate filling.
Green band round centre of shell.	Indicates a filling of 80/20 amatol. If filled with other proportions of amatol, a green band and also a fraction will be stencilled below the band.
Green band round centre with TROTYL overprinted.	The shell is filled trotyl, or a mixture of trotyl and beeswax.
Black band above driving band on gun shell.	A gun and howitzer of the same calibre exists. Shell suitable to gun.
(Two vertical strips, diametrically opposite, from driving band to shoulder.)	(Indicates "Economy driving bands.")
Two black bands round body.	Practice shell filled H.E.
Yellow band on shrapnel.	Practice shrapnel.
Yellow band on body of powder shell.	Practice projectile.
Yellow band on smoke shell.	Smoke target shell.
Two yellow bands.	Practice projectile special.

Yellow band on body.	Practice projectile, filled inert and practice shot.
White band below driving band.	For star shell when suitable for firing with full charges.
Black band round body.	Empty H.E. shell used for drill purposes.

Stencilling.—The following general stencilling will be found :—

1. Calibre.
2. III or IIIA, etc. Indicating the Mark of shell ; the letter denoting the approximate c.r.h.
3. GUN or HOW., or GUN & HOW as applicable.
4. The letters H or L as applicable.
5. The monogram of the filling firm or filling station, date of filling (day, month and year) and a series number in a ring to distinguish the filled Lot are stencilled on all projectiles on the reverse side to that of the calibre, numeral, etc., except in the case of smoke shell, on which the markings relating to the body charging and to the head filling are stencilled below the calibre and numeral.

Markings on the shell relating to the *filling* :—

1. Fraction 70/30, proportion of mixture for shellite.
2. Fraction 40/60 or 50/50, proportion of mixture for amatol.
3. Fraction 93/7, proportion of mixture for trotyl/beeswax.
4. Design No. of method of filling.
5. The letters PHOS when charged with phosphorus.

Markings for presence of *smoke box* or *flash producers*.

1. Two green discs indicating smoke box in small shell.
2. Two aluminium discs indicating flash pellets in powder shell.

Markings on body of shell relating to the *exploders*.

1. EXPR BAG with weight and Lot No. of picric powder in the exploder in the case of C.P., S.A.P., A.P., and A.P.C. shell filled lyddite or shellite.
2. EXPR C.E. when shell are fitted with C.E. exploders.

   NOTE.—When trotyl exploders are fitted, no exploder markings will appear on the shell.

Markings on body of shell relating to *gaines*.

1. G.9 . . . IIZ . . . 5. Which markings indicate the inclusion of Gaine, Mark and Lot. No. respectively.

   NOTE.—If FOR G.9 appears on shell, it implies absence of gaine.

Markings on body of shell relating to the *fuze*.

1. When A.P. and C.P., etc., shell are fuzed, the letters FZD are stencilled together with the Lot No. of the fuzes, *e.g.*, FZD LOT 6.

In addition to the above markings the bases of A.P. and C.P. shell filled and fuzed are painted as follows :—

Cover plate painted red ... Non-delay fuze fitted.
,, ,, blue ... Delay fuze fitted.

2. Smoke shell issued fuzed with the No. 106 type fuze, have the body of the fuze painted green.
3. Shell suitable for one type of fuze only (other than No. 101 fuze) have the words USE . . . FUZE.
4. Shell with G.S. fuze-holes fitted with exploders for the No. 13 fuze in fixed armaments, have the words—USE SHORT FUZE.
5. On 6-pr. steel and 3-pr. A.P. shell fitted with fuze, percussion, base Hotchkiss, the fuze particulars are stencilled as follows :—

(Typical) MK. X FUZE. LOT 6.

On 3-pr. A.P. shell fitted with tracer fuze the Lot No. of the fuze only is stencilled.

Markings on shell body relating to *tracers*.

Projectiles prepared for or fitted with a tracer, in addition to the usual distinctive markings, have the following symbols stencilled on the head :—

 ... ... When prepared for tracer.

 ... ... When fitted with tracer.

 ... ... When fitted with tracer fuze.

Miscellaneous markings :—

A on 3-pr. and 6-pr. steel shell which have been annealed.

 denotes the centre of gravity on 9·2-inch gun shells.

LIM projectiles which are limited as to the full charge with which they may be fired. It is stencilled on the base and body of shell.

B B stencilled on the base of a shell filled with amatol and indicates the presence of a brass nose-bush.

R. PRAC 1934. the central tube of unboxed shrapnel shell affected by rust. When the powder is caked (central tube choked or not choked).

R. when the powder is free, but central tube rusted.

(year as applicable). Amatol filled shell with rusty fixed containers

$\frac{\text{R.F.C.}}{\text{B.M.1934}}$ which have been authorized for firing at practice.

$\frac{4976}{4}$ With date and monogram of station, evidence of repair or conversion. For H.E. the fillings of which have been repaired, have the design number of the method of rectification stencilled on them as above.

§5.37. Weight markings on projectiles.

(*a*) **Fixed Armaments :** All shell in fixed armaments whose weight filled and fuzed is within a limit of $\pm$ 0·1 per cent. of mean weight are stencilled DEAD WT.

Those which vary from the above are stencilled with their actual weights as follows :—

9·2-inch	To the nearest 1 lb.
Under 9·2-inch to 6-inch ...	To the nearest ¼ lb.
5-inch and under	To the nearest 1 oz.

(*b*) **Field Service :** All shell 3·7-inch and above for Field Service (except shrapnel) are weight marked on the " unit system," by which variations above and below the mean weight of a filled and fuzed shell are indicated.

The value of the units in lbs. and the calibres to which they are applicable are as follows :—

¼ lb. for 3·7-inch to 4·5-inch inclusive.
½ lb. for 60-pr.
1 lb. for 6-inch guns and howitzers.
2 lb. for 8-inch howitzers and 9·2-inch guns and hows.
5 lb. for 12-inch howitzers.
10 lb. for 14-inch gun.
15 lb. for 18-inch howitzer.

On every shell the number of units variation in weight from the mean nominal weight is marked, preceded by + or – to show whether it is over or under weight, thus :—

+ 3 on a 3·7-inch shell means that it is not more than ¾ lb. over weight.

– 1 on a 12-inch shell means that it is not more than 5 lb. under weight.

0 on any shell indicates nominal weight.

This marking is based on the actual weight of the filled shell fitted with the " Standard " fuze.

Position of symbols.

(*a*) Shell above 6-inch, in line with the calibre and Mark of shell—on the nose—in three places.

(*b*) Shell 6-inch to 3·7-inch, in line with the calibre and Mark of shell—on the nose—in two places.

(*c*) All smoke and target shell, fixed ammunition have the actual weight, filled and fuzed, to the nearest $\frac{1}{8}$ lb. stencilled on the shoulder.

(*d*) Practice projectiles 3-inch 20-cwt., the weight is marked to the nearest $\frac{1}{2}$ lb.

(*e*) All shrapnel shell are " dead weight."

(*f*) Star shell have no weight markings.

Position of markings on projectiles.

(*a*) Unboxed projectiles.—Markings are confined, as far as possible, to the ogive, and care is taken to avoid the shoulder, where most rubbing is likely to take place.

(*b*) Boxed projectiles.—As far as possible, markings are not placed on parts of projectiles which are liable to come in contact with fittings of boxes.

Markings on exploders.

Exploders are distinguished as follows :—

(*a*) By alphabetical series.—The empty bags are marked with the numeral, contractor's initials or recognized trade mark, and the word " Trotyl " or the letters " C.E." as applicable.

At the time of filling the bags are marked on the reverse side by the filling factory with the Lot No. of the explosive, date of filling and the initials of the filling firm or monogram of the filling station. The identification letter is added above the word " Trotyl " or the letters " C.E."

(*b*) By weight.—Exploders not in the alphabetical series, either bag or wrapped pellet, are distinguished by the weight and nature of the filling and are marked as in (*a*) above, with the exception that the weight replaces the identification letter.

(*c*) All trotyl exploders inserted in shell since April, 1929, are marked with a green bar across the plain ends.

(*d*) In the past, exploders were marked with the following particulars, viz. the numeral and weight, Lot No. of picric powder or trotyl, date of filling, monogram of the filling station and the letters " P.P." (except in the case of 7-dram exploders) or " T.N.T." as applicable, both on the bag and the paper cylinder.

CHAPTER VI

FUZES AND GAINES

INTRODUCTION

§6.01. General remarks.

The bursting charge of a shell is ignited or detonated by means of a fuze designed to act upon or after impact, or on graze or at any particular instant during flight.

Fuzes may thus be divided into two distinct classes :

1. Percussion fuzes.
2. Time, and Time and Percussion fuzes.

Percussion fuzes function as a result of the projectile in which they are fitted striking any object offering sufficient resistance.

There are two types of percussion fuzes :—

(*a*) Those which function as a result of a direct blow on the nose, a hammer or needle being forced on to a detonator. These fuzes are known as direct action or direct action impact fuzes.

(*b*) Those which depend on the forward movement of a " graze pellet " relatively to the main body of the fuze. These fuzes are known as graze fuzes.

The action of time fuzes is initiated by the shock experienced on the discharge of the gun, but the explosion of the filling does not occur until the expiration of a definite period of time, dependent upon the position in which the fuze is set before firing.

There are two types of time fuzes :—

(*a*) Those which depend for their timing action on the burning of composition, usually pressed into rings. These fuzes are known as combustion time fuzes.

(*b*) Those which depend for their timing action upon a clockwork or other mechanical device. These fuzes are known as mechanical time fuzes.

Time and percussion fuzes are time fuzes in which a percussion mechanism is embodied.

§6.02. Metals used in manufacture of fuzes.

Both ferrous and non-ferrous metals are used in fuze construction.

The ferrous metals used are :—

(*a*) Steels of different qualities, according to the employment of the component concerned which includes set-screws, washers, collars, safety caps of Fuze 106 type, needles, etc.

(*b*) Grey cast iron, the use of which was confined to bodies of Fuze, time, No. 79.

For needles, rustless steel is generally employed. Hard drawn steel containing not less than 0·8 per cent. of carbon may be used for needles for Fuze, Base, Hotchkiss type.

For springs, tinned steel piano wire, brass and phosphor bronze wires are used. Piano wire has a very high breaking (tensile) strength and is more extensively used than brass or bronze wire.

The non-ferrous metals used are :—

(*a*) Copper alloys—known for identification purposes as classes " A," " B," " G," " G.Pb " and " G.Y."

(*b*) Hard rolled brass.

(*c*) Aluminium alloy.

(*d*) Phosphor bronze.

(*e*) Manganese bronze.

(*f*) Aluminium bronze.

The copper alloys are in more general use for fuze bodies and components. Metal, class " A," having the highest tensile property, is used for small and medium type base fuzes and such fuze components which are required to withstand severe stresses. Metal, class " B," is the medium between the " A " and " G " classes and is used for certain small components such as pea-balls and plugs for base fuzes.

Metals, class " G " and " G.Pb." are in general use for all nose fuzes and their components. Metal, class " G.Pb," is " lead restricted " and is not to contain more than 0.1 per cent lead. Metal, class " G.Y," is also " lead restricted " up to a maximum of 2 per cent. lead, and, in certain circumstances, is allowed as an alternative to metal, class " G.Pb."

Fuze bodies and those components which come into contact with certain H.E. fillings, *e.g.*, lyddite, amatol, etc., are made of metal containing not more than 0·1 per cent. lead, in order to avoid the danger of the formation of sensitive salts. Metal containing up to 2 per cent. lead may be used in an emergency.

During the Great War, fuzes of the " lead restricted " class made of metal containing an excess of lead above 0·1 per cent. were tin or nickel plated all over.

It is now the practice for the external surfaces of fuzes and gaines, and those parts which come into contact with the fuze and gaine filling, such as magazine chambers, fire channels, grooves in time fuze rings, to be coated with fuze lacquer. In addition, the plain (unthreaded) surfaces of the fuzes and gaines, base caps, adapters, etc., which may come into contact with the shell filling are coated with copal varnish, stoved on.

Copper alloys are employed because of—

(i) Their superiority over ferrous metals in regard to tarnishing and corrosion.

(ii) Their ease of procurement and of mechanical treatment.

Hard rolled brass is employed because of its adaptability for springs, such as stirrup springs for time pellets, and for the smooth surface properties obtainable where required for shutters.

Aluminium alloy is employed owing to its low density. It has the disadvantage, however, of being easily corrodible and direct contact with gunpowder is avoided as far as possible :—

Examples are Fuze, T & P, No. 80, Mk. IVA.
Fuze, percussion, D.A., No. 132.

Bronzes are employed because—

(i) They have the same advantages as copper alloys.

(ii) They are superior to copper alloys in strength and resistance to shock.

Example.—Fuze, percussion, Base, Large.

The plates which are included in this chapter are coloured to denote the type of metal employed in their manufacture ; thus :—

Yellow	denotes	copper alloys and brass.
Purple	,,	steel.
Red	,,	copper.
Light blue	,,	aluminium and tinned plate.
Dark blue	,,	wrought iron.
Brown	,,	bronzes.

The following table shows the mechanical properties of metals used in the manufacture of fuzes.

TENSILE STRENGTH

Metal	Tenacity tons to the square inch		Elongation in a test piece conforming to British Standard Formula :— $\frac{\text{Length}}{\sqrt{\text{Area}}} = 4$
	Yield	Break	
Metal Class " A " rolled or extruded	20	30	20 per cent.
Metal Class " A " after stamping	20	28	13 ,,
Metal Class " B " rolled or extruded ...	12	20	30 ,,
Metal Class " G," " G.Pb " and " G.Y " (either bar, casting or stamping) ...	8	20	12 ,,
Aluminium alloy (either bar, casting or stamping)	7·5	12	7 ,,
Aluminium bronze rolled or extruded	20	42	18 ,,
Aluminium bronze after stamping ...	20	40	12 ,,
Phosphor bronze, rolled	20	30	20 ,,
Manganese bronze, rolled	18	34	20 ,,
Brass, hard rolled sheet	6	12	10 ,,
Wire, piano, tinned steel	—	160 to 200	—
Grey cast iron	—	9 to 10	—

COMPRESSION STRENGTH

Dimensions of test piece—Length equal to diameter

Material	Permanent depression after a load of 8 tons to the square inch not to be greater than—	Minimum depression without serious cracking	Minimum crushing load a square inch
Metal Class "G," "G.Y." and "G.Pb." ...	1 per cent.	35 per cent.	60 tons.
Aluminium alloy	1 ,,	50 ,,	60 ,,

§6.03. Safety arrangements.

Arrangements are incorporated in the design of fuzes to ensure that they shall be safe (i) for storage and transport, (ii) during the shock of discharge from the gun, and (iii) during flight.

Arming of the fuze. This is a mechanical event which is designed to take place as the shell leaves the muzzle of the gun. Before the fuze is "armed" it should not be capable of being put into operation by any conceivable rough usage, or by any drop in any position which is likely to occur in the Service. Generally speaking, the moving parts of a fuze are securely locked together by a number of devices, and the particular combination of forces to which it is subjected when being projected from a gun are used to unlock them. Any other combination of forces would not have the required effect. It is in fact a form of combination lock in which firing from a rifled gun is the key. The mechanical details of the locking and arming arrangements vary with the fuze and will be given separately with each type.

Safety shutters.—Modern fuzes for H.E. shell also have a shutter which isolates the detonator from the magazine until the projectile is clear of the bore. The detonator is the most sensitive part of the fuze, and however rigorous its conditions of manufacture and examination, there is always a slight danger of its functioning on its own account even with all the moving parts properly locked. The shutter prevents such an accident from having serious effects.

In addition, safety devices may be provided which can be dispensed with immediately before loading. For instance, D.A. fuzes generally have removable caps, and some fuzes have been provided with safety pins, inserted from the exterior, and locking a moving part. The latter have the disadvantage of allowing the ingress of moisture to the interior of the fuze, and are now seldom used.

The above methods are employed to ensure safety during storage and transport and while the shell is in the gun.

During flight, the shell encounters the resistance of the air, which as far as the fuze is concerned has one of two effects: (*a*) With a direct action type the hammer or needle of the fuze tends to get pushed in by the air, so firing the fuze, or (*b*) with a graze type the retardation of the shell caused by the resistance of the air tends to cause the graze pellet to move forward by its own inertia until the needle is in contact with the detonator. These two effects are prevented (*a*) by holding the needle with a spring, shearing wire, or disc, or by placing a light cover over the needle to keep the rush of air away from it ; and (*b*) by interposing a light " creep spring " in front of the graze pellet of just sufficient strength to prevent its creeping forward during flight. It is evident that the weaker the spring or other device that is used, the more sensitive will be the fuze. In the design of graze fuzes the strength of the spring is determined by what is known as a " creep spring factor " which is—

$$\frac{\text{Creep force on pellet}}{\text{Strength of spring when needle is touching detonator.}}$$

§6.04. Dynamical factors governing design.

Four dynamical factors are available for the purpose of arming and functioning the several types of fuzes. These are :—

1. The rapid acceleration of the projectile in the bore of the gun.
2. The pressure of the propellant gases on the base of the projectile.
3. The rotation of the projectile.
4. The retardation which occurs on impact of the projectile on its objective.

In addition, consideration must be given to the " set forward " force due to the retardation of the projectile caused by the resistance of the air.

Dealing with these factors separately, we have :—

1. *The rapid acceleration of the projectile in the bore.*

The shell has an acceleration a in the bore, which is a maximum at the point of maximum chamber pressure P, that is to say, fairly early during shot travel. Under this acceleration anything loose inside the fuze tends to be left behind. This " set-back " tendency is used with various devices such as " pellets," " ferrules," " arming sleeves " and " detents," in order to cause the fuze to function or arm.

Pellets are little loose cylinders of metal carrying (usually) a detonator. For example, a " lighting pellet " in a time fuze is caused to fly back on to a needle by the " set-back " force, and so light the time ring of the fuze. The pellet must be prevented from setting back prematurely, due to accidental shocks, and this is done generally by a spring, either in the form of a stirrup or a spiral. We have to be able to calculate the necessary strength of these springs.

Again, in graze fuzes, a pellet is also used, this time to set forward owing to the shock of impact on arrival at the target end. Here also a stirrup spring is sometimes used for safety purposes and this is put out of action, on the arming of the fuze, by a *ferrule*, a hollow metal cylinder which sets back over the spring and forces itself down around the graze pellet. The ferrule and pellet become jammed, forming in effect, one pellet, which will now be separated from the needle by a creep spring only. Again the strength of the stirrup spring is needed.

Detents are little bolts lying longitudinally in a fuze as part of the locking arrangement. They are also kept in place by a spiral spring.

The force of set-back.—If the pellet, ferrule or detent has a weight of m lb. the set-back force $= a\frac{m}{g}$ lb. weight. The spring must therefore yield under this force for the fuze to be effective.

Since $$a = \pi \frac{d^2 P}{4M} g \times 2240 \text{ ft/sec/sec. approximately,}$$

where P = maximum chamber pressure in tons a square inch,
M = weight of shell in lb.
d = calibre in inches,
the set-back force on the pellet, ferrule or detent

$$= \pi \frac{d^2 m}{4M} \text{P tons.}$$

This formula neglects certain factors. In calculating the acceleration, the effect of friction of the driving band and the fact that part of the weight of the charge is also accelerated, have been omitted. The value obtained, therefore, is slightly too high. The " set-back " force formula takes no account of friction between the moving part and the recess in which it slides. In the case of a pellet (say) fitted eccentrically from the axis of the fuze, this friction is increased by the centrifugal force acting on the pellet. Since, however, the maximum pressure, and hence the maximum " set-back " force, occurs within a very short distance of the start, the rotational velocity is small and the consequent centrifugal force can usually be neglected.

Example.—In Fuze, T & P, No. 80, the weight of the time pellet and stirrup spring is 189 grains and the stirrup yields and allows the pellet to set back under a force of 165 lb. weight. What is the least pressure at which the fuze can be used in an 18-pr. gun, assuming a " factor of certainty " of 2 ?

Let p = the required minimum pressure.

Then " set-back " force on pellet

$$= \pi \frac{d^2}{4} \frac{m}{M} \frac{p}{7000} \times 2240 \text{ lb.}$$

and this must be equal to or greater than 165 × 2, *i.e.*, 330 lb.

Therefore $$p = 330 \times \frac{7000}{2240} \times \frac{4M}{\pi d^2 m}$$

$$= \frac{330 \times 7000 \times 4 \times 18{\cdot}5}{2240 \times \pi \times (3{\cdot}3)^2 \times 189}$$

$$= 11.8$$

Therefore, the least chamber pressure is 11·8 tons a square inch.

2. *Gas pressure.*

The pressure of the propellant gases on the base of the shell is used to crush in a pressure plate in a base fuze. The pressure plate is usually attached to a spindle, and the forward movement of the spindle unlocks the moving parts of the fuze, thus arming it.

If d_1 = effective diameter of pressure plate in inches, the force acting on the pressure plate will be

$$\frac{\pi P d_1^2}{4} \text{ tons}$$

Against this there is :—

(1) The resistance of the pressure plate and spindle to movement, irrespective of the motion of the shell.

(2) The " set-back " force on the pressure plate and spindle due to the acceleration of the shell.

(1) Can be measured by finding the hydrostatic pressure required to arm the pressure plate at rest. This will be

$$\frac{\pi P_1 d_1^2}{4} \text{ tons,}$$

where P_1 is the hydrostatic pressure in tons a square inch.

(2) Can be found as already explained in the previous example, and is equal to :—

$$\frac{\pi P d^2 w}{4W} \text{ tons.}$$

The minimum total pressure necessary to arm the pressure plate in the gun is therefore the sum of (1) and (2), and is equal to

$$\frac{\pi P_1 d_1^2}{4} + \frac{\pi P d^2 w}{4W} \text{ tons weight.}$$

This must be less than $\frac{\pi P d_1^2}{4}$ or the fuze will not arm.

Example :—In a 6-inch B.L. gun, P = 18 tons a square inch, d = 6 inches, and W = 100 lb. In a No. 16 base fuze, assume P_1 = 4 tons a square inch, d_1 = 0·6 inch, and w = 6½ drachms.

Then the minimum total pressure on the pressure plate necessary to arm the fuze is :—

$$\frac{\pi \times 4 \times (\cdot 6)^2}{4} + \frac{\pi \times 18 \times (6)^2 \times 6\cdot 5}{4 \times 100 \times 16 \times 16} = 1\cdot 13 + \cdot 129 \text{ tons weight,}$$

and so the total pressure necessary to arm the fuze in the gun is 1·26 tons weight.

The actual pressure on the pressure plate is given by the formula already quoted $\left(\frac{\pi P d_1^2}{4}\right)$ and equals 5·08 tons weight. This is therefore more than sufficient.

3. *The rotation of the projectile.*

In this case centrifugal force is used to withdraw safety bolts, open shutters, unwind tapes, or to remove generally any components necessary to arm the fuze.

To obtain a measure of the centrifugal force.—When a projectile travels up the bore of a gun after firing, two opposite sets of forces are engendered by the rotation of the shell. One, centrifugal force, tending to withdraw the bolt, unwind the tape, open the shutter, or otherwise arm the fuze ; the other neutralizing this effect.

Centrifugal force is directly proportional to the square of the angular velocity, and, therefore, to the square of the axial velocity of the shell. This force increases during shot travel, and attains a theoretical maximum value at the muzzle where the highest velocity is reached.

The forces opposing the action of centrifugal force and keeping the bolt or shutter in its initial position are :—

(1) Friction (*a*) due to the moment of inertia of the moving part round the fuze axis, and
(*b*) due to longitudinal inertia (set-back) owing to the linear acceleration of the projectile.

(2) The pressure or torque of the spring inserted behind the bolt or in the shutter.

Of these, the force due to friction varies directly with the axial acceleration, which is itself directly proportional to the pressure on the base of the shell, while the force due to the spring is constant. Once the point of maximum pressure is passed, therefore, the rate of acceleration decreases and begins to die away. On the projectile leaving the muzzle, deceleration sets in, and a point must exist where there is neither acceleration not deceleration, *i.e.*, where the velocity of the projectile is constant. At this neutral spot forces produced by the moment of inertia of the moving part and the set-back are eliminated, and the spring alone is operative against the action of centrifugal force.

Similarly a point must exist between the point of maximum pressure and this neutral region (assumed for convenience at the muzzle where it theoretically exists), where the two opposing sets of forces have an equal and opposite value.

The fuze, therefore, should arm after the shell has just passed that point, provided the spring of the shutter or bolt has already been overcome.

The problem is consequently to find where this takes place, so as to ascertain whether a certain design of fuze is suitable for a particular gun.

In the case of shutters, a further complication is introduced owing to the fact that shutters may be designed to open either with or against the twist of the rifling.

The centrifugal force.—If the rifling of the gun has a twist of 1 turn in n calibres, the angular velocity of the projectile in radians per second is

$$w = \frac{24\pi v}{nd}$$

where v is the linear velocity in feet per second, and d the calibre in inches.

The centrifugal force F, acting along a bolt of weight m lb., whose centre of gravity is distant r feet from the axis of the shell, is equal to

$$\frac{m}{g} r\omega^2 \text{ lb. wt.}$$

$$= \frac{(24)^2\pi^2 v^2 mr}{2240 n^2 d^2 g} \text{ tons.}$$

This force as before stated reaches a maximum at the muzzle.

Force due to the moment of inertia of the moving part round the fuze axis.

Let k = radius of gyration of the bolt in feet.

Then $\frac{m}{g} k^2$ is its moment of inertia.

And if Fk = turning moment exerted by the fuze body on the bolt, we have

$$Fk = \frac{m}{g} k^2 \dot{\omega}$$

where $\dot{\omega}$ = angular acceleration.

If μ = coefficient of friction, the force opposing the setting out of the bolt due to centrifugal action is

$$\mu F = \mu \frac{m}{g} k\dot{\omega} \text{ lb. wt.}$$

Force due to friction between the bolt or shutter and its seating owing to the acceleration of the projectile.

In this case the retarding force will be

$$\mu \frac{m}{g} a \text{ lb. wt.}$$

where a = axial acceleration in the bore.

Force due to spring.—Let x lb. wt. be the force exerted by the spring.

Total forces.—Therefore the total forces preventing the bolt withdrawing under the action of centrifugal force are :—

$$x + \mu \frac{m}{g}\left(k\dot{\omega} + a\right) \text{ lb. wt.}$$

As, however, in practice the frictional force due to the moment of inertia of the moving part is extremely small in comparison with that of the set-back force, it may be neglected. The above expression, substituting the value of a, already found, then becomes

$$\frac{x}{2240} + \frac{\mu \pi d^2 m}{4\text{M}} p \text{ tons.}$$

(p in this case being the pressure at any selected point in the bore)

This force will obviously be at a maximum when p is a maximum and will diminish after the point of maximum pressure is passed, Therefore, in order that the fuze may arm as the shell reaches the muzzle, the value of this force should be less than the centrifugal force at the muzzle,

i.e.,
$$\frac{m}{g} r\omega^2 > \frac{x}{2240} + \frac{\mu \pi d^2 m}{4\text{M}} p$$

at the muzzle in order to allow the fuze to function correctly.

Example.—In a 6-inch gun, M.V. 2400 f./s., the twist of rifling is 1 in 30, and the muzzle pressure is 5·8 tons a square inch. Determine whether the centrifugal bolt of a base fuze will withdraw correctly, and thus whether the fuze is suitable for use in that gun, assuming that $\text{M} = 100$ lb., $r = 0{\cdot}1$ inch, $n = 30$, and $\mu = 0{\cdot}07$. There is no spring in these fuzes.

For the fuze to function correctly, the bolt should withdraw at a point where the pressure just exceeds the muzzle pressure.

To find that pressure p :—

In the first place a base fuze has no spring behind its centrifugal bolt, so the term $\frac{x}{2240}$ can be omitted.

We then have
$$\frac{(24)^2 \pi v^2 r}{2240 n^2 d^2 g} = \frac{\mu d^2}{4\text{M}} p$$

$$\therefore \frac{9}{35} \times \frac{\pi \times (2400)^2 \times 0{\cdot}1}{(30)^2 \times 36 \times 32{\cdot}2 \times 12} = \frac{{\cdot}07 \times 36}{400} p$$

$$\therefore \frac{0{\cdot}63}{100} p = {\cdot}0372$$

$$\therefore p = 5{\cdot}9 \text{ tons a square inch}$$

The bolt will therefore withdraw at a point just before the muzzle is reached.

The bolt is therefore suitable.

4. *The retardation due to the impact of the projectile on its objective.*

This deceleration is utilized either to cause pellets held back by creep springs to set forward owing to their inertia and to function the fuze, or to crush in some portion of the fuze when the shell strikes the target.

For example :—

(*a*) In the No. 80 type of fuze, after the rapid acceleration of the projectile has caused the ferrule to set back over the percussion pellet, the latter is held back by a creep spring, which, on the shell striking, is compressed and causes the fuze to function by the detonator striking the needle.

(*b*) In the No. 44 type of fuze, the needle disc is crushed in, and the needle strikes the detonator.

" Set-forward " forces.—With reference to (*a*) above, the creep spring must be strong enough to prevent the pellet moving forward when the shell is retarded by the resistance of the air. Consistent with this condition, the weaker the spring, the more sensitive the graze action of the fuze.

The retardation due to air resistance is greatest at or just beyond the muzzle. Its value " a' " can be obtained from the ballistic tables when the ballistic coefficient and muzzle velocity are known. If the range table only is available, an approximation to the initial retardation may be obtained by dividing the loss in muzzle velocity over the first 100 yards by the time of flight.

The forces acting on the pellet of weight w lb. is, therefore,

$$\frac{m}{g} a' \text{ lb. wt.}$$

Example.—The weight of the percussion pellet of a fuze T. & P. No. 80 is 156 grains. What must be the least power of the creep spring to prevent the fuze functioning on the shell leaving the muzzle of an 18-pr. gun ?

From the range table we find " a' " to be approximately 195 feet per sec. per sec.

Therefore the force acting on the pellet is

$$\frac{156}{7000} \times \frac{195}{32{\cdot}2} \text{ lb. wt.} = {\cdot}135 \text{ lb. wt.}$$

$$= 2 \text{ oz. wt. approximately.}$$

The creep spring must therefore resist a force of at least 2 oz. wt.

Since, however, a shell on first leaving the muzzle may be unsteady owing to various causes, such, for instance, as a worn gun, and this unsteadiness causes increased resistance to its passage through the air, it is necessary to make the creep spring stronger than the figure obtained from the range table, if premature action of the fuze is to be avoided.

It is customary to increase the creep spring to a strength which will allow the shell to yaw up to 10 degrees before the pellet can move forward.

§6.05. Detonators for fuzes and gaines.

There are two classes of detonator used in fuzes and gaines :—

(1) *Disruptive* detonators to initiate a wave of detonation in H.E. filled fuzes and gaines and to break through the diaphragm leading to the powder filling in Fuze, No. 45P, type.

(2) *Igniferous* detonators to ignite powder filling by means of a flash or to fire another disruptive detonator.

(1) Disruptive detonators are usually fired by a direct blow from a hard sharp needle, an essential to obtain best results. The degree of detonation is also dependent to some extent upon the velocity of the blow, the shape of the needle and the pressure under which the charge is pressed into the detonator shell. All these considerations are taken into account in the design of the fuzes.

All modern H.E. filled fuzes and gaines, as well as No. 45P, fitted with shutters, now contain detonators filled three grains lead azide and a " topping " of two grains detonating composition " A " mixture. Fuzes and gaines fitted with this detonator are distinguished by the letter " Z " *after the Mark.*

Unshuttered H.E. fuzes and gaines are fitted with detonators filled fulminate of mercury.

Lead azide has better keeping qualities than fulminate of mercury.

With the exception of Fuze, No. 18, which is fitted with a four-grain fulminate detonator, all modern D.A. and D.A.I. nose detonating fuzes, also gaines, Nos. 8 and 9, are fitted with five-grain detonators.

A ten-grain fulminate of mercury detonator is used in gaine, No. 2.

(2) Igniferous detonators are in three classes :—

(*a*) Filled detonating composition " A " mixture.
(*b*) Filled detonating composition " B " mixture.
(*c*) Filled detonating composition " C " mixture.

There are four types of detonators filled " A " mixture.

(i) 1·5 grains used for time detonator of Fuzes, Nos. 88, Mk. III*, and 197, Mk. I.
(ii) 1·7 grains used in Fuzes, Nos. 101E, 106PE Mk. I and as percussion detonators in Fuze, No. 88.
(iii) 2 grains used in Fuze, Base, Hotchkiss and No. 280.
(iv) 3 grains used in Fuzes, Base, Percussion, large and medium, also Fuzes, Nos. 106PD, 132 and mechanical fuzes.

There are three types of detonators filled " B " mixture.

(i) 1·62 grain, composed of 0·75 grains " B " mixture and 0·87 grains R.P. or G.20 gunpowder priming. Used for time detonators of Fuzes, Nos. 80 and 88 types.

(ii) 2 grains, composed of 0·85 grain " B " mixture and 1·15 grains R.P. or G.20 gunpowder. Used for time detonator of Fuze, No. 199.

(iii) 2·5 grains, composed of 1·1 grains of " B " mixture and 1·4 grains of R.P. or G.20 gunpowder. Used for the time and percussion detonators of fuze T. & P. No. 220.

There is only one type of detonator filled " C " mixture.

This is used for the percussion detonator of Fuze, T. & P., No. 80, and contains 2·99 grains composed of 1·39 grains " C " mixture and 1·6 grains R.P. or G.20 gunpowder.

§6.06. Gauge of fuzes.

Fuzes are designed to screw into the nose or base of a shell.

Direct action, direct action impact, time, and time and percussion fuzes screw into the nose.

Graze fuzes may be designed to screw into the nose or base of the projectile.

The two normal sizes of nose fuze-holes in the service are the general service (G.S.) 1-inch gauge and the 2-inch gauge ; a special gauge (1·2-inch) is used for the Q.F. 1½-pr. and 2-pr.

The gauge for the 3-inch mortar is 1·375 inch.

There are three sizes of gauge for base fuzes :—

A large base gauge (2-inch) for 5·5-inch shell and above.
A medium base gauge (1·6-inch) for 4·7-inch shell and below.
A special base gauge (0·775-inch) for Q.F. 1½-pr., 2-pr., and Hotchkiss 3-pr. and 6-pr.

§6.07. Nomenclature of fuzes.

The general principles on which fuzes are numbered are :—

(*a*) According to gauge of screwed part of body.
(*b*) ,, action.
(*c*) ,, position in shell (*i.e.*, nose or base).
(*d*) ,, filling of magazine.
(*e*) ,, physical characteristics.

The system of group numbering of fuzes according to their design was introduced many years ago, but owing to the fact that the groups provided by the originators were not sufficiently extensive to cope with the great expansion of fuze design, various cases of apparently inconsistent numbering will be met with. In modern practice, however, the system will be found to be applied as consistently as possible, *e.g.* :—

Time mechanical fuzes of 2-inch gauge begin with No. 200.
Small percussion fuzes, *i.e.*, of gauge smaller than G.S. 1-inch begin with No. 130.
Small time fuzes are numbered No. 120 and onwards.
Nose percussion fuzes 2-inch gauge begin at No. 100.
Fuzes, T. & P., of 2-inch gauge begin at No. 80.
Fuzes, T. & P., of G.S. 1-inch gauge begin with No. 50.
Time fuzes of 2-inch gauge were introduced with No. 30, although congestion subsequently necessitated the use of numbers 29 and 28 for this type of fuze.
G.S. gauge time fuzes of fairly modern description are numbered upwards from No. 25, although earlier patterns of this class of fuze bore earlier numbers.
G.S. gauge percussion fuzes of the D.A. and D.A.I. types begin with No. 1 and continue after No. 19, owing to congestion, with No. 44.

The remaining nose and base percussion groups bear various numbers out of the general order ; this is the result of the lack of numbers available.

In certain cases (viz. Nos. 65A, 80, 80B, 83, 88) time and percussion fuzes have had their number advanced by 100, *i.e.*, becoming 165A, 180, etc., upon conversion to time fuzes only. This indicates that the percussion portion has been rendered inoperative by removal or other means.

Fuzes bearing fractional numbers, such as 80/44, have also had their percussion portion rendered inoperative in a similar manner to Nos. 180, 188, etc., but differ therefrom in having the first two graduations on the ring obliterated.

The fractional number given to these fuzes was originally used to denote the fact that they were time fuzes of the type designated by the numerator adapted for use with a modified No. 44 fuze, used as a gaine.

§6.08. Blinds and prematures.

Blinds and prematures due to faults in the shell have already been enumerated in Chapter V, page 139 ; they may also be attributed to faults in the fuze as follows :—

Blinds.—

(1) Faulty preparation for action, such as the gun numbers failing to withdraw safety pins, remove safety caps, etc.
(2) Fuze used in ordnance unsuitable for it, *i.e.*, using a fuze designed for use in a gun of high chamber pressure and M.V. in a howitzer of which the chamber pressure and M.V. are insufficient to enable the safety arrangements to be overcome and allow the fuze to function.

(3) Shell fitted with fuze of the "graze" percussion type being insufficiently checked on impact. This is most likely to occur with a heavy shell striking at a small angle of impact, since a heavy body moving at a high rate of speed is more difficult to check than a light one.
(4) Fuzes of the D.A. type may give blinds due to too small an angle of impact, owing to the shell not getting a direct blow on the nose.
(5) The wrong type of fuze being used.
(6) Failure of the mechanism of the fuze to "arm" when discharged from the gun.
(7) Mechanical defects in the fuze caused by rough handling in storage or transport.

Prematures.—A premature is a most serious defect, and one considered of such importance that, if a case occurs at original proof, the whole lot is rejected.

Prematures may arise from—

(1) Fuze improperly prepared for action. This is most likely to occur with time and time and percussion fuzes, *i.e.*, the fuze not being clamped up, thus allowing the fuze to flash over or the ring to slip.
(2) Defective fuze. This may be due to fuzes being tampered with by unauthorized persons to satisfy curiosity as to internal construction. This is not liable to occur under peace conditions but should be guarded against in time of war. It is especially necessary with the No. 106 type fuze where the sealing wire is broken, when care must be taken to see that the tape and shearing wire are correct before loading. A fuze found with a broken sealing wire should never be fired from a gun.
(3) With graze fuzes, a premature may be caused by an obstruction in the bore.

This may occur when using cartridges such as were used with the 15-inch howitzer which had millboard strengthening pieces, unless the bore is always examined before reloading.

Discs of primers blowing out and remaining in the bore may also be a cause.

Shell should also be properly rammed home, as otherwise there is an increased jar on the shell on firing, and increased pressure from the propellant gases.

(4) A fuze used in a gun for which it was not intended. That is to say, a fuze intended for a howitzer having a low chamber pressure and M.V. being used in a high pressure, high velocity gun, thereby imposing stresses on the fuze it was not designed to withstand.
(5) An over-sensitive detonator combined with shutter failure.

(6) With base fuzes a premature may occur owing to the propellant gases penetrating into the interior of the fuze or the shell.

(7) In some cases rough handling in transit may place the fuze in a sensitive condition.

The above are some of the more common causes of blinds and prematures, but the latter, owing to rigorous inspection and proof, are of very rare occurrence in the Service. Blinds may occur, not through any fuze defect, but owing to the natures of soil on which the shell may occasionally land.

Other causes capable of producing blinds and prematures are :—

Blinds.—Fuze magazine left unfilled ; absence of detonator or detonator needle ; fuzes damp or in a bad condition, etc.

Prematures.—Absence of creep springs, safety tapes and half collars, centrifugal bolts, shearing wire, etc.

Such causes can be summed up under :—

(*a*) Faulty manufacture.
(*b*) Bad storage conditions.

All fuzes before being accepted for service are subjected to a rigorous examination during and after manufacture and filling, and a proportion of each lot are proved for results. Fuzes already accepted are always re-examined and reproved, if necessary, before being issued for service.

Besides these safeguards, there are periodical inspections of all fuzes held on charge by units.

§6.09. Marking of fuzes.

All fuzes are stamped with their distinguishing number and Mark, the maker's initials, the filler's initials, the date of filling, and the filled lot number.

Lot Number.—Time and T. & P. fuzes are manufactured in lots of 2,000.

Percussion fuzes and gaines are manufactured in lots of 1,000.

Certain letters placed after the number of the fuze denote a standard alteration to the original parent design. These are :—

" D " ... indicating that the fuze is fitted with a delay pellet.

" E " ... indicating that the fuze is fitted with a safety shutter.

" P " ... indicates that the magazine of the fuze is filled powder.

Painting and stencilling of fuzes.—Delay or " D " fuzes can be easily identified by the colour marking on the caps of the fuzes, and in the case of Fuze, No. 16D, by the painting on the base.

Fuze, No. 16D, is marked on the base thus : A blue circle on a red ground.

Fuze	*Painting*	*Meaning*
106PE Mk. II 139P 45P	Cap lacquered red	Denotes magazine filled with powder.
101E	Cap painted blue ...	Denotes gaine fitted with delay.
132	Head painted white	Denotes powder filled shell only.
197	Cover painted white	To distinguish from 80/44, 83 or 88, from which fuzes it is converted.
General	Blue 1-inch T on fuze	Denotes time fuze only.
,, ...	Time rings lacquered red.	Denotes filling of long burning R.D.202 composition.

In the following paragraphs a brief description is given of all current service fuzes and gaines, together with plates of latest Marks of such stores. In order to avoid unnecessary repetition, the class of material from which the various components are made is not given. This can be ascertained by reference to the various colourings on each plate, the key to which is given on page 149.

The current Marks of the fuzes and gaines are given, particulars of earlier Marks are already described in previous publications.

Where recent modifications in structure or filling have necessitated an advance in Mark, details of such modifications are given.

The nature of round for which the fuzes and gaines are to be used is given in Table 6.41, which will be found at the end of this chapter.

PERCUSSION FUZES

§6.10. Direct action (D.A.) fuzes.

A direct blow on the nose forcing the hammer or needle on to the detonator is necessary to cause these fuzes to function.

The D.A. type do not require such a heavy blow as the D.A.I., as the hammer or needle is less rigidly supported. For example, in the D.A. (No. 44) fuze, the needle is suspended on a copper disc, and only a moderately light blow is necessary to actuate it ; in the D.A.I. (No. 45) fuze, on the other hand, a steel hammer is suspended over the detonator by means of a strong steel shearing pin, and before the fuze can act, the head of the fuze must be crushed in and the shearing pin broken.

The No. 106 fuze is even more sensitive and rapid than the D.A. (No. 44) type, owing to the protrusion of the hammer from the nose of the fuze when armed. The hammer thus receives its blow before the shell proper actually touches the ground, while the copper disc bearing the needle being recessed in the No. 44 fuze, the shell has to strike the ground squarely before the fuze can operate.

These types of fuzes are usually fitted with a disruptive detonator carrying a charge on pure fulminate, or lead azide, and a C.E. pellet. Other fuzes of the same nature filled with powder are, however, found. These may have an igniferous detonator filled with what is known as detonating composition "A" mixture, though in some cases, notably the No. 45P, a disruptive detonator is used in order to break through the thin metal skin closing the hole leading to the magazine.

Safety arrangements in D.A. and D.A.I. fuzes.—Such arrangements are designed :—

(*a*) To protect the head of the fuze prior to loading.
(*b*) To seal the flash of the detonator from the magazine of the fuze should the detonator ignite or detonate prematurely.

They are effected by introducing :—

(*a*) A cap.
(*b*) A shutter, which is arranged to open under the action of centrifugal force set up by the rotation of the projectile.

This shutter is held in position by a spring until that spring is overcome by centrifugal force, and also in certain cases, *i.e.*, in Fuzes, Nos. 44 and 45, by a safety pin passing through the shutter.

The safety cap is removed and the safety pin withdrawn on loading.

Uses.—Direct action (D.A.) fuzes, owing to their rapid action, are suitable for use with H.E. shell where splinter effect is required above ground, such as in the attack of personnel and wire cutting, and for bursting smoke shell; the No. 106 fuze is the most effective.

No. 44 fuze in comparison with No. 106 is more suitable for H.E. shell in heavy guns for the purposes for which they are used, as it enables the shell to penetrate slightly before bursting.

During the Great War the No. 44 fuze with small modifications was used beneath a No. 80/44 time fuze for bursting time H.E. This modified fuze was known as the No. 44/80; it is now being replaced in the land service by a No. 8 gaine which is similar in construction. (*See* also page 187.)

Direct action impact (D.A.I.) fuzes are used in H.E. shell for naval and coast defence guns against lightly armoured vessels, or the upper works of those heavily armoured.

Cause of blinds.—Certainty of action with these types of fuzes depends upon the angle of impact.

Thus the No. 106 fuze will not function with certainty on hard ground at angles below 6 degrees, or on meadowland below $2\frac{1}{2}$ degrees.

For angles of impact of less than 8 degrees a graze fuze may be employed against land targets to obtain greater certainty of action.

This uncertainty of action is greater with shell with a large calibre or a small c.r.h., as such shell may ricochet from the shoulder without the fuze being touched.

As examples of direct action Fuzes, Nos. 44, 106E and 117 are described in the following pages.

§6.11. Fuze, percussion, D.A., No. 44, Mk. X.

Use.—This fuze is used with H.E. shell having a G.S. gauge fuze-hole, or fitted with an adapter of that gauge and having trotyl or C.E. exploders.

The various components and general structure of the fuze are shown on the diagram. The body and base plug are of lead restricted (class G.Pb.) metal. The filled shutter is designed to open fully when the fuze is spun between 1,200 and 2,100 revolutions a minute.

The fuze is fitted with a five-grain detonator and is filled with C.E.

The threads of the detonator plug, holder, screwed collar, base plug, and the top and bottom of the fuze are coated with R.D. cement.

Earlier Marks of this fuze, other than those which are obsolete, differ but slightly from that shown in the diagram. For certain Marks, the securing pin is secured to the cap by means of a brass clip arrangement instead of a whipcord becket. The shutter and skin of metal over the fire channel in the body have been thickened slightly as an extra precaution against prematures.

Fuze, Percussion, No. 44/80.

May be obtained by conversion of Fuze, No. 44, and differs only in that the shutter springs are adjusted so that the shutter will open when the fuze is spun between 6,000 and 8,000 revolutions a minute. As they are for use under time fuzes they are without safety pin and safety cap, but are not issued direct to the service.

Action of the Fuze, No. 44.

1. On loading, the locking pin, which withdraws the safety pin with it, and safety cap are removed.

The safety cap has prevented the needle disc from receiving an accidental blow. The safety pin has held the shutter in the closed position so that, even if the detonator did fire accidentally, the wave of detonation is prevented from reaching the magazine by the solid portion of the shutter. Also should the C.E. in the shutter ignite, the fuze would be safe owing to the metal of the solid diaphragm intervening between it and the magazine.

2. On firing, nothing happens while the shell begins to accelerate up the bore, when the shutter, owing to its inertia, sets back. This prevents it being opened by centrifugal force, and seals the fire channel while the shell is in the bore.

After the shell leaves the muzzle, acceleration ceases; the centrifugal force, therefore, overcomes the shutter spring. The shutter then opens, and the channel filled with C.E. now comes into line with the fire channel, thus forming a train of C.E. to the magazine.

3. On impact of the shell, the needle disc is crushed inwards and the needle pierces the detonator. The detonator fires and detonates the C.E. pellet.

Action of the Fuze, No. 44/80.

Is similar to that described of Fuze, No. 44, except that the needle disc is crushed inwards by the exploding of the time fuze magazine.

§6.12. Fuze, percussion, D.A., No. 106E, Mk. VIIIz.

This fuze is employed when rapidity of action is required, such as wire cutting, against personnel in the open, or in smoke shell.

It is intended for use in shell having the 2-inch fuze-hole gauge and fitted with trotyl or C.E. exploders.

The various components and general structure of the fuze are shown on the diagram. The body, magazine and bottom cap are of lead restricted (class G.Pb.) metal.

The filled shutter is designed to open fully when the fuze is spun between 1,300 and 1,700 revolutions a minute.

The fuze is fitted with a five-grain detonator and is filled with C.E.

The hammer has a steel stem and head, the latter having an aluminium centre inserted to give a cushioning effect. It is assembled into the fuze as follows:—

The copper tape is passed half round the hammer stem, the retaining pin in one half-collar being passed through the hole in the end of the tape into the hole in the stem of the hammer. The other half-collar is then placed in position with the tape leading off between the division between the half-collars. The tape is then wound tightly round the outside of both the half-collars in a clockwise direction, looking at the hammer from the needle end. The brass weight soldered to the end of the tape must be close against the outer turns of the tape.

The steel washer is then threaded on to the stem of the hammer and the hammer placed through the hole in the top of the fuze, so that the shearing wire and guide pin can be inserted through the body of the fuze passing through the slot in the hammer stem. The ends of the shearing wire are next pressed into grooves. The leather washer is then placed in position and the safety cap screwed home in order to make an airtight joint. The cap is secured by a brass wire threaded singly through the eye on the cap. The wire is then bent double and twisted. The loose ends are then passed in a clockwise direction when viewed from the top of the cap, through the eye in the fuze and back again over the eye. The ends are then inserted into the sealing hole and fixed in position by a lead plug.

The threads of the guide pin are coated with R.D. cement before insertion.

The threads of the detonator plug and magazine and the whole of the base of the fuze is coated with R.D. cement. The head of the set-screw is covered by waterproofing composition.

1. Before loading, the sealing wire is broken and the cap removed. The nose of the fuze should then be inspected to see that the shearing wire and tape are in the correct position. If either are incorrect the fuze should on no account be fired.

2. On acceleration in the bore the hammer sets back and prevents the tape from unwinding during the time that acceleration is taking place, *i.e.*, while the shell is travelling up the bore.

A similar action causes grip to take place between the shutter and diaphragm, retarding the action of centrifugal force in tending to open the shutter.

After the shell has left the muzzle, acceleration ceases and deceleration sets in. The tape therefore unwinds under the action of centrifugal force, and allows the half-collars to fly outwards, leaving the hammer suspended solely on the shearing wire. The guide pin prevents any rotary motion of the hammer which might shear the wire prematurely.

At the same time, the shutter, under the action of centrifugal force, overcomes its spring, and the hole filled with C.E. comes into alignment with the fire channel and magazine.

3. On impact, the hammer is forced in, firing the detonator and actuating the fuze

Adaptations of Fuze, No. 106, filled powder, are known as Fuze, Nos. 106PE and 106PD. The 106PE, Mk. II, is similar to the 106E, Mk. VIIz., except that the magazine is filled with gunpowder and the bottom cap has a hole drilled through it which is closed by a brass disc. Fuze, No. 106PD, is not fitted with a shutter and has a three-grain detonator filled " A " mixture.

The safety cap of both fuzes is painted or lacquered red.

Notes on the design of the No. 106 *Fuze.*—This fuze has undergone more modifications, radical and slight, than any other fuze ever introduced into the service. Including the Marks of No. 106E, the design of which is adapted from No. 106, there have been altogether some fifty-five Marks contemplated or adopted.

In the original design of No. 106 there was no shutter. The base was closed by a plug instead of a cap, there was no disc over the detonator, and the shearing wire had no visible ends. Fuzes fitted with base plugs on conversion to the base cap type are distinguished by the letter " A " after the Mark.

An alteration was effected in some designs (indicated by having an S placed after the numeral of the Mark) by instituting an all-steel hammer on the score of ease of manufacture. It was not found, however, to function so efficiently as the steel hammer with the aluminium pad in the head. When a shell, fitted with this fuze carrying an all-steel hammer, strikes a hard surface at a very small angle of arrival, the shock of impact tends to revolve the hammer momentarily round the shearing wire as centre. The stem of the needle has, therefore, a tendency to tilt towards the wall of the body, affecting in this manner the speed of action of the fuze: whereas in fuzes containing a composite needle, the aluminium pad acts as a shock absorber and allows the needle to function correctly.

During the Great War prematures occurred, it was surmised, owing to the stem of the hammer breaking off near the head, with the result that during acceleration in the gun, the needle stem being left unsupported, set back and fired the detonator. As an additional safeguard, therefore, the detonator was covered by a brass disc. These fuzes are identified by the letter " D " after the Mark.

To ensure greater safety in handling and firing the fuze, the No. 106E was introduced. This type effected the introduction of the shutter, with consequent alteration to the magazine and base of the fuze.

The No. 106E fuze has not proved altogether satisfactory; the shutter, while undoubtedly ensuring safety to personnel, has introduced some slight delay effect, with the result that craters are formed on impact. It has therefore proved less efficient than the shutterless No. 106.

There are also objections to the stemmed in C.E. in the fire channel and in the hole in the shutter :—

(*a*) Owing to the difficulty of filling.

(*b*) Owing to the fact that, if the paper tabs retaining the stemmed C.E. under the detonator and in the hole in the shutter become detached, the stemmed C.E. will become displaced into the shutter recess during travelling of the fuze.

This will cause discontinuity and a consequent decrease in the pressure wave. The result is a lower order of detonation, an explosion only, or even in certain cases a total failure.

The Mk. VIIIz fuze depicted on the diagram is designed to overcome certain of the objectionable features mentioned above.

The introduction of the detonator plug with an enlarged though shallow straight through flash channel, the altered shape of the shutter, together with the enlarged diameter and lengthened flash channel beneath the shutter will facilitate filling, produce a better order of detonation and give a more instantaneous effect. The altered shape of the shutter and larger throw-back, due to the increased diameter of the recess in the magazine, is a further assurance against blinds as there is a suspicion that failures of this nature may be caused by the shutter returning to the partly closed position due to loss of spin immediately prior to or on moment of impact. The Mk. VIIz fuze is identical with the Mk. VIIIz except that the former is obtained by conversion from Fuze, No. 106.

As neither of the above marks are in general use in the service at present, a diagram showing a part-sectioned view of the Mk. VIz fuze, which is typical of the Marks in current use, appears below the view of the Mk. VIIIz fuze.

Fuze, Percussion, D.A., No. 115E, Mk. III.

The internal arrangement of this fuze is identical with that of Fuze, No. 106E, Mk. VI, but the shape of the body is more elongated to preserve contour of the streamline shell with which it was designed for use. It is now obsolete.

§6.13. Fuze, percussion, D.A., No. 117, Mk. IIIz.

This fuze is approved for use in 6-inch howitzer streamline, 60-pr. 56-lb., and 18-pr. streamline H.E. shell, and 18-pr. streamline smoke.

The various components and general structure of the fuze are shown on the diagram. The body, magazine and bottom cap are of lead restricted (class G.Pb.) metal.

The filled shutter is designed to open fully when the fuze is spun between 1,800 and 2,200 revolutions a minute.

The fuze is fitted with a five-grain detonator and is filled C.E.

The striker is retained in position by four segments, the segments being held in position by means of the striker and arming sleeves.

The threads of the guide bush, magazine and set-screws are coated with R.D. cement ; the whole of the bottom of the fuze and bottom cap are similarly coated.

The Mk. IIz differs from Mk. IIIz only that the latter has a small recess formed at the base of the upper cavity in the body to provide a seating for the bottom of the arming spring. This was introduced to prevent the end of the spring fouling the striker. Mk. I design is cancelled.

Action.

(1) *On loading,* the safety cap is removed.

(2) On discharge taking place, the arming sleeve, which has been kept in position by the arming spring, sets back and compresses that spring.

Assisted by the action of centrifugal force the four segments fall out and become displaced.

The striker spring then expands from its state of compression, and the movable end, acting against the striker head, withdraws the striker until the striker sleeve impinges upon the guide bush.

The point of the striker is thus drawn up clear of the shutter, which, however, is still held in the closed position owing to the action of friction, due to set-back, while the projectile is accelerating up the bore.

After the shell has left the bore, deceleration sets in and the shutter is free to open under the action of centrifugal force. When the shutter is fully open the detonator is swung under the point of the striker and over the fire channel. The striker is then held off the detonator solely by means of the striker spring.

(3) *On impact*, the striker is forced on to the detonator, causing the functioning of the fuze.

The *No.* 230 *fuze* differs from the No. 117 in that it has a shorter magazine and the C.E. pellet is kept in position by a screwed plug instead of a cap. It is used in conjunction with No. 9 gaine.

§6.14. Fuze, percussion, D.A., impact, No. 45, Mk. IXz.

This fuze is used with H.E. shell having G.S. fuze-hole gauge.

The various components and general structure of the fuze are shown on the diagram. The body and base plug are of lead restricted (class G.Pb.) metal.

The shutter is designed to open fully when the fuze is spun between 1,200 and 2,000 revolutions a minute.

The fuze is fitted with a five-grain detonator and is filled with C.E.

The threads of the detonator plug, holder, set-screw and base plug, also the whole of the top and bottom surfaces of the fuze are coated with R.D. cement.

There has been little structural alteration in this fuze since its inception. Earlier Marks were fitted with a detonator plug which contained no C.E. filling and which screwed down on to the detonator. These fuzes were also fitted with "plunger" and "thin" type shutters. Fuzes of this type are suitable for conversion to No. 45P (powder filling). For certain Marks the securing pin is secured to the cap by means of a brass clip arrangement instead of a whipcord becket.

The shutter and the skin of metal over the fire channel have been thickened slightly as an extra precaution against prematures.

Action of the fuze.

Is similar to that described for Fuze, No. 44.

Adaptations of Fuze, No. 45, are known as Nos. 45P, 450 and 450P.

Fuze, No. 45*P*, is for use in shell filled lyddite and is identical with Fuze, No. 45, except that the former is filled with G20 gunpowder in the hole in the diaphragm and G12 gunpowder in the magazine, the detonator plug and shutter contain no filling and the base plug has a hole drilled through the bottom, the hole being closed by a brass disc coated with R.D. cement and spun in. The top of the fuze is painted red, over the cement ; the safety cap is also painted red.

Fuzes, No. 450 and 450*P*.

These fuzes are intended for proof purposes.

Fuze, No. 450, is filled with C.E. as for Fuze, No. 45, and Fuze, No. 450P, with powder as for Fuze, No. 45P. The former is painted green on the top of the fuze and the whole of the safety cap, the latter is painted black.

They are converted by removing the hammer, modifying the holder and fitting a needle with disc similar to that for Fuze, No. 44.

§6.15. Graze-action fuzes.

In the British Service graze-action fuzes are either of the nose or base type ; they may also be fitted with a delay, if required.

This nature of fuze differs chiefly from the direct-action fuze in that it comes into action when the forward velocity of the shell receives a sudden check instead of depending on a blow on the fuze.

This principle involves the use of a comparatively heavy pellet carrying a detonator (or needle), loosely enclosed in a chamber usually concentric with the axis of the fuze, and capable of longitudinal movement to the front. The pellet with detonator (or needle) is positively prevented from making contact with the needle (or detonator) until after the act of firing has withdrawn the mechanism immobilizing the pellet.

On graze the shell loses velocity, but the pellet, by reason of its inertia, continues on with negligible loss of velocity, and the detonator, striking the needle, is fired, and causes the fuze to function.

The percussion portion of a time and percussion fuze is generally of the graze-action type.

Safety arrangements.—As the action of the graze fuze is very sensitive, measures are taken to render innocuous accidental or premature ignition of the detonator before the gun has been fired by closing the flash channel with a shutter, which opens only when the shell has acquired a considerable velocity of rotation.

In graze percussion fuzes such devices as centrifugal bolts, detents, and pressure plates are usually employed to prevent longitudinal movement of the graze pellet until the fuze has been armed.

Another method, adopted in the Fuze, Base, Hotchkiss, consists of a graze pellet made in two parts, namely, a needle-holder embedded in lead surrounded by brass tubing. Before firing, the needle point is masked by the surrounding lead and tubing, but on firing, the latter sets back on the needle-holder, unmasking the needle, so that it can penetrate the detonator when the graze pellet goes forward on impact.

Premature ignition of the detonator during the flight of the projectile, due to the " creeping action " of the pellet, is prevented in the following manner :—

Once the shell has left the bore it decelerates due to air resistance. Since the loose pellet is enclosed in its chamber, it is not acted on by the resistance of the air, and therefore has a tendency to creep forward.

Usually, the creeping action is prevented by what is termed a " creep spring," though " anti-creep spring " would be a more correct name. Before the needle can pierce the detonator this pellet has to compress the creep spring. The strength of the spring is, however, so arranged as to retain the pellet on its seating until the forward movement of the projectile is sufficiently checked. The greater degree of sensitiveness required in the fuze, the more closely is the strength of the creep spring adjusted.

Comparison between Graze and Direct-action Fuzes.—1. Owing to the interval of time taken by the pellet to compress the creep spring and move forward on to the needle (or detonator, as the case may be), the graze fuze is distinctly slower on its action than the D.A. or D.A. impact fuze.

2. Since graze fuzes depend for their action on the checking of the forward velocity of the shell, and not on direct blows on the nose, there is less liability to blinds occurring at low angles of arrival than with D.A. fuzes, especially in the smaller calibres of shell.

Delay action.—Where the shell is required to penetrate before bursting, the delay action mentioned in (1) above is increased by the introduction of a delay pellet in the filling of the fuze or gaine.

Such graze-action fuzes are termed " delay " fuzes; those without a delay pellet " non-delay."

The terms " delay " and " non-delay " are used by way of contrast, and not in comparison with D.A. fuzes, which are practically instantaneous in their action.

§6.16. Graze percussion nose fuzes.

This type of fuze is not capable by itself of initiating detonation in an H.E. shell, as at present designed in the British Service.

It is used in conjunction with a gaine screwed into an adapter in the base of the fuze.

The gaine, ignited by the flash of the detonator, is designed to initiate the detonating wave.

Gaines.—Gaines were introduced into the Service for use under graze fuzes to initiate the detonating wave necessary for detonating a H.E. shell.

In the Fuze, No. 101E, at present in use, the delay composition (if present) and the powder pellet in the gaine are directly ignited by the flash from the fuze detonator.

The following are the types of gaines to be found in the service:

1. No. 2 Mk. IV used in conjunction with No. 101E fuze in H.E. shell, also with an adapter for smoke target shell for anti-aircraft equipments.

2. No. 8 and No. 9 used by anti-aircraft equipments under a No. 199, 80/44 or other suitable fuze for H.E. shell in place of Fuze, No. 44/80. They are safer than a No. 2 gaine as they have a shutter interposed between the detonator and the C.E. pellet.

The No. 8 gaine is similar in construction and action to the No. 44 fuze which has been described previously, with the exception that the safety pin and cap is omitted, a stronger shutter spring is provided, and the external shape of the body is cylindrical, with screw threads on the upper portion.

The shutter is adjusted to open between 6,000 and 8,000 revolutions a minute.

3. No. 9, Mk. II (*see* Fig. 6.16) is similar in construction and action to the No. 8 except that it is of larger diameter and has a different type of safety shutter. Premature movement of the shutter assembly is prevented by a detent similar to that in Fuze, No. 101E, and a blast channel is formed in the body on one side of the magazine to allow for expansion of gas in the case of premature explosion of the detonator. The filled shutter must open fully when the gaine is spun between 1,800 to 2,200 revolutions a minute.

It will be seen that gaines are mainly used for:—

(1) Bursting H.E. shell fitted with a graze percussion fuze.
(2) Bursting H.E. shell fitted with a time fuze.

§6.17. Fuze, percussion, No. 101E, Mk. IIM and Gaine No. 2, Mk. IV, with delay.

This fuze, being typical, will be described with the No. 2 gaine.

The general design and components of the fuze and gaine can best be seen from the diagram.

The body and adapter are of lead restricted (class G.Pb.) metal.

The shutter is designed to open fully when the fuze is spun between 2,000 and 3,000 revolutions a minute. The fuze is fitted with a 1·7-grain detonator filled detonating composition "A" mixture.

The threads of the graze pellet plug, and the needle, also the top face of the body, prior to screwing home the top cap, are coated with R.D. cement.

Certain of the earlier Marks of this fuze were fitted with a needle spun into the cap ; these are now obsolete.

Gaine, No. 2, Mk. IV.

The gaine depicted on the diagram is filled " delay." When filled " non-delay " a longer perforated powder pellet is used in the upper cavity in the gaine.

The gaine is fitted with a 10-grain fulminate of mercury detonator and the magazine is filled with C.E.

The exterior surface of the gaine is coated with copal varnish, lead free, and the screw threads of the gaine, prior to assembly into the adapter and screwing home the cap, are coated with R.D. cement.

Previous Marks of this gaine were fitted with an " open " type detonator filled 10 grains fulminate of mercury covered with 4·5 grains of R.P. or G.20 gunpowder. This method of filling is now obsolescent.

N.B.—It will be observed that the large fulminate detonator in the gaine is not cut off from the shell filling by any safety shutter.

Action of fuze and gaine.—Before discharge, safety is ensured by the centrifugal bolt holding the pellet and detonator away from the needle ; in addition, the fire channel between the fuze detonator and the gaine is blocked by the shutter.

On discharge, the detent of the centrifugal bolt of the fuze sets back, compressing the spring. The stalk of the detent is thrown outwards by centrifugal force and catches under the shoulder of the recess ; it cannot return to its original position The rotation of the shell then causes the centrifugal bolt, now free, to move outwards clear of the graze pellet, which is held back by the creep spring.

While acceleration is taking place, the shutter sets back, retarding the action of centrifugal force, the shutter consequently remaining in the closed position while the shell is in the bore.

When the shell leaves the bore, deceleration sets in, and the shutter overcoming its spring, allows its detent to free itself ; the stem of the detent swings to one side of the slot and holds the shutter in the open position.

On graze or impact, the pellet overcomes the creep spring and goes forward carrying the detonator on to the needle. The flash passes down the passage in the pellet and fuze, igniting the delay composition or powder pellet in the gaine. The flash from the pellet ignites the 10-grain fulminate detonator, which in turn detonates the C.E. pellets in the body of the gaine.

Distinguishing marks.—If the gaine is fitted with a delay, the cap of the fuze is painted blue, and a blue band is painted round the body of the gaine.

Shutters for No. 101 *type fuze.*—The shutters belong to the adapter rather than to the fuze proper and are only found in Fuzes, No. 101B and No. 101E.

Fuze, No. 101, was without a shutter.

There are two kinds of safety shutter arrangements in use with this type of fuze.

(1) A shutter made apart from the fuze adapter and fitted into the adapter recess—known as the " Pill-box " or self-contained type, lettered " D."

(2) Shutters contained in the adapter—known as the " Adapter-slot " type and lettered G, J, K, and M.

As neither of these types has given satisfactory sealing against an over-sensitive detonator a new and improved shutter will be fitted in the future.

Fuzes converted to the new shutter will be designated 101E Mk. IIIC.

Of the above, D, G and M were approved for use with Nos. 101E and 101EX fuzes—the latter fuze being now obsolete.

Shutters D and G were approved for No. 109E fuze—this fuze is also now obsolete.

Shutters J and K for the No. 101B fuze.

In design and function all these shutters are generally similar, differing mainly in dimensions of spring and weight of shutter, and method of accommodation in the " Adapter for gaine."

The following table shows the number of revolutions a minute at which the shutters are designed to open.

Type I	D	2,000–3,000 r.p.m.
	G	2,000–3,000 r.p.m.
Type II	J	1,300–1,700 r.p.m.
	K	1,300–1,700 r.p.m.
	M	2,000–3,000 r.p.m.

Shutters J, K have weaker springs, in order to enable them to open correctly, when fitted to the No. 101B fuze which is used for heavy howitzers. Owing to the comparatively low rotational velocity of the shell, the centrifugal force tending to open the shutter is small.

§6.18. Graze, base, percussion, fuzes.

Base fuzes at present extant in the service are Fuze, percussion, base, Hotchkiss, and Fuzes, percussion, base, Nos. 11, 12, 15, 16, 346, 480, 500 and 501. The letter " D " is added to the nomenclature of Fuze, No. 16, when the fuze is fitted with delay.

The Hotchkiss fuze is quite distinct in type. Of the remainder No. 16 may be taken as typical.

If detailed descriptions of Nos. 11, 12 and 15 should be required, reference should be made to the handbook of the particular gun in which it is used.

Detailed descriptions of Fuzes, Nos. 346, 480, 500 and 501 are not available.

§6.19. Fuze, percussion, base, Hotchkiss, Mk. X.

This fuze is used with small natures of A.P. shell.

The various components and general structure of the fuze are shown on the diagram. The body and cap are of lead-restricted metal, the former is of class " A " metal and the latter of class " B " metal. The external threads of the body are left-handed. The lip formed on the body acts as a gas-check.

The percussion pellet consists of a brass cylinder, filled with an alloy of lead and tin, in which is embedded the needle-holder with needle. The needle-holder is enlarged at the lower end and has three rows of crimps cut in it to grip the lead alloy. The needle must not protrude beyond the top of the percussion pellet. A pressure of 120 to 220 lb. is required to force the needle forward.

The fuze is fitted with a two-grain detonator, filled detonating composition " A " mixture. A fire hole is bored through the centre of the cap.

The Mk. X fuze differs from its predecessor (Mk. VII) only in regard to certain manufacturing tolerances. There are no Mks. VIII and IX in the land service.

Action of the fuze.—On discharge, the percussion pellet sets back, but the needle-holder, owing to its enlarged end resting on the inside base of the fuze, cannot set back. The needle-holder is therefore forced through the lead alloy and the needle protrudes beyond the edge of the brass casing. The creep spring prevents the pellet from moving forward before impact occurs.

Fuze, Percussion, Base, Hotchkiss, Tracer, No. 280, Mk. I.

This fuze is identical with the Base Hotchkiss Fuze, except that the body is extended beyond the base to provide a separate cavity or central tube to receive tracer composition. It is threaded at the end to receive a steel cap. A hole is drilled through the base of the cap, which is closed by a brass disc sweated into a recess formed at the bottom of the cap.

The filling of the tracer portion consists of approximately 100 grains of illuminating composition, with a layer of **5** grains of priming composition and **5** grains S.F.G.2 gunpowder ; the whole being pressed in.

The tracing composition is fired by the flash from the cartridge and the burning should be visible for a period of not less than five seconds.

§6.20. Fuze, percussion, base, large, No. 16, Mk. IV.

This fuze is used with large natures of A.P.C. and C.P. shell.

The various components and general structure of the fuze are shown on the diagram. The body is made of bronze and the cap of class " A " metal, both of which metals are lead restricted. The body is screw-threaded externally with a left-handed thread ; the flange below the threads provide a seating in the shell.

The fuze is fitted with a detonator filled three grains of detonating composition " A " mixture, and filled with R.P. or G.20 gunpowder.

The vertical channel in the body is filled with five perforated R.P. or G.20 gunpowder pellets and the circular ring in the groove at the top of the body is made of compressed G.20 gunpowder.

Action of the fuze.—The fuze is specially designed to avoid any premature action.

If the detonator be accidentally fired, the flash is masked :—

(1) By the centrifugal bolt.

(2) By the flange on the bottom of the pellet fitting into the groove in the body.

(3) By the pea-ball closing the fire channel.

As a further precaution against gas entering the fuze from the rear, a copper gas-check and steel cover plate are fitted to shell which use this fuze.

On discharge, the propellant gases penetrate the perforations in the steel cover plate fitted to the base of the shell, and exert a pressure on the copper gas-check plate covering the fuze. That portion of the copper gas-check plate which covers the pressure plate of the fuze is so shaped as to fit inside the steel protecting ring, and the gas pressure is thus communicated through the gas-check plate to the pressure plate of the fuze, pushing it and its spindle forward. This releases the retaining bolt which moves outwards under the action of centrifugal force, the grooved portion of the retaining bolt moving on to the reduced portion of the spindle.

The centrifugal bolt is, in turn, released and its smaller end is freed from the recess in the body, bringing the flash hole in the stem in line with the flash hole in the pellet.

The centrifugal bolt is prevented from turning by the small pin on the underside of the head working in its recess.

At the same time the retaining bolt of the pea-ball compresses its spring under the action of centrifugal force, and leaves room for the pea-ball to enter the recess.

The detonator pellet is prevented from turning by the small brass guide pin working in its groove, and any creep action is checked by the creep spring.

During acceleration, the detonator pellet sets back, thus preventing the pea-ball from entering its recess, also bringing the end of the inclined channel in the detonator pellet hard up against the fuze body, and the flange on the base of the detonator pellet well into the annular groove on the body of the fuze. The pea-ball will also set back on its seating. On deceleration, the detonator pellet moves forward a sufficient amount to allow the pea-ball to escape into its recess.

On impact, the pellet, overcoming its inertia, moves forward, compresses its spring, and carries the detonator on to the needle. It is locked in this position by the locking pellet engaging the recess in the body, and the upper coned portion of the detonator pellet wedging itself in the cap ; this allows the flash from the detonator to pass down the flash hole in the pellet and centrifugal bolt, through the hole left open by the pea-ball, thence to the perforated powder pellets, thus firing the magazine. The flash from the magazine passes through the six fire holes in the cap into the shell.

Delay action.—This may be provided in Fuze, No. 16, in order to give the shell sufficient time to penetrate heavy armour. The fuze is then called No. 16D.

Comparison between Fuze, No. 16, *and Fuzes, Nos.* 11, 12, *and* 15.—The main points of difference between Fuze, No. 16, and Fuze, Nos. 11, 12, and 15 are :—

(1) The body and cap of the No. 16 fuze are made of " lead-restricted " metal, whereas Nos. 11 and 15 are not so restricted. It is essential, therefore, that A.P. or C.P. shell filled H.E. should always be fuzed No. 16, the Nos. 11 and 15 being reserved for powder-filled shell.

(2) No. 11 is made of a weaker material than No. 15 and is used with uncapped shell, No. 15 being used with capped shell.

(3) The copper pressure plate of No. 16 is retained in position by a steel protecting ring instead of being covered by a steel protecting plate.

(4) The No. 12 *Special* and 12F *Special* are made of " lead-restricted " metal.

§6.21. New base fuzes.

Fuzes, Base, Nos. 159, 346, 480, 500 and 501 have been introduced into the land service. They are for use with pointed shell.

TIME AND TIME AND PERCUSSION FUZES

§6.22. Introduction.

Definition.—A time fuze is a fuze which can be set to burst a shell at a certain predetermined time after firing.

A time and percussion fuze embodies in addition a graze percussion mechanism which will cause the shell to burst, should—

(*a*) the shell strike the target before the time at which the fuze has been set, has elapsed, or

(*b*) the time mechanism fails in its functioning.

As before stated, time fuzes may be divided into two main classes :—

1. Burning or composition fuzes.
2. Mechanical fuzes.

The time portion of T. and P. fuzes is at present of the burning or composition type.

Burning or composition fuzes.—This type of fuze consists generally of a body containing a lighting mechanism for igniting the time rings, a magazine for igniting the bursting charge of the shell, and, around the stem of the fuze, a time ring (or rings) capable of being rotated for setting purposes.

The time ring has an annular groove on its underside filled with time composition, which is pressed in solid at a load of about 20 tons to the square inch and faced off by machinery. This composition burns at a relatively regular rate.

In the early types of time fuze only one time ring was employed ; in order, however, to obtain a greater length of composition without an appreciable increase in the diameter of the fuze, two rings are now employed. The rings are superimposed and communicate one with the other. The top ring is fixed and the lower is capable of rotation, in order to secure the different lengths required in the time of burning to suit differing ranges.

Compositions or "fuze powders."—Prior to the introduction of A.A. artillery and long-range shrapnel the standard composition was a fine-grain gunpowder normally composed of—

Potassium nitrate	74 to 76 per cent.
Sulphur	8 to 10½ per cent.
Charcoal	by difference.

The powders are manufactured in batches, which are blended together to produce given times of burning when ignited in their compressed condition in a time-ring channel.

In order to obtain increased times of burning for longer ranges, variations are made in the method of manufacture as follows :—

(1) Different classes of wood are used for the charcoal.
(2) Variation in the temperature and duration of charring when the wood is being converted into charcoal.

Charcoals made of a hard wood burnt at a high temperature, give longer times of burning than those made from the lighter classes of wood charred at a lower temperature.

By the blending of a slow-burning powder with a fast-burning powder, any total rate of burning between the two limits may be obtained.

It was formerly supposed that fuze powders so made only differed from one another in their time of burning, at rest, under standard conditions ; later it was discovered that, in addition, the effect of pressure, temperature, and spin differs with each fuze powder. This fact was elucidated in an investigation into the cause of the erratic burning of anti-aircraft fuzes.

The anti-aircraft gun has a higher muzzle velocity than the field gun of the same calibre, and the spin is consequently higher, *i.e.*, with a 3-inch 20-cwt. anti-aircraft gun, M.V. 2,500 f/s,* the spin is 19,980 revolutions a minute, and with an 18-pr. field gun, M.V. 1615 f/s, the spin is 12,000 revolutions a minute.

The effects of increased spin are twofold, and will be dealt with in greater detail later. They are :—

(1) The ring vents becoming choked with slag.

(2) Mechanical disintegration of the fuze powder at the burning surface.

In order to prevent the formation of slag, a powder has been produced by the Research Department, Woolwich, which is slagless. This powder is called R.D. composition No. 202, and is composed of—

Ammonium perchlorate	77 per cent.
Charcoal	20 ,,
Starch	3 ,,

Rings filled with R.D. 202 composition are lacquered with red lacquer.

The following table shows the powders used for filling land service fuzes.

Powder giving time of burning Secs.	Fuzes
22	No. 80 type.
25	,, 199 ,,
30	,, 83 ,,

R.D. 202 composition, used in the lower ring of the Fuze, No. 88, in conjunction with powder giving 30 seconds time of burning in the upper rings, gives a time of burning of 48 seconds.

§6.23. Influences affecting rate of burning.

Three general sets of factors influence the rate of burning of fuze powders. They are :—

(1) The chemical factors.

(2) The age and climatic factors.

(3) The physical factors.

Chemical factors.—The effect on the time of burning of variations in the quality of the ingredients and the method of manufacture.

Age and climatic factors.—The time of burning of a fuze differs with age according to the conditions under which it has been kept. If the moisture in the powder increases, the time of burning

---

* *i.e.* with 12½ lb. projectile.

increases, and the reverse is the case when the moisture content is lowered.

Alternate subjection to high and low temperatures will therefore tend to give variable results.

Physical factors.—These may be considered under three headings :—

(1) The rotational velocity or spin of the fuze.
(2) The effect of pressure on the burning face of the composition.
(3) The effect of the temperature of the composition.

The effect of these three factors on the burning of fuzes used for short and medium ranges and low trajectories is of small account and can be easily allowed for, but, with the introduction of high-angle fire and long-range shrapnel, the physical effects become very marked, resulting in erratic burning and blinds.

It must be realized that these effects are largely inter-dependent ; for instance, the effect of spin, which is normally negligible at speeds less than 12,000 revolutions a minute, will become quite perceptible at low speeds if the atmospheric pressure is down to 15 inches.

However, it has been shown that, if the performance of the gun is normal, the errors due to physical effects are sufficiently small to allow of their being treated as additive, *i.e.*, assuming the resultant effect to be the sum of the effects due to each cause when acting alone, all the other factors being normal. Therefore, a constant percentage correction, which will give fairly accurate results, can be calculated. There are, in addition, random errors in the time of burning which are not apparently due to any regular cause in the fuze itself or in external conditions, just as there are random errors in range. It is the object of experiments and manufacture to reduce the percentage zone of random errors to a minimum.

(1) *Effect of spin.*—This effect is due to centrifugal action. When the spin is high, the composition warmed, softened, and just ignited by the combustion of the previous layer is " spun " outwards to the outer edge of the groove, before it has had time to burn properly and ignite the next layer. Consequently combustion is slower and may fail altogether.

In fuzes filled with ordinary gunpowder this effect is negligible at spins below about 12,000 revolutions a minute, but since centrifugal force varies directly with the square of the rotational velocity, the effects increase very rapidly with the spin, so that at 30,000 revolutions a minute or over, the burning may cease altogether, and in any case will be most irregular. This effect is particularly noticeable with fuze powders having a high sulphur content, so that great care is now taken to ensure that the sulphur content does not exceed 10 per cent.

Another effect of high rotational speed is to cause the blocking of the ring vents with slag, when ordinary composition is used. This causes a banking-up of pressure inside the ring, until the pressure is sufficient to blow out the obstruction. This succession of explosive bursts, accompanied by pulsations of pressure, is equivalent to a reduction of pressure on the burning surface with consequent prolongation of the time of burning.

(2) *Effect of pressure.*—This effect may be considered under two headings :—

(*a*) The static pressure of the atmosphere in which the fuze is burnt.

(*b*) The dynamic pressure due to the motion of the shell through the air.

An increase of pressure on the burning surface of the composition increases the rate of burning and so decreases the time of burning.

(*a*) The atmospheric pressure may vary from slightly above normal atmospheric pressure at sea level when the barometer is high, to less than half an atmosphere (15 inches) at a height of 20,000 feet.

Each diminution of atmospheric pressure of 1-inch of the barometer from a normal height of 30 inches increases the time of burning of Fuzes, time, Nos. 31, 80/44, 180, 180B, 183; Fuzes, T. and P., Nos. 65A, 80, 80B, 197 by 1/44th (0·023) of the mean time of burning of each fuze, and Fuze, time, No. 188; Fuzes, T. and P., Nos. 88 and 220 by 1/60th (0·0166) of the mean time of burning of each fuze. As regards Fuze, No. 199, a tentative figure of 0·015 is given. For Fuze, time, No. 79 no correction is necessary.

(*b*) The dynamic pressure is a maximum at the nose of the fuze, and then falls off rapidly to a pressure below the atmospheric line before the cylindrical part of the shell is reached. The pressure varies with the velocity, but the general shape of the pressure curve remains the same. At velocities below 800 f/s, the pressure on the nose is small and does not vary much, but at higher velocities it increases rapidly, and the region of positive pressure is extended further back from the nose.

It follows that, for a given fuze and shell, the higher the velocity the greater will be the pressure. Also for the same fuze the pressure on the escape hole will be greater in a large shell than in a small one, other conditions of spin, etc., being the same, since the escape hole will be relatively further forward when the fuze is fitted to a large shell than to a small one.

The contour of the fuze has considerable effect on the shape of the pressure curve; a fuze with a smooth contour running into the lines of the head is to be preferred, as it conduces to a more regular and more gradual decline of pressure. Projecting portions immediately in front of an escape hole are objectionable, as they may lead to a region of diminished pressure.

(3) *Effect of temperature.*—This effect may be considered under three headings.

(*a*) The initial temperature of the fuze.

(*b*) The gain of heat by conduction through the body of the fuze from other parts of the composition as they burn.

(*c*) The loss or gain of heat by contact with the external air.

An increase in temperature increases the rate of burning, and so decreases the time of burning.

(i) The initial temperature of the fuze depends on the atmospheric temperature of the chamber, etc.

(ii) The gain of heat by conduction is only noticeable when a fuze is burnt at rest. When the fuze is in rapid motion, the body of the fuze is kept at practically the same temperature as the surrounding atmosphere by the cooling action due to its flight through the air.

(iii) Loss or gain of heat will occur when the fuze comes into contact with the external air, and is due to three causes :—

(*a*) Initial difference in temperature between the fuze and the external air.

(*b*) Fall of atmospheric temperature at increasing altitudes.

The following table gives the normal temperature for various altitudes, calculated at half saturation :—

Height	*Temperature*
Sea Level	60° F.
5,000 ft.	50·4° F.
10,000 ft.	39° F.
20,000 ft.	10·4° F.
30,000 ft.	– 31·2° F.

(*c*) The effect of the rapid motion in compressing the air round the head. The compression is rapid and consequently is followed by rise of temperature.

§6.24. Burning or composition fuze mechanisms.

If the fuze is a time fuze only, it naturally has only a time mechanism ; if a time and percussion fuze, it may be divided into two portions—

1. The time mechanism.
2. The percussion mechanism.

1. *Time mechanism.*—The essential qualities of a time mechanism are that—

(*a*) It should be capable of withstanding rough usage, without becoming dangerous.

(*b*) At the same time it should " arm " and fire the detonator at low orders of acceleration.

Time mechanisms may be divided roughly into three classes according to the matter in which the igniting pellet is supported—

(i) By a shearing wire.
(ii) By a stirrup spring.
(iii) By a spiral spring.

Method (i) is practically obsolete, but may still be met with in Fuze, T. & P., No. 65A. The objections to it are that, firstly, a safety pin is essential to guard against accidental ignition of the detonator before firing owing to rough usage; and, secondly, there is very little scope for making the type more sensitive if it should be required to do so.

Method (ii) is met with in No. 80 fuze. A safety pin is not necessary provided high accelerations are available, as the detonator can be held at some distance from the needle, and the stirrup spring made fairly strong. The effects of successive jars which the fuze may receive, however, tend to become cumulative and so may cause it to arm.

Method (iii) is found in No. 199 and 220 fuzes. This method has not the objections referred to in the other two methods. Unless made excessively sensitive, there is no necessity for a safety pin, owing to the long distance between the detonator and the needle: also the effect of any jars will not be cumulative.

2. *Percussion mechanisms.*—The percussion mechanism of all T. & P. fuzes is on the graze principle, but various devices have been employed to " arm " the fuze. These are:—

(*a*) A safety pellet with shearing wire and ball combined with a centrifugal bolt in the percussion pellet.

This method of controlling the percussion pellet is *practically obsolete*; it may be found in the older types of fuzes such as the No. 65A.

(*b*) A centrifugal bolt alone.

Also *practically obsolete*, but may still be met with in some older types of T. & P. fuzes such as No. 84 used in the naval service.

(*c*) The stirrup spring and ferrule type with or without a ball. This principle is adopted in Nos. 80 and 88 fuzes.

(*d*) The centrifugal bolt and detent type. This exists in the 220 fuze.

Method (*a*) depends firstly on the " set-back " force on acceleration in the gun causing the pellet to shear the shearing wire and drop into a recess, leaving the ball free to move outwards. A safety pin is, therefore, essential to prevent accidental shearing of the wire.

Secondly, it depends on centrifugal force due to rotational velocity acting on a retaining bolt in such a manner that the percussion pellet is free to move forward on impact. The objections to this method are the same as those for the shearing wire type of time mechanism which have been previously mentioned.

Method (*b*) depends solely on centrifugal force. The rotation of the shell causes centrifugal bolts to fly out of recesses in the percussion pellet. The springs for the pellet are weak, and there is also the great objection that a severe jar may cause the mechanism to " arm " after the safety pin has been withdrawn, while the shell is travelling up the bore.

Method (*c*) depends on the " set-back " on discharge, and in the case where a ball is employed in addition, on centrifugal force which removes the ball. There are the same objections in this method as in the " stirrup " type of time mechanism. With the addition of the ball, these objections are to some extent minimized.

Method (*d*) depends on both " set-back " and centrifugal force. In this case, if the detent spring be properly designed, the fuze is capable of arming at a low order of acceleration, and, as the travel of the detent is long, there is no necessity for a safety pin.

Safety pins.—Safety pins are embodied in fuzes whenever the safety arrangements do not ensure that the time or percussion mechanisms are incapable of being armed by rough usage before the fuze is fired from the gun.

The safety pin passes from the exterior of the fuze through the movable pellet, or between that pellet and either the needle or detonator, with the object of ensuring that no movement of the pellet can take place before the pin is withdrawn.

In modern fuzes in the land service they are dispensed with as far as possible, consistent with safety, as they are liable to become difficult to withdraw ; the rate of fire is thus apt to become reduced ; moreover, they form a point of entry for moisture to the interior of the fuze.

Tensioning and clamping of time rings.—The lower time rings of fuzes are prevented from movement, subsequent to setting, in handling, loading and during flight, by two methods :—

(1) Tensioning.
(2) Clamping.

Tensioning consists in screwing down the top cap to such an extent that the time rings can only be moved by the application of a certain definite force. The force usually employed is 250 inch-ounces ± 20 (applied by the torque due to the movement of a 25-oz. weight on the extremity of a 10-inch lever). When correctly tensioned, it is impossible to turn the rings with the fingers. The cap is then fixed by a set-screw.

Occasionally, owing to various causes, this tension may be lost —cases have occurred where the box-cloth washers between the rings have perished, producing a loss of tension—and the rings become loose. This may lead to a " flash-over " or premature, as the tensioning is the means by which the flame from the composition is confined; it also provides a seal against the entry of the flash of any propellant gases that may pass over the shell (*see also* § 6.33, page 197).

Further, if the rings be loose, they may move relatively to one another and alter the setting, and so lead to either a blind, a burst out of the calculated position, or possibly a premature.

For field-gun and anti-aircraft equipments, where a very rapid rate of fire is required, some method of tensioning is essential.

Clamping consists in screwing down the top cap by means of a key subsequent to the rotation of the ring to the required graduation in the operation of setting. This method is more definite, and is suitable for the heavier guns and howitzers whose rate of fire is comparatively slow. A typical " tensioned " fuze is T. & P. No. 80, and T. & P. No. 88 is a fuze whose rings are fixed by clamping.

Setting of time fuzes.—Setting may be done by fuze keys, or fuze-setting machines. " Tensioned " fuzes may be set by either method; " clamping " fuzes by hand only.

Mechanical fuze-setters are rapid and accurate in action, and are particularly suitable when accuracy combined with rapidity of fire are required, as a number of fuzes can be set to the same length.

Setting studs.—However well designed and accurately made the setter may be, it must be borne in mind that good results depend largely on the accuracy of the setting studs or slots in the fuze.

When the fuze is manufactured the setting studs are very accurately inserted, one in the movable ring, the other in the body of the fuze. They are usually made of rustless steel.

The following points should be observed :—

(1) They should project sufficiently to ensure good engagement with the stops of the setter.
(2) They should be so placed as not to interfere with a cover, if provided.
(3) The material used for the studs should not chip or flake and leave pieces of metal in the slots of the setter.
(4) They must not affect the flight of the projectile.

After issue to the service, the greatest care must be taken not to damage or distort the studs. They must be free from rust and burrs, and must not be used for any other purpose such as a hold for a fuze key, etc.

Escape holes.—This term refers to the method of providing an escape for the gases evolved by the combustion of the powder in the composition time rings.

Radial channels are provided in the time rings themselves, so that the gas generated may escape through them into the atmosphere.

In the past, considerable difficulty was experienced in closing the escape holes to seal the ingress of muzzle flash, and at the same time to open, as soon as the adjacent composition was ignited. This difficulty has now been overcome as follows :—

Escape hole discs are lightly held in position in the escape hole recess in the time rings by stabbing ; and powder is pressed into the escape holes. These discs are placed in position with sufficient security to withstand the action of centrifugal force due to rotation, and, by pressing in the powder behind them, are relieved of the force exerted by this powder when the fuze is spinning. In the case of the top ring, the pellet in the escape hole is ignited by the flash from the detonator, and in the case of the bottom ring, the powder pressed in is fired by the descending flash from the composition in the top ring.

Any blocking of the exhaust channels by slag, etc., resulting from the combustion of the fuze powder, seriously affects the regularity of burning. Therefore, one of the important points to be borne in mind is to arrange a free passage for the ejection of the slag.

Time fuzes used with H.E. shell.—During the Great War, a composite arrangement of fuzes was introduced in order to obtain a " time " burst with H.E. shell.

For example, a modified Fuze, D.A., No. 44, is used immediately below a specially prepared No. 80 fuze, and is called Fuze, No. 44/80. The special preparation of the No. 80 fuze consists in removing the percussion portion, and replacing it by a wooden plug and in blotting out that part of the graduated scale below the figure 2 to obviate any danger from accidental short setting. It is known as Fuze, time, No. 80/44 (*see* Fig. 6.25(*b*)). Fuze, time, No. 199, may be used for the same purpose.

Fuzes, No. 44/80 will generally be replaced by gaines Nos. 8 or 9 in the land service.

Description of typical time and time and percussion fuzes.—It is not proposed to describe any time fuze in detail, as, in almost every case, the time fuzes at present extant in the land service are simply time and percussion fuzes with the percussion arrangement replaced by a wooden plug.

§6.25. Fuze, time and percussion, No. 80, Mk. XI.

This fuze is used with the light group of 18-pr. shell and is made of unrestricted (class " G ") metal. It is similar in design to the Mk. V, with the exception that the latter is a lighter fuze, made of aluminium. The Mk. V fuze is used with the heavy group of 18-pr. shell.

Fuze, T. & P., No. 80, Mk. XI, consists of various components which can best be identified by reference to the diagram.

Two time rings superimposed fit over the upper portion of the body, the lower one resting on the platform. This platform on the body is graduated from 0 to 22. Each ring has a groove on its underside forming a circle broken by a bridge of metal. The fuze composition is pressed into the groove.

Boxcloth washers are placed between the rings, and paper rings cover the composition in order to exclude damp.

A hole, fitted with an escape hole disc, is cut in each ring at the beginning of the powder channel to act as an exit for the products of combustion. These brass escape-hole discs are most necessary to ensure correct functioning of the fuze. If any of these are missing in any particular fuze, that fuze should not be used.

The upper ring is prevented from rotating on the body by pins; the lower ring can be rotated for setting. For this purpose a line is engraved on the lower time ring so as to be above the scale of 0 to 22, agreeing with the nominal time of burning, inscribed upon the platform. A red cross is also engraved upon the body, which, in conjunction with the setting mark on the ring, denotes the position of safety.

Lighting mechanism.—Immediately below the cap the mechanism for lighting the time rings is situated. This consists of a time pellet fitted with a detonator and supported by a stirrup spring over the needle.

Percussion mechanism.—The percussion mechanism is fitted in the body above the magazine. It consists of a pellet, fitted with a detonator, on which rests a stirrup spring. The creep spring and ferrule rest on the stirrup spring with a cap to hold them in position. R.P. or G.20 powder is then poured into the magazine, through a hole in the base plug which is then closed by a screwed plug.

Fuze cover.—When in the shell, the fuze is protected from damp by a sealed cover (*see* page 198).

Action of the time mechanism.—On acceleration in the gun, the pellet sets back, straightening out the arms of the stirrup spring, and carrying the detonator on to the upper point of the needle.

The flash from the detonator passes through an opening in the stem and lights the composition in the upper time ring and the powder pellet in the escape hole. This blows out the escape hole disc.

The flame burns round the upper ring in the direction in which the shell is rotating until, after an interval of time determined by the setting, it reaches a passage communicating with the lower time ring provided with a powder pellet to ensure certain ignition.

The composition in this ring is thus lighted, the escape hole disc is blown out, and the flame travels in the reverse direction until after an interval of time, also determined by the setting, it reaches the powder pellets in a passage leading to the magazine which is then ignited.

Action of the percussion mechanism.—On discharge, the ferrule sets back, straightening the arms of the stirrup spring, and travelling on, forces itself over the pellet to a rim at the base ; the creep spring prevents any forward movement of the percussion pellet during flight.

On graze or impact, the percussion pellet flies forward, overcoming the creep spring, and carries the detonator on to the point of the needle. The flash passes through the pellet and fires the magazine.

Safety arrangements.—When the setting on the lower time ring is opposite the red cross on the body, *i.e.*, when the fuze is set at safety, the flash holes to the lower time ring and to the magazine are masked by the bridges of the upper and lower time rings, respectively. This provides a double safety against ignition of the magazine, should the lighting mechanism act prematurely.

Fuzes should invariably be set safe during transport of ammunition. If rounds, of which the fuzes have been set, have to travel, then the fuzes must be re-set to safety, *not to zero*, in order to ensure safe transport of the ammunition.

The lighting pellet is protected by the cap of the fuze. Although this protection is sufficient for ordinary usage, yet it is possible for the fuze to be fired if the round is dropped heavily on to the cap, particularly if it be a fuze of which the cap is made of aluminium.

It is important, therefore, that the rounds should be handled with care, and that the fuzes should be set at safety during the transport of the ammunition.

§6.26. Fuze, time and percussion, No. 88, Mk. VI.

In general this fuze is similar to No. 80 fuze previously described.

The main differences lie in the mechanism of the lighting and percussion arrangements.

The time rings are similar except that the lower ring is filled with R.D. 202 composition, which gives the fuze a total time of burning of 48 seconds. The graduations, however, remain the same, 0 to 22. The lower ring is lacquered red.

Time mechanism.—The pellet with the detonator is supported by a coiled spring, which is strong enough to ensure safety during transit. There is a considerable distance between this detonator and the needle.

Percussion mechanism.—The percussion pellet with detonator occupies a central recess in the lower part of the body.

A stirrup spring, fitting over the top of the pellet, supports a ferrule, which keeps the detonator at a distance from the needle.

A metal ball is interposed between the pellet and the top of the recess.

A creep spring rests on the flange round the base of the percussion pellet.

Action of the time mechanism.—On discharge, the pellet sets back, overcoming the coiled spring, and carries the detonator on to the needle. The further action of the time mechanism is similar to that of the No. 80 fuze.

Action of the percussion mechanism.—The acceleration on firing causes the ferrule to set back over the pellet, straightening out the arms of the stirrup spring.

The rotation of the projectile causes the metal ball to fly out into a recess in the body.

" Creeping " during flight is prevented by the creep spring.

On impact, the pellet flies forward, compressing the creep spring, and carries the detonator on to the needle.

Safety arrangements.—The safety arrangements of the time mechanism are similar in principle to those of the No. 80 fuze.

The percussion mechanism has an additional safeguard in the metal ball, which ensures that the detonator cannot be touched by the needle until the ball has been moved clear by the action of centrifugal force.

In the earlier Marks of the fuze, the percussion needle was spun in, but in this Mark the needle is screwed in from the top. By this construction the needle is more firmly held and is capable of withstanding the forces set up on firing.

In the Mk. III, the time mechanism was placed eccentrically with regard to the axis of the fuze. There were, in consequence, two needles.

§6.26 (*a*). Fuze time and percussion, No. 220 Mk. I.

The fuze is a tension type of 2-in. gauge.

The time portion is similar to that of the No. 199 fuze, but is located to one side of the centre line of the fuze.

The percussion arrangement is in the centre line of the fuze, and resembles that of the No. 101E fuze. The needle pellet is provided with a flash groove passing round the sides and bottom, so that on graze the flash has easy access to the magazine.

The dome shaped cap is fixed after tensioning by two set-screws.

When in the shell, a copper asbestos washer fits under the flange.

The top ring is filled with 30-second powder whilst the bottom contains R.D.202 ; the latter is therefore lacquered red.

The Mark II fuze differs only in the graduations, which are from 0 to 22.

§6.27. Fuze, time, No. 199, Mk. III.

This fuze has been introduced for anti-aircraft guns.

The various components and general structure of the fuze are shown on the diagram. The body and base plug are of unrestricted (class " G ") metal.

The method of manufacture and filling is similar to that described for Fuze, T. & P., No. 88, except that there is no percussion arrangement; the detonator is filled two grains detonating composition " B " mixture and G.20 gunpowder and the time rings are filled with a fuze powder of new composition to give a time of burning of 25 seconds. The upper ring is prevented from rotating by a single brass pin. Both rings have a setting slot.

There is a two-way channel leading from the lighting pellet chamber to the upper ring to allow the flame and gas from the detonator to circulate in the body of the fuze and so cause the ignition of the time ring to be less violent.

The markings round the fuze body are graduated 0 to 30.

Mk. I has small setting slots in top and bottom ring and the graduations are round the top ring.

Mk. II has widened setting slots in both rings and certain minor structural differences, otherwise as Mk. I.

Mk. III is similar to Mk. II except that the graduations are transferred from top ring to flange round the body.

Mks. I and II are now obsolescent.

Action of fuze.

Similar to that described for time mechanism of Fuze, No. 88.

§6.28. Mechanical fuzes.

Mechanical fuzes are time fuzes in which the time to burst is controlled by mechanical means in place of the burning away of a train of composition.

They may be divided into two distinct types :—

(1) Mechanical time fuzes, such as No. 203.
(2) Mechanical distance fuzes.

Advantages of mechanical fuzes.—In general the advantages of mechanical time and mechanical distance fuzes over the ordinary burning or composition time fuzes are :—

(*a*) The absence of irregularity in time due to inherent variation in the time of burning of batches of powder, and the detrimental action of the slag produced by combustion.

(*b*) The absence of variation caused by dynamic pressure at the escape holes of ordinary time rings and the rarification of the atmosphere at high altitudes. This is of importance in the case of fuzes fitted to shell fired at high angles of elevation against aircraft flying at considerable altitudes.

(*c*) The absence of variation in the rate of burning at high altitudes owing to the lowering of atmospheric temperature.

(*d*) The effect of spin is not so great at high spins as it is in the case of combustion fuzes.

(*e*) In the case of mechanical time fuzes, it is possible to test each individual fuze for time. This can only be done in the case of a combustion fuze by destroying it.

(*f*) Age should not affect a mechanical fuze, and in any case it can be tested for time-keeping at any period in its life and re-regulated if necessary. Nothing can be done with aged combustion fuzes, which burn irregularly.

Disadvantages of mechanical time fuzes.—

(*a*) They are, at present, expensive to manufacture, but this may be more than compensated by the life in store.

(*b*) The mechanism is affected by spin when the spin is high, but perhaps not to the same extent as a combustion fuze.

(*c*) Possible loss of tension in springs in storage, but there is not yet sufficient data on this point. The Germans experienced no loss of tension after ten years.

(*d*) Possible deformation in high velocity guns, but this point awaits proof.

§6.29. Mechanical time fuzes.

In this type the mechanism is designed to run at a predetermined rate after the fuze has been armed, this rate being little affected by the rotational velocity of the projectile in which it is fired.

The time of running or time of burst is, therefore, practically constant in any type of gun or howitzer. Consequently, the design has the advantage over others in that it is capable of being used in nearly all equipments, provided the time of flight is known, even though the range table does not include a scale for the actual fuze.

There are several different mechanical time fuzes in existence in the service, differing only slightly in their general make-up.

No mechanical types are at present in general use in the land service. A general description of Fuze, No. 203, is given. A list of the others is appended showing their essential differences from No. 203 fuze.

§6.30. Fuze, time, No. 203, Mk. I.

This fuze, designed to burst a shell at any interval of time up to 60 seconds after the firing of the gun, consists primarily of three parts :—

(*a*) The body.

(*b*) The magazine in the base of the body.

(*c*) The clockwork mechanism.

(*a*) The body consists of a metal base piece, screw-threaded to the 2-inch fuze-hole gauge, and provided with a flanged platform on which rests the dome with steel cap attached. The cap is conical and brought up almost to a point in order to continue the streamline contour of the shell into which the fuze fits.

At the top of the dome is a race-way, known as the " hand-race," which consists of a perfectly flat, smooth band of metal in true vertical relationship with the platform of the base-piece. This hand-race is slotted to accommodate the " hand."

Inside the dome, below the hand-race, a locking ring is held in position by means of three copper wire rivets passing right through the wall of the dome and the thickness of the ring. The lower edge of the ring is recessed so as to leave only a thin rim of metal. Five hollow-edged pins, which come immediately under the thin rim of the locking ring, are secured in the platform of the base-piece.

The dome is held in position by means of a screwed collar, which screws down into the flange of the base-piece, housing a tensioning wire between it and the flange provided at the base of the dome. The screwing down of the ring on to the tensioning wire regulates the turning movement necessary to rotate and set the dome. This turning movement is fixed at 250 inch-ounces plus or minus 25 inch-ounces.

(*b*) The underside of the body is hollowed out to form the magazine, and a hole through the diaphragm thus formed between the magazine and platform is bored and screw-threaded to take the detonator plug. Holes are also drilled to allow of the passage of the holding or anchoring screws of the clockwork mechanism.

The detonator is an ordinary igniferous detonator filled three grains Detonating Composition " A " mixture, and the magazine is powder filled and closed by a base plug in the usual manner.

(*c*) The " action " is a piece of unjewelled clockwork mechanism. It consists of a train of wheels operating a hand, the motive power being supplied by a mainspring. The hand itself is capable of rising out of its normal position under the influence of a spring, and is the main factor in the functioning of the fuze. The hand is double-ended, and is mounted on a hollow centre, the rim of which embraces a lip on the end of the striker lever. The striker lever is released by the rising of the hand. The shape of the hand is such as to allow it to rise vertically when it reaches a definite angular position of coincidence with the recesses in the hand race.

The method of regulating the movement is based on the present-day watch-making practice, but the mode of application is different. In this mechanism a pallet and straight length of steel spring or phosphor-bronze ribbon take the place of the balance wheel and coiled hair-spring.

The pallet consists of four arms, two upstanding in such an angular position as to engage the teeth of the 'scape-wheel, the other two being flat and in a straight plane with the base plate of the movement. A small brass weight is placed at each end of the latter pair of arms in order to regulate the rate of oscillation of the pallet by the adjustment of the measure of weight.

The rate of oscillation is controlled by the straight hair-spring, which is secured centrally through the arbor carrying the pallet. The hair-spring has one end housed so as to be free to move radially in a saw-cut in the bottom plate of the movement, and the other end similarly secured in a radially adjustable block, which can be adjusted until the free length of hair-spring is such as to ensure the correct rate of oscillation of the pallet and therefore the correct rate of movement of the rotating hand.

The normal rate of oscillation is 87·98 complete beats a second.

The mechanism is anchored to the platform of the base piece by holding screws, and also embraces a number of safety devices described below.

Safety arrangements.—The main safety arrangements employed are as follows :—

(*a*) The safety catch (or centrifugal safety catch) to maintain the striker in the unarmed position.

(*b*) Trigger safety catch to prevent premature arming.

(*c*) A shearing pin to maintain setting at safety in storage and transit.

(*a*) The centrifugal safety catch is housed under the cam on the striker and serves a dual purpose—

(i) In the event of the movement being accidentally set in action, the striker, owing to its being upheld by the interposing catch, is prevented from reaching the detonator.

(ii) If a fuze in this condition were loaded into a gun and the gun fired, the catch would be prevented from swinging out from under the striker, because the latter would immediately jamb down on to a step cut in the catch for that purpose, and hold it firmly in place.

(*b*) The hand is secured against movement by means of a trigger pivoted on the top plate of the movement at one end, the opposite end being provided with a lip of overhanging metal, which engages the hand and retains the trigger in position. The trigger, being free to move up and down, is controlled by means of an eccentric screwed to the top plate and overlapping the edge of the trigger, thus preventing any tendency of the trigger to rise under the influence of the hand spring. It has, of course, to descend in order to free the hand, but once down it must not be allowed to reassert itself due to rebound. This is countered by a small locking bolt which functions under the action of a spring immediately on descent of the trigger.

A trigger safety catch is housed under the trigger in order to prevent its being prematurely functioned. The catch is withdrawn immediately the fuze is set away from the safety mark.

(*c*) The shearing pin of copper passes through a small hole in the flange of the base piece and dome, and is for the purpose of preventing the possibility of the dome being accidentally moved in transit after being issued set at safety.

Assembled fuzes.—The clock is fully wound before being secured in position in the body. The striker point is immediately above the centre of the detonator.

The dome and base piece carry setting dots for use with a fuze setter, the setting depending upon the relative positions of the hand of the movement and the dome hand race slots.

The fuze is waterproof to prevent ingress of moisture, and is secured against interference as regards its tension by locking the screwed collar in position after tensioning the fuze.

Action.—On firing, the locking ring in the dome shears its suspending wires and sets down on to the steel pins in the platform of the base piece, thus locking itself and preventing any shifting of the dome from the position it was set in, by reason that a feather or guide pin in the dome engages a slot or featherway in the locking ring. The trigger descends and is caught by the locking bolt, releasing the hand: during acceleration in the bore it is unlikely that the pallet will oscillate owing to the force of " set-back "; when the muzzle is reached the hand begins to rotate under the influence of the mainspring. After acceleration ceases, the centrifugal safety catch, flying out under the influence of centrifugal force, leaves the striker supported only by its cam resting on the sloping surface of the pillar provided for the purpose.

The fuze having reached its setting, the hand is brought immediately under the slots in the hand race into which it rises under the influence of its spring, and releases the lever controlling the striker. The lever flies outwards owing to centrifugal force, and the action of the striker spring turns the cam of the striker off the pillar, the sloping surface of the latter facilitates this action, and the striker fires the detonator and the powder magazine.

The other mechanical fuzes are :—

§6.31. Fuze, time, No. 200.

This fuze differs from No. 203 in respect of the dome-locking device, and in being without a safety catch under the trigger.

The clock is contained in a clock case having a flanged base, and the pins are affixed in the locking ring instead of in the fuze platform. The locking ring is also suspended by springs instead of shearing wires.

On firing, the locking ring sets out of the springs, the pins puncture the clock case rim, and thereby lock the dome, as in the case of No. 203.

Fuze, time, No. 201.

This fuze is identical with No. 200 except that it is timed to run 85 seconds instead of 60. So far, none have been made and formal approval is still withheld.

Fuze, time, No. 202.

This fuze is identical with No. 203, but is for use in naval service only.

Fuze, time, No. 204.

This fuze consists of the No. 200 mechanism made in France, fitted with centrifugal safety catch as in No. 203 (*i.e.*, locking step) and fitted to bodies pertaining to the No. 203 design.

Fuze, time, No. 205.

Is similar to No. 203, but with contour for use with 3-inch 20-cwt. 16-lb. shell.

Fuze, time, No. 206.

Differs principally from the No. 202 fuze in being fitted with a modified clock mechanism giving a running time of 45 seconds. The bottom ring is graduated in seconds and marked 0–225, the figures representing the number of 1/5th seconds of time. The fuze is fitted with a cover similar to that for the No. 202. It is used in the naval service.

*Fuze, time, German, converted, Dopp Z*16.

It was from this fuze that the No. 200 was developed. It is identical for all practical purposes with the No. 200, except as regards the fuze-hole gauge. To render it suitable for use with the 2-inch fuze-hole an adapter is necessary. It is also fitted with centrifugal safety catch of the No. 203 type (*i.e.*, locking step).

§6.32. Mechanical distance fuzes.

Under this heading are included those fuzes in which the time burst depends on the distance travelled, or more strictly on the number of revolutions the fuze makes about a spindle which passes through its axis. The spindle is held against rotation by means of vanes, or by the attachment of some pendulous weight or similar " inertia " member.

This type of fuze is therefore inoperative until fired. Its time to burst is dependent upon and varies with the muzzle velocity, the pitch of the rifling, and the calibre.

This design requires no self-contained driving force such as a mainspring.

On the other hand, the necessity for the projecting vanes or other exterior inertia member introduces, in the majority of cases, a grave risk of failure due to injury during loading.

No fuzes of this type yet exist in the land service.

PRESERVATION OF FUZES

§6.33. Effect of climate.

Fuzes deteriorate when exposed to damp, and climatic conditions affect fuzes adversely, especially abroad in hot moist climates.

The injury is permanent and tends to increase with time, although there may be no further exposure.

Moisture acts on the detonators, and, if they become affected, blinds or unsatisfactory bursts may result. Damp also acts on the composition of time fuzes. It may lengthen the time of burning, prevent ignition, lead to a premature functioning of the fuze, or cause the rings to become so tight that they cannot be set by hand. This latter defect is frequently met with in time and time and percussion fuzes of war-time manufacture. It is caused by the composition in the time rings swelling and thereby taking a firm grip on the boxcloth washers.

Dampness also corrodes the bodies of fuzes, especially those parts made of aluminium. It rusts or corrodes safety pins, making them difficult to withdraw.

In mechanical fuzes moisture disturbs the harmonious working of the clock. Precautions have, therefore, to be taken to protect fuzes, particularly time fuzes, from damp.

§6.34. Fuze Cylinders.

Fuzes, when issued separately from shell are packed in tinned-plate cylinders. The lids are hermetically sealed by a tin strip soldered on, and the cylinders are vacuum tested before issue.

In certain cases where fuzes are only to be issued to home units, the lids are tape-banded, *i.e.*, sealed by a piece of tape shellaced on.

Each cylinder has a label on the top, showing the number, nature, Mark, filled lot number, date of filling and packing, and the filler's initials.

Painting of fuze cylinders.—Fuze cylinders containing time and time and percussion fuzes having the 2-inch gauge are painted green. Those containing detonating fuzes are painted yellow. All other cylinders are painted black.

§6.35. Waterproofing and sealing of fuzes.

All openings in percussion fuzes are coated with R.D. cement or waterproofing composition to prevent the ingress of damp.

In all the latest time and time and percussion fuzes, the spaces between the cap, time rings, and body, also the set-screw recess of the cap, and the escape hole discs in the time rings, are now waterproofed with a composition of beeswax, mineral jelly, and french chalk.

The base plugs are waterproofed by having the threads coated with R.D. cement before being screwed in, and then the whole of the base is covered with R.D. cement.

§6.36. Fuze covers.

Time and 80 type T. & P. fuzes are, with minor exceptions, further protected by sealed covers. The covers may be made of brass, tinned-plate, or lead foil. Only brass and tinned-plate covers now exist for the No. 80 fuze.

The cover is cup-shaped to fit over the fuze, and is secured by soldering a brass band to the junction of body of fuze and cover. It forms an integral part of the fuze.

Lead foil covers are not now fitted.

Metal covers with tear-off bands once removed cannot be replaced, that of the 199. however, is replaceable.

All covers should be fitted in a dry atmosphere.

The chief essentials of a cover, neglecting questions of manufacture, cost, etc., are :—

(*a*) Their capacity of withstanding rough treatment, without rendering their removal difficult, or losing their air-tight property.

(*b*) Their ease of removal without the use of special tools, and without injury to the operator.

(*c*) Their freedom from projections or fringes left on the fuze after stripping, which might influence the subsequent flight of the projectile.

MORTAR FUZES

§6.37. Introduction.

As mortar fuzes differ essentially in some respects from fuzes used in gun ammunition, it has been considered preferable to treat them separately.

All general remarks on fuzes, their function, design, waterproofing, etc., naturally apply to these mortar fuzes, so that the description of the types that follow must be read in conjunction with any statements of a general nature already made.

Mortar fuzes may be divided into two classes :—

1. Percussion fuzes.
2. Time fuzes.

Percussion fuzes.

There were three distinct types of percussion fuze :—

(*a*) *The direct action type.*

This type is similar in principle to an ordinary D.A. gun ammunition fuze. Its functions on impact by having the needle forced home on to the detonator.

(*b*) *The all-ways type.*

This type has the capability of functioning as soon as any part of the cylindrical bomb, to which it is fitted, strikes the ground. The fuze itself need not be touched by the impact. This type is now obsolete.

(*c*) *The pistol type.*

This type only contains the mechanism of a fuze; it is used in conjunction with a detonator carried separately in transport, and fixed into the bomb itself when preparing for action. It is now obsolete.

Three percussion fuzes exist for mortars at present. These are :—

Fuze, Percussion, D.A., Nos. 138, 139 and 139P, and are used in the vaned type of 3-inch mortar bombs.

§6.38. Fuze, percussion, D.A., No. 138, Mk. I.

The construction and action of this fuze is generally similar to the No. 117 described on page 169.

In the design of mortar fuzes which are to be fitted with shutters to render them safe in transport and during acceleration in the bore, the following points should be considered.

(i) In a mortar the chamber pressure available for arming a fuze is relatively of a low order—usually under 2 tons to the square inch—and, further, the acceleration period is of short duration.

(ii) As the mortar is smooth bore, no spin can be imparted to the bomb; neither is it possible for practical considerations to twist the vanes, as stability in flight and air resistance must be taken into consideration.

The general differences from the No. 117 fuze are given below (*see also* Fig. 6.13).

The shutter containing a five-grain fulminate detonator is held against the shutter spring by the point of the striker.

The arming sleeve surrounds the striker and retains six steel balls in a seating, these together with the sleeve and arming spring retain the striker, and therefore the shutter, in the safe position.

Above the guide bush is the striker head, striker spring, split pin, two locking pins and safety cap.

The striker head on its top surface is shaped to form a key; the interior of the safety cap is cut to form a key-way. The striker head is extended to form a sleeve which has two semi-circular recesses; the guide bush is also prepared in a similar manner. Into these recesses fit the two locking pins. Thus if the fuze should accidentally arm, and the cap be turned, the feather on the striker head enters the featherway inside the safety cap, thereby locking the cap, striker head and striker to the fuze body, through the medium of the two locking pins and guide bush. The two locking pins allow vertical but not rotary movement of the striker.

Action of the fuze.—The cap is removed before loading. If the cap is difficult to remove by hand, it must not be forced, as the fuze may have become armed through rough usage and consequently be in a dangerous condition.

On acceleration, the arming sleeve sets back, compressing the arming spring until it is tripped by the tongues of the retaining sleeve; the balls are now free to fall out of their seating. The striker spring under the head now forces up the striker, the point of which leaves the hole in the shutter. The shutter spring forces the shutter containing the detonator into the armed position.

On impact the striker is forced in and the needle enters the detonator initiating detonation.

§6.39. Fuzes, percussion, Nos. 139 and 139P.

These are D.A. fuzes, and the construction of the No. 139 can be seen from Fig. 6.39. It is generally similar to Fuze, Percussion, D.A., No. 44, but the safety shutter is omitted, since there is no rotational velocity to open such a fitting.

The D.A., No. 139P, fuze is similar to the No. 139, but is filled with gunpowder instead of C.E. The cap of this fuze is painted with red lacquer as a means of identification.

§6.40. Rust-proofing.

In all mortar fuzes, those components made of iron or steel must be efficiently rust-proofed, in order to avoid deterioration setting in owing to damp, the usual process being electro-tinning.

TABLE 6.41

FUZES FOR LAND SERVICE

FIELD AND MOBILE ARTILLERY

Percussion fuzes.

(*a*) Direct-action fuzes.

Fuze	Projectile	Equipment
No. 44 ...	H.E. Shell	B.L. 9·2-inch gun. ,, 6-inch gun.
	Smoke Shell	Q.F. 4·5-inch how.
No. 106 ...	H.E. Shell	Q.F. 18-pr. (Service only). Q.F. 13-pr. (Service only).
No. 106E ...	H.E. Shell	B.L. 18-inch how. ,, 14-inch gun. ,, 12-inch how. ,, 9·2-inch gun and how. ,, 8-inch how. ,, 6-inch gun and how. ,, 60-pr. Q.F. 4·5-inch how. ,, 3·7-inch how. and mortar. ,, 18-pr. ,, 13-pr.
	Smoke Shell	B.L. 6-inch how. Q.F. 4·5-inch how. ,, 3·7-inch how. or mortar. ,, 18-pr.
No. 106P.E.	Practice Projectile... ...	B.L. 6-inch how. (*z*) ,, 60-pr. (*z*). Q.F. 4·5-inch how. (*z*). ,, 3·7-inch (*z*). ,, 18-pr. (*y*).
No. 106P.D.	Practice Projectile, powder filled.	B.L. 6-inch how. ,, 60-pr. Q.F. 4·5-inch how. ,, 3·7-inch ,,
No. 117 ...	H.E. Shell	B.L. 6-inch how. (86-lb. shell). ,, 60-pr. (56-lb. shell). Q.F. 18-pr. (streamline shell).
	Smoke Shell	Q.F. 18-pr. (streamline shell).

(*z*) Powder filled.
(*y*) H.E. filled.

Percussion fuzes—continued.

(*b*) Graze-action fuzes.

Fuze	Projectile	Equipment
No. 101B ...	H.E. Shell	B.L. 18-inch how. ,, 12-inch ,, ,, 9·2-inch ,, ,, 8-inch ,,
No. 101E ...	H.E. Shell	B.L. 6-inch how. ,, 60-pr. B.L. 2·75-inch. Q.F. 4·5-inch how. ,, 3·7-inch how. ,, 18-pr (not for Imperial service). ,, 13-pr. ,, ,, ,, ,,
No. 109E ...	Practice Projectile, filled H.E. (with picric powder exploder).	B.L. 6-inch how. ,, 60-pr. Q.F. 4·5-inch how. } India only.

(*c*) Graze-action base fuzes.

Fuze	Projectile	Equipment
No. 12 Special	A.P. Projectile	Q.F. 18-pr.
No. 16 ...	C.P. Shell, filled Shellite ...	B.L. 18-inch how. ,, 12-inch ,, ,, 9·2-inch ,, ,, 8-inch ,, ,, 6-inch gun. ,, 6-inch how.

Time and percussion fuzes.

Fuze	Time of Burning	Shell	Equipment
No. 80 No. 80B*	22 secs. ...	Shrapnel Shell ...	Q.F. 3·7-inch how. (H fuze). ,, 18-pr. (*x*). Q.F. 13-pr. (*x*). B.L. 2·75-inch (*x*).
No. 88	48·5 secs. ...	Shrapnel Shell ...	B.L. 9·2-inch gun. ,, 6-inch gun. ,, 60-pr.
No. 220	4.5 secs. (approx.)	Shrapnel Shell ...	60-pr. (practice reduced charge).

(H) Heavy (brass or steel body) fuzes.

(L) Light (aluminium body) fuzes.

(*x*) The various marks of shell, without fuze, are not all of the same weight, and (H) or (L) fuzes are used as required to produce correct "filled and fuzed" weight for these calibres. In Imperial service, the latest marks of shell require (H) fuzes only.

* Obsolescent.

Time fuzes.

(*a*) Combustion type.

Fuze	Time of Burning	Shell	Equipment
No. 183	30 secs. ...	Star Shell ...	B.L. 6-inch how. Q.F. 4·5-inch how.† ,, 3·7-inch ,,
No. 188	48·5 secs. ...	Star Shell ...	B.L. 6-inch how. Q.F. 4·5-inch how.† ,, 3·7-inch how.†

(*b*) Mechanical type.

Fuze	Time of run	Shell	Equipment
No. 203	60 secs. ...	Shrapnel Shell ...	B.L. 9·2-inch gun.

Note.—In shell filled H.E. only shuttered fuzes will be used for practice.

† No. 188 fuze to be used when proving these shell, but if No. 188 is not available, the service fuze (No. 183) may be used for that purpose.

ANTI-AIRCRAFT ARTILLERY

Time fuzes.

(*a*) *Combustion Type.*

Fuze	Time of Burning	Projectile	Adapter (if used) and Gaine	Equipment
No 80/44 (H) *No. 80B/44 (H)	22 secs.	H.E. Shell ...	Adapter, No. 9 and Fuze, No. 44/80. Adapter, No. 16, and Gaine, No. 8. Gaine, No. 9.	Q.F. 3-inch 20-cwt.
		Practice Projectile		Q.F. 3-inch 20-cwt.
	22 secs.	H.E. Shell (for Canada)	Adapter, No. 9, and Fuze, No. 44/80 Adapter, No. 16, and Gaine, No. 8	Q.F. 13-pr. 9-cwt.
No. 80/44 (L) *No. 80B/44 (L)	22 secs.	H.E. Shell (for Canada)	Adapter, No. 11, and Gaine, No. 2	Q.F. 13-pr. 9-cwt.
No. 180 (H) *No. 180B (H)	22 secs.	Target Shell charged phosphorus	Adapter, No. 16, and Gaine, No. 8 Adapter, No. 15, and Gaine, No. 2 Gaine, No. 9.	Q.F. 3-inch 20-cwt.
		Practice Projectile, Mk. IIB		Q.F. 3-inch 20-cwt.
No. 180 (L) *No. 180B (L)		Shrapnel Shell (for Canada)		Q.F. 13-pr. 9-cwt.
		Target Shell charged phosphorus (for Canada)	Adapter, No. 15, and Gaine, No. 2	Q.F. 13-pr. 9-cwt.
No. 197	22 secs.	Practice Projectile in burst short rounds		Q.F. 3-inch 20-cwt.
No. 199	25 secs.	H.E. Shell ...	Adapter, No. 16, and Gaine, No. 8, Gaine, No. 9	Q.F. 3-inch 20-cwt.
		Shrapnel Shell, Mk. IIBT		Q.F. 3-inch 20-cwt.
		Practice Projectile		Q.F. 3-inch 20-cwt.

* Obsolescent.

(H) Heavy (brass or steel body) fuzes. (L) Light (aluminium body) fuzes.

Time fuzes—continued.

(*b*) *Mechanical Type.*

Fuze	Time of Run	Projectile	Adapter (if used) and Gaine	Equipment
No. 205	60 secs.	H.E. Shell	Adapter, No. 16 and Gaine, No. 8, Gaine, No. 9	Q.F. 3-inch 20-cwt.

COAST DEFENCE ARTILLERY

Percussion fuzes.

(*a*) Direct-action impact fuzes.

Fuze	Shell	Equipment
*No. 13	H.E. Shell except 6-inch Mk. XXVI/XXB. *See* note (*w*)	B.L. 9·2-inch. ,, 7·5-inch (India) ,, 6-inch. Q.F. 6-inch. ,, 4·7-inch.
*No. 18	H.E. Shell except 6-inch Mk. XXVI/XXB. *See* note (*w*)	B.L. 7·5-inch (India). B.L. 6-inch. Q.F. 6-inch. ,, 4·7-inch. ,, 4-inch. ,, 12-pr. 12-cwt.
No. 45	H.E. Shell except 6-inch Mk. XXVI/XXB. *See* note (*w*)	B.L. 9·2-inch. ,, 7·5 inch (India). ,, 6-inch. Q.F. 6-inch. ,, 4·7-inch. ,, 4-inch. ,, 12-pr. 12-cwt.
No. 45P	H.E. Shell. *See* note (*w*).	B.L. 9·2-inch. ,, 7·5-inch (India). ,, 6-inch. Q.F. 6-inch. ,, 4·7-inch. ,, 4-inch. ,, 12-pr. 12-cwt.

Note.—No. 18 Fuze will not be used for practice. * Obsolescent.

(*w*) *Filling, Exploders, Gaines and Adapters for H.E. Shell with D.A. Impact Fuzes.*

(i) Shell, B.L. 6-inch, Mk. XXVI/XXB :—

Lyddite filling with trotyl exploder / Trotyl ,, C.E. ,, — {Fuze, No. 45P in Adapter, No. 23, over Gaine, No. 9.}

(ii) New filling of shell other than (i), 2-inch fuze-hole :—

Lyddite filling with picric powder exploders. Fuze, No. 45P, in Adapter, No. 2.

(iii) Existing filled shell, other than (i) :—

Lyddite filling with picric powder or trotyl exploders. Fuzes to be allotted are No. 45P for shell 9·2-inch to 4·7-inch, inclusive, and No. 18 or 45 to 4-inch and 12-pr. Alternatively, 4·7-inch shell may have Nos. 18 or 45 in lieu of No. 45P fuze. Adapter, No. 2, required for shell with 2-inch fuze-hole. This allocation of fuzes will be adopted in replacement of the No. 13 fuze issued for use with lyddite filled shell 4·7-inch and above, fitted with picric powder or trotyl exploders.

Amatol or trotyl filling, with trotyl or C.E. exploders, No. 13, 18, 45 or 45P fuzes may have been issued for use with these shell, but in future and until the shell have been used up, Fuzes No. 18 or 45 only will be allotted.

Percussion fuzes—continued.

(*b*) Direct-action fuzes.

Fuze	Shell	Equipment
No. 44	H.E. Shell filled trotyl	Q.F. 6-pr. (obsolete for Imperial service).
	,, ,, lyddite	6-pr. 10 cwt. on twin mounting.
No. 117	H.E. Shell filled lyddite with trotyl exploders or filled trotyl with C.E. exploders (for 4 c.r.h. shell when firing landwards)	B.L. 9·2-inch.
No. 230	H.E. Shell, Mk. XXVI/XXB, with lyddite filling and trotyl exploder or trotyl filling and C.E. exploder over Gaine, No. 9, when firing landwards	B.L. 6-inch.

(*c*) Graze-action base fuzes.

Fuze	Shell	Equipment
Hotchkiss ...	Common shell ... ... Steel shell	Q.F. 6-pr. and 3-pr. (obsolete for Imperial service).
*No. 11	A.P. Shell, filled powder	B.L. 9·2-inch. ,, 6-inch. Q.F. 6-inch.
	C.P. Shell, filled powder	B.L. 6-inch. Q.F. 6-inch.
No. 12	C.P. Shell, filled powder	Q.F. 4-inch and 12-pr. Q.F. 4·7-inch.
No. 15 or 15C ...	A.P. or A.P.C. Shell, filled powder ... A.P.C. Shell, filled powder ... ... A.P. or A.P.C. Shell, filled powder ... A.P. Shell, filled powder	B.L. 9·2-inch. ,, 7·5-inch (India). ,, 6-inch. Q.F. 6-inch.
	C.P.C. Shell, filled powder ... ... C.P. or C.P.C. Shell, filled powder ... ,, ,, ,, ,, ...	B.L. 9·2-inch. B.L. 6-inch. Q.F. 6-inch.
No. 16	A.P.C. Shell, filled lyddite or shellite...	B.L. 9·2-inch. ,, 6-inch.
*No. 16 (with delay) No. 16D	A.P.C. Shell, filled shellite or lyddite...	B.L. 9·2-inch. ,, 6-inch.
No. 12F special... No. 500 ... ... No. 501 ... ... No. 480 ... ... No. 346 ... ... No. 159	S.A.P. Shell, filled lyddite. *See* note (*v*) S.A.P., filled trotyl C.P.B.C. Shell, filled trotyl/beeswax ... A.P.C. shell, filled trotyl/beeswax ...	Q.F. 4·7-inch. B.L. 6-inch. B.L. 9·2-inch. B.L. 15-inch

Note (*v*).—Shell taken over from N.S. were fitted with No. 12N fuzes and shell so fuzed may be met with in the service. Shell fitted with No. 12F special fuzes (converted from 12N) have also been issued. All lyddite filled S.A.P. shell will eventually be re-explodered and fitted with No. 500 fuzes in replacement of No. 12 type.

* Obsolescent.

Percussion fuzes—continued.

(*d*) Time and percussion fuzes.

Fuze	Time of Burning	Shell	Equipment
No.65 A...	20·1 secs....	Common nose-fuze shell ...	Q.F. 6-pr. (obsolete for Imperial service).
No. 88 ...	48·5 secs....	Shrapnel shell	B.L. 9·2-inch. ,, 6-inch. Q.F. 6-inch. ,, 4·7-inch.

(*e*) Time fuzes, combustion type.

Fuze	Time of Burning	Shell	Equipment
No. 188	48·5 secs....	H.E. Shell, Mk. XXVI/ XXB over No. 9 Gaine (for anti-beach landing operations)	B.L. 6-inch.

(*e*) Time fuzes, mechanical type.

Fuze	Time of run	Shell	Equipment
No. 203	60 secs. ...	Shrapnel Shell	B.L. 6-inch.

TANK EQUIPMENT

(*a*) Percussion fuzes.

Fuze	Shell	Equipment
Base Hotchkiss ...	A.P. Shell, Mk. I ... ... Practice Projectile, Mk. I	Q.F. 3-pr. 2-cwt.
	Common Shell ... ... Steel Shell	Q.F. 6-pr. 6-cwt.
No. 280 (fuze and tracer)	A.P. Shell, other than Mk. I Shell	Q.F. 3-pr. 2-cwt.
Not decided ...	Smoke Shell	Q.F. 6-pr. 6-cwt.
No. 106E	H.E. and Smoke	Q.F. 3·7-inch mortar

M.L. MORTAR EQUIPMENT

(*a*) Percussion fuzes.

Fuze	Bomb	Equipment
No. 146	H.E. Bomb, Cylindrical Type Bomb	M.L. 3-inch.
No. 138 (shuttered)	Streamline H.E. Bomb	M.L. 3-inch.
No. 139	Streamline Smoke Bomb	M.L. 3-inch.
No. 139P	Streamline Practice Bomb	M.L. 3-inch.

AIRCRAFT EQUIPMENT

(*a*) Percussion Fuzes.

Fuze	Shell	Equipment
No. 131	H.E. Shell	Q.F. 1½-pr. and 2-pr.
No. 240	H.E. Shell	Q.F. 1½-pr.
Base Hotchkiss ...	C.P. Shell	Q.F. 1½-pr.

CHAPTER VII

IMPLEMENTS FOR SETTING TIME FUZES

§7.01. Introduction.

To set a time fuze, whether a fuze with a powder train that burns at a definite rate, or a mechanical fuze, the fundamental quantity required is the time of flight to the desired point of burst. The subject is discussed in Artillery Training and in the Manual of Anti-Aircraft Defence, and we find that two implements are used. For use in the field these are :—

1. A *fuze indicator* which finds the fuze length to suit the conditions of the moment.
2. A *fuze setter* which adjusts the fuze (usually by turning the time ring).

For anti-aircraft defence a height-fuze indicator takes the place of the fuze indicator. It gives the fuze length in terms of the height of the target and the angle of sight.

The *fuze indicator* performs three functions :—

(*a*) Gives the length required for a given range, instead of for a given time of flight, or for a given gun elevation.

(*b*) Corrects the standard length for range to allow for the factors which affect the time of burning of a powder train fuze.

(*c*) Corrects for any change in the time of flight due to variation in the muzzle velocity from that in the range table.

§7.02. Theory of the corrector.

The correctors in use are :—

1. Proportional corrector to the fuze.
2. Non-proportional corrector.
3. Muzzle velocity corrector.

The speed at which fuze composition burns is affected by a number of factors such as barometric pressure, temperature, velocity, age of the fuze, and others ; the length of the fuze required, therefore, is not constant but varies with each of these conditions. The variation is proportional to the length required, as calculated for standard conditions in the fuze scale tables.

1. *Proportional corrector.*—The function of the proportional corrector is to determine the increase or decrease in setting from the

normal for all times of flight, on the assumption that the amount of increase or decrease is proportional to the time of burning of the fuze. In other words, it gives the correct fuze setting for effective burst at any range when once the indicator has been adjusted for one range, so long as the conditions remain unchanged.

Such a corrector is usually made in the form of a slide rule. Suppose f_0 to be the length of fuze required for a given time of flight under standard conditions, f_1 the length required for the same time of flight under the condition of the moment, and x the length required to be added or subtracted from each unit length to give a burst at the required spot under the condition of the moment. Then, since the corrector is to be proportional,

$$f_1 = f_0 \pm f_0 x$$
$$\text{or } f_1 = f_0(1 \pm x)$$
$$\text{and } \log f_1 = \log f_0 + \log (1 \pm x)_0$$

If a slide rule be constructed with one of its fixed scales graduated proportionally to the logarithm of the time of flight, the other proportionally to log $(1 \pm x)$, and the slider to the logarithm of the lengths of fuze, then by setting the slider so that the fuze length corresponds to the time of flight under standard conditions, the reading of the scale log $(1 \pm x)$ should be zero. If the scale be displaced by an amount of $(1 \pm x)$, the fuze lengths brought opposite the graduations of the time of flight scale will be corrected lengths for all times of flight, and the reading of the scale log $(1 \pm x)$ will be the correction of the moment (assuming the gun is shooting to the range table).

The *Service fuze indicator* is based on these principles with certain modifications.

(i) The logarithms of the time of flight are replaced by a scale of ranges. The graduations on this scale are proportional to the times of flight to graze.

(ii) The divisions of the fuze scale are modified to give effective height of burst at the ranges on (i). To simplify the work of the battery commander an approximation is arrived at for each gun, estimated to give a constant angular height of burst at all ranges.

NOTE.—The even graduations of the corrector scale give a slight disproportionality in the percentage time corrections. This error is negligible for small variations.

(iii) The scale log $(1 + x)$ is replaced by an empirical even scale graduated from 0 to 200 or 300, and giving approximately the right proportional corrections. To avoid the inconvenience of the double sign the corrector scale is so placed that the graduation giving no correction is 150 instead of 0. The graduations are such that 10 divisions alteration of corrector setting at a certain range (3,000 yards in the case of the 18-pr. gun) alters the mean angular height of burst by 10 minutes.

Fig. 7·02(a)

INDICATOR, FUZE, No 10 MARK I.

SCALE = 1/3.

The figures 150, the line round them, & the 150 line on Corrector Scale together with the arrow heads on the fuze Scales to be filled with red wax. All other engraving to be filled with black wax.

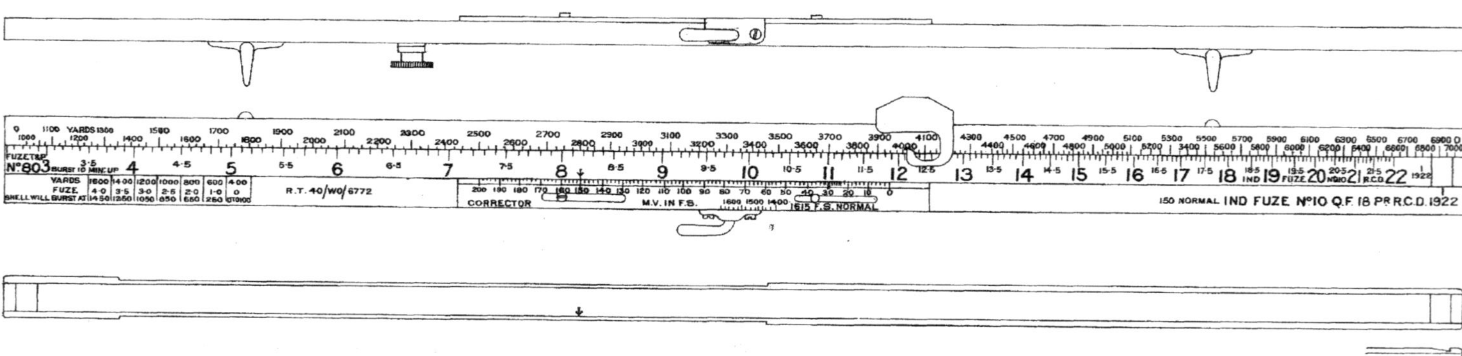

These logarithmic graduations have the disadvantage that the divisions are closer together as the range increases, while they spread out more as the range shortens. Consequently, if the higher ends of the scales have their divisions wide enough to be easily set, the whole indicator would be of an unwieldy size. The difficulty is met by making the lower end of the scale non-proportional; the range at which proportionality ceases is fixed for each design of indicator.

In the latest pattern of indicator for the 18-pr. the change is made at 2,500 yards, below which the scales give a non-proportional correction. Fig. 7.02(*a*) shows this pattern.

With corrector normal the scale gives a height of burst 10 minutes up throughout. Ten divisions on the corrector correspond to 10 minutes change in height at 3,000 yards; above this range the value of 10 corrector, expressed in height, increases gradually; below it the value decreases gradually to 2,500 yards where it is about 8 minutes. It is then kept constant at this value on the non-proportional part of the scale.

On the lower left-hand side will be seen a table of fuze settings for short ranges, 1,600 to 300; these are used instead of the readings given by the corrector scale. The reference number of the range table is also engraved on the bar; this is the range table from which the divisions of the scales are computed.

If it is desired to use the indicator with any other ammunition than that marked on it, *e.g.*, a reduced charge or a different type of shell, a table of corrections must be employed. It can, however, be used with more than one pattern of fuze and the slider is usually engraved with a scale for fuze on each side.

2. *Non-proportional corrector.*—In a non-proportional corrector the range scale is evenly divided, and the fuze scale is graduated in accordance with the range table. The increase and decrease in fuze length, obtained by altering the corrector setting, are no longer proportional to the time of flight, and the fuze indicator must be readjusted when the range is altered.

3. *Muzzle velocity corrector.*—The object of the muzzle velocity corrector is to enable the normal fuze scale to be used with a gun whose muzzle velocity differs from normal and to obtain a correct burst without calculation. Such a corrector can be applied to the indicator described above, and can be used whether the carriage has calibrating sights or non-calibrating sights. The M.V. scale will, however, be different in the two cases, the reason being as follows:—

If the elevation required for a given range R be ϕ with normal M.V., then for a fall in M.V. an elevation $\phi + d\phi$ is required to cause the trajectory to intersect the line of sight at the target (*i.e.*, to obtain the same range). If the gun be fitted with a calibrating sight the range on the drum will be R, while the elevation will be $\phi + d\phi$. But if the fuze reading be taken for the range R from the indicator the shell will burst short, since it will take a longer time to traverse the higher trajectory due to the elevation $\phi + d\phi$. Therefore it is necessary to adjust the indicator to give a longer fuze setting for range R. This will be done if we move the corrector scale to the left.

On the other hand, if the sight is not provided with a muzzle

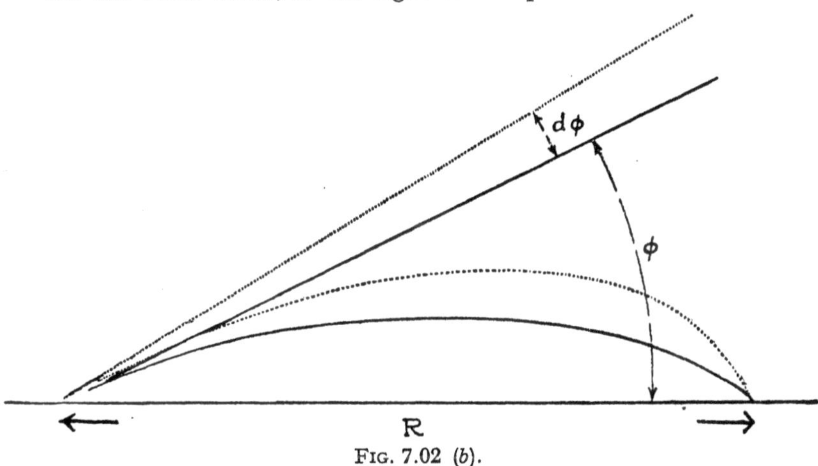

FIG. 7.02 (*b*).

velocity corrector, the elevation $\phi + d\phi$ required to allow for loss in M.V. will give a range on the drum of R + dR. If the fuze setting for R + dR be taken from the indicator, the burst will be beyond the required point, for it can be demonstrated that it would take longer to travel R + dR with elevation $\phi + d\phi$ at normal M.V. than it would take to travel R with elevation $\phi + d\phi$ at a

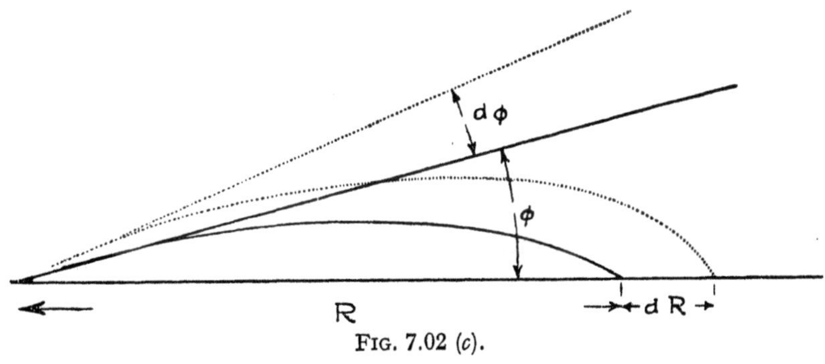

FIG. 7.02 (*c*).

lower M.V. Therefore, a shorter fuze is required. This will be obtained if the corrector be moved to the right.

A change in the muzzle velocity of the gun involves a variation in the spin of the shell. This has an effect on the time of burning of the fuze; a fall in M.V. means an increase in the rate of burning, and *vice versa*.

For example, in the 3-inch 20-cwt. gun (M.V. 2,000 f/s) a fall of 100 f/s in M.V. decreases the spin by 800 r.p.m. The resulting increase in the rate of burning of No. 80 fuze is of the order 0·5 per cent. at most, and not more than 1 per cent. at the vertex of a high trajectory. Even at extreme fuze settings the correction for this will not be more than 0·2.

The fuze indicator shown in Fig. 7.02(*a*) has a M.V. scale for use with a calibrating sight engraved on a movable plate, which also has on it the corrector scale. This plate replaces the fixed proportionate corrector scale and is capable of limited movement right and left, as it is secured by a clamp. Fixed to the side of the indicator is an index which is opposite the 1615 on the M.V. scale when no correction for M.V. is required.

If the plate be moved to the right or left, it will be seen that the sliding fuze scale must be moved also, in order to bring the reading 150 on the corrector to zero again. Consequently, a movement of the plate to the right gives a shorter fuze, while a movement to the left gives a longer fuze for the same corrector setting.

§7.03. Height-fuze, indicators for high-angle fire.

Since the time of flight to the future position of the target will vary according to the height as well as the angle of sight, the fuze setting must be corrected accordingly. The time of burning of the fuze is also affected by the height. The construction of the height-fuze indicator is explained in the Text Book of Gun Carriages and Gun Mountings and in the Manual of Anti-Aircraft Defence.

In its simple form it consists of a vertical plate with fuze curves engraved on one side. A straight edge graduated in heights is pivoted at the origin of the curves and can be directed in the future line of sight. The height is set on the straight edge by a cursor and the appropriate fuze length may be read. The fuze curves are calculated to correct for the two variables mentioned above, and there is also a corrector to allow for the error of the moment.

§7.04. Fuze-setting instruments.

The implements actually used for setting the fuzes are :—

Fuze keys.
Fuze-setting machines.

§7.05. Fuze keys.

Are wrenches suitably shaped for the fuze ; they are described in the various handbooks.

When using a fuze key it is necessary to read the scale on the fuze every time.

There are in this method two personal errors possible, that of the man setting the fuze and that of the man working the indicator. This is, nevertheless, the service method of setting, both for fire on the flat and for high-angle fire.

§7.06. Fuze-setting machines.

The search for a satisfactory design of a fuze-setting machine was begun many years ago, and a large number of patents have been taken out. Some of these never reached the stage of a thorough trial, while others have proved satisfactory in some respects but possessing disadvantages in the requirements of accuracy and facility in use. The only setting machine adopted for use in the field is the *débouchoir* of the French artillery. The fuze used with it is, however, not a ring fuze ; it has a tube of burning composition wound spirally round the exterior, and the action of the machine is merely to punch a hole between the burning composition and the flash composition at the required distance. The mechanism is, therefore, more simple than that required to turn the time ring of a tensioned fuze such as is used in the British Service.

The object of a fuze-setting machine is to render the setting of time fuzes more rapid and accurate than the same operation performed by means of a fuze key, and unless such a machine does fulfil these essentials, there is no advantage in its use, more especially in the field, where weight is important. It should be easier to set the dial of a machine than to set the small scale on the time ring of the fuze. On the other hand, if the machine has no corrector the fuze indicator must still be used, and the number of operations is not reduced. To include a proportional corrector in the mechanism renders it more complicated.

If any machine is to set fuzes accurately it must be given a fuze made to great accuracy. In particular, the slot in the time ring of the fuze in which the turning member of the machine engages must be accurately and clearly made. This has proved one of the chief difficulties in the design of fuze-setting machines in the past. The problem would be easier were a projection allowed on the time ring to engage against a suitable surface in the turning member of the machine ; but such a projection would be objectionable ballistically and has been definitely excluded from all modern fuze designs. The machine itself must therefore have a projection to engage in the small slot in the time ring of the fuze. This projection is, of course, very subject to wear, thus causing inaccuracy.

It will be seen, therefore, that the introduction of a machine for use with most existing fuzes is almost out of the question. With the introduction of new types, such as the Fuze, No. 199, and the mechanical fuze, which give results of improved regularity, the provision of some form of setting machine is desirable. Experimental mechanical fuze setters for use with Fuze, time, No. 199, are now under trial.

CHAPTER VIII

SMALL ARM AMMUNITION

§8.01. Introduction.

Small arm ammunition is used with the rifle, machine gun, revolver and pistol. The term *S.A. Cartridge* is held to mean the complete round, and includes the cartridge case, percussion cap, propellant charge and bullet.

The *Cartridge Case* is normally made of cartridge brass (70 per cent. copper and 30 per cent. zinc), solid drawn, and its design and manufacture follow the same lines as a Q.F. case for a gun. The base may be formed with a rim for purposes of extraction and to position the cartridge in the chamber, or with the rimless type the case may be provided with a groove for the extractor to catch in, the case being then positioned in the chamber from the shoulder. The first type gives better positioning of the cartridge in the rifle, the latter is more convenient for loading in a magazine and for automatic weapons.

In place of the primer of a Q.F. case, a cap chamber is usually formed in the base of the case, and connected by fire holes to the interior. In the centre of the cap chamber is a metal projection or anvil on which the cap composition is crushed by the blow of the striker.

The *Percussion Cap* must be made to fit accurately in the cap chamber, or there will be an escape of gas between it and the cartridge case known as a " blowback," and the cap would be liable to come out.

The cap is made of copper or of brass, and is sometimes secured in the cap chamber by " ringing," *i.e.*, the metal of the cap chamber is pressed over the edge of the base of the cap.

Irregularity in the priming results in uneven ignition and hang-fires, while indifferent quality leads to instability and deterioration with resulting missfires. The composition must be well protected from moisture ; this is accomplished by covering the composition with a varnished lead-tinfoil disc and varnishing after insertion of the disc.

The *Propellant* varies according to the nature of the cartridge. The propellant most commonly used for S.A. ammunition is cordite, though various types of nitrocellulose powder have been used. Each type of propellant has advantages, though for most cartridges either type is suitable.

Bullet.—The design of a S.A. bullet differs from that of a gun projectile in the following particulars :—

(i) Except in the case of special types, such as tracer, it is to all intents and purposes solid, so questions of strength of wall do not arise.

(ii) It has no driving band, but is made to engage the rifling by " set-up " (*i.e.* expansion of the base of the bullet on firing) and by the bullet being made slightly larger in diameter than the bore of the weapon.

In other respects, questions of shape and balance of bullet are the same as those dealt with under shell design (Chapter V). Bullets being produced by mass production, it is, however, of great importance that the machines used shall be capable of ensuring great accuracy in such matters as symmetry. If the centre of mass of a bullet is only one-thousandth of an inch away from the axis of the bullet it will cause an error of between 13 and 14 inches at a range of 600 yards.

The ordinary British rifle bullet is not streamlined. For the reasons given in Chapter V, streamlining makes little difference to the flatness of the trajectory at the most effective ranges for rifle fire (up to about 600 yards) where the velocity is high, and also makes it more difficult to obtain satisfactory set-up. It is, however, possible to get considerably increased ranges by the use of streamlined bullets.

In general, the modern rifle bullet consists of an envelope and a core. The envelope is usually made of fairly soft metal, that will be engaged by the rifling without undue strain on the bullet or on the rifling. Cupro-nickel (80 per cent. copper, 20 per cent. nickel) or soft steel are both used for this purpose.

The quality of the cupro-nickel must be good. If it is too hard or brittle, the bullet envelopes will split and break up in flight ; if too soft, the metal is liable to strip off ; if the quality of the metal is bad, there may be metallic fouling.

The core depends on the nature of the bullet. It should be sufficiently heavy to give the bullet the momentum necessary for long ranges.

Certain revolver bullets and some old type rifle bullets have no envelope, but are made of solid lead alloy, the acceleration of the bullet being too small to cause the lead to strip off.

Types of modern cartridge.—There are several types of modern cartridge for the rifle or machine gun, viz :—

1. Service ball.
2. Armour-piercing.
3. Tracer.
4. Observing.

§8.02. ·303-inch service ball cartridge.

The service ball cartridge is Cartridge, S.A., Ball, ·303-in., Mk. VII.

The Mk. VII is used for cordite-loaded ammunition. If loaded with nitro-cellulose the Mk. is VIIz.

The case is of solid-drawn brass, and has a rim or flange at the base end, by which the cartridge is positioned in the gun and extracted. It is bottle-necked near the front end to allow the mouth of the case to fit over the base of the bullet.

The base of the case is recessed to form a cap chamber and an anvil ; two fireholes connect the cap chamber to the interior of the case.

The percussion cap is of copper-zinc alloy and is pressed into the cap chamber of the case and ringed in ; the joint between the cap and the case is varnished.

The priming of the cap consists of about six-tenths of a grain of cap composition, which is pressed into the cap, covered with a disc of varnished lead-tin foil and then varnished over the disc.

Charge.—For this cartridge there are two types of charge, viz., cordite (about 37 grs) and nitrocellulose powder (about 41 grs).

The cordite is used in the form of tubular sticks (M.D.T. 5–2).

The nitrocellulose powder is in the form of chopped tube, and is of the "progressive" type, *i.e.*, it is impregnated to a certain extent with dinitrotoluene, which has the effect of slowing down the rate of burning during the early stages of combustion.

A glazedboard disc is provided as a wad, and is placed between the propellant and the bullet.

The material of which the wad is made has to be carefully selected, and must be chemically neutral, since the presence of either acid or alkali is liable in time to affect the quality of the cordite or corrode the case.

The bullet, weighing 174 grains, is pointed, and has a lead-antimony core (2 per cent. antimony) enclosed in a cupro-nickel envelope. The point is struck to a radius of about 8 calibres.

In the pointed end of the envelope in front of the lead is an aluminium tip. This is required partly to bring the bullet to the correct weight for the required ballistics, and partly to balance the bullet suitably for accurate shooting. Fibre and compressed paper tips were used during the Great War as alternatives to aluminium.

Near the rear end of the bullet is a cannelure which is filled with beeswax, partly for lubrication and partly to waterproof the joint between the bullet and case.

The bullet is secured into the case by three indents of the metal of the case into the cannelure, and by pressing the case with a cone-shaped die on to the parallel portion of the bullet.

The weight of the complete cartridge is about 386 grains.

Marking.—The base of the case is stamped with—

(*a*) The contractor's initials or recognized trade mark.
(*b*) The last two figures of the year of manufacture.
(*c*) The numeral, *i.e.*, VII or VIIz, as applicable.

The annulus of the cap is varnished with dark purple varnish as a means of identification, and also to assist in waterproofing.

§8.03. ·303-inch armour-piercing cartridge.

At the most favourable range the service bullet will not penetrate more than 5 mm. of modern armour. For dealing with armoured vehicles and aircraft, loophole plates, etc., a specially designed armour-piercing projectile and heavier charge are required.

The armour-piercing cartridge is designated "Cartridge, S.A., Armour Piercing, ·303-in., W, Mk. I" when cordite loaded; when nitrocellulose loaded the Mark is Iz.

The case and cap are the same as in the Mk. VII cartridge.

The charge for the Mk. I cartridge consists of about 37 grains of cordite M.D.T. 5–2 and for the Mk. Iz cartridge about 43 grains of N.C. powder.

The bullet consists of a soft steel envelope coated with cupro-nickel and a hard steel core. Between the core and the envelope is a sleeve of lead-antimony alloy.

A glazedboard disc is used as a wad.

A hardened steel projectile would give the greatest penetration, but as it would not take the rifling, the softer envelope is provided. As it is, the " set-up " (or expansion of the base due to gas pressure) obtained is very small, and although the bullets are made larger in diameter than the ordinary service bullet, the accuracy obtained is not as good.

On the bullet striking an armour plate, only the core penetrates; the envelope and sleeve flatten out and support the point of the core during the first instant of penetration.

The penetrating power, therefore, depends almost entirely on the striking energy of the core. Both the weight of the core and its velocity should consequently be as great as possible. This means that the thickness of the sleeve has to be cut down to the lowest practicable limits.

Marking.—The base is stamped similarly to the service cartridge, except that W I (or W Iz, as applicable) replaces the VII.

The annulus of the cap is varnished with green varnish.

§8.04. ·303-inch tracer cartridge.

Tracer cartridges are so called because they leave a visible wake or " trace " behind them so that the trajectory can be seen.

They are used to ascertain visually where the bullets are going, with the object of correcting the aim. It is, therefore, essential that they should have approximately the same trajectory as other service bullets. Unfortunately the trajectories of the service and armour-piercing bullets differ somewhat, so it is not possible to imitate exactly the trajectories of both.

The " trace " is obtained by filling the bullet with some chemical which burns during flight, with the result that from the time the bullet leaves the muzzle until the chemical is completely consumed, the bullet is continually becoming lighter. The shape of the trajectory is consequently not the same as that of an ordinary service bullet, whose weight does not alter.

Nevertheless, it is possible to make a tracer bullet which up to a range of 600 yards has a trajectory sufficiently close for practical purposes to those of both the ordinary service and armour-piercing bullets, the trajectories of which do not differ much up to this range.

Beyond this range, the path of the tracer bullet diverges considerably, but this is of no great consequence, as the bullet does not trace much further than this.

The tracer bullet is very destructive to the weapon in which it is fired. The barrel rapidly becomes fouled, and there is a tendency for the bore to become coated with a hard metallic deposit, which is not easy to remove.

When firing tracer cartridges, it is desirable to fire ordinary or armour-piercing cartridges at the same time, using alternate rounds of the different types, as this reduces the deposit considerably. The best way of getting rid of this deposit is to fill the rifle barrel with a strong solution of ammonia and leave it to soak overnight.

The service tracer cartridge is designated " Cartridge, S.A., Tracer, ·303-in., G, Mk. I " when cordite loaded; when nitrocellulose loaded the Mark is Iz.

The case, percussion cap, wad and charge are identical with the Mk. VII or VIIz cartridges, respectively.

The bullet consists of a cupro-nickel envelope, and a core in two parts; the front portion of the core is of lead-antimony alloy, the rear portion consisting of a solid drawn copper cylinder, open at the rear end, and filled with tracer composition.

Marking.—The base is stamped similarly to the service cartridge, except that G I (or G Iz, as applicable) replaces the VII. The annulus of the cap is varnished with red varnish.

§8·04 (*a*). ·303-inch observing cartridge.

This cartridge is used only as a training store. Its purpose is to assist in the observation of fire and it is designed to make a small puff of smoke on the bullet striking the ground or the target. The cartridge is designated :—" Cartridge, S.A., Observing, ·303-in., O, Mk. I."

The case, percussion cap, wad and charge are the same as in the Mk. VII cartridge.

The bullet consists of a cupro-nickel envelope, in the nose of which a hole is bored which is closed with a plug of fusible metal; the front end of the envelope contains phosphorus and powdered aluminium, behind which is a lead core, and the base is soldered.

Marking.—The base is stamped similarly to the service cartridge except that 0 I replaces the VII. The annulus of the cap is varnished with black varnish and the tip of the bullet is lacquered black.

This cartridge is in Explosive Group XII, whereas ball, A.P., tracer and blank S.A. cartridges are in Group VI.

§8.05. Special ammunition for the Royal Air Force.

Ammunition which is used by the R.A.F. in controlled guns is not of special design, but it is made within closer limits of overall length than ammunition for the army.

The term " controlled " gun refers to a machine gun, mounted in an aeroplane, of which the rate of fire is synchronized with the rotation of the propeller. The gear employed ensures that, whatever may be the speed of the propeller within certain limits, the gun will only fire when there is a clear space between the blades in front of the muzzle.

The safety of this device also depends on the correct functioning of the ammunition. Any appreciable " hang-fire " of the cartridge at high speeds would be a grave source of danger, as it might cause the bullet to strike one of the propeller blades of the aeroplane.

Although the ammunition is of the same quality as army ammunition, precautions are taken during manufacture and inspection to ensure as far as possible a high standard of regularity of ignition, and particular attention is paid to immunity from hang-fires and functioning defects in machine guns.

Earlier supplies of this special ammunition had wads of glazed-board, but the ammunition is now being supplied with strawboard wads, to avoid damage to the blades of the propeller.

The designation of this special ammunition (ball, tracer and armour-piercing) is the same as the corresponding types of army ammunition, but with the addition of the words " Special for R.A.F. Red Label." The service letter is /A/.

Marking.—For purposes of identification, this special ammunition is distinguished by a label printed in red and bearing the words " Red Label—Synchronized guns, R.A.F.," which is pasted on the box in which the ammunition is packed. The year of manufacture is stamped in full on the base of the ammunition, instead of the last two numerals only. The remaining marking is as for other ammunition of the corresponding types.

Ammunition thus marked will be found, however, among ordinary stocks, as batches made for the R.A.F. are relegated from time to time to ground service use. When this is done the " red label " is removed from the package.

§8.06. ·303-inch blank cartridge.

The chief essentials in a blank cartridge are that it must make as much noise as possible and must be perfectly safe in use.

The principle danger in the use of the cartridge is that, when fired, some portion may be ejected from the barrel in the form of a dangerous missile.

There must be no risk of confusion between blank and ball cartridges capable of functioning in the same gun. It is also necessary that blank ammunition should function satisfactorily in all types of service weapons, and, if any such weapons should function automatically with ball cartridges, it is desirable that it should do the same with the blank cartridge.

FIG. 8.06.—Cartridge, S.A. Blank, ·303-in. L, Mk. V.

The service blank cartridge is designated " Cartridge, S.A., Blank, ·303-in., L, Mk. V."

The case and cap for this ammunition are the same as for the Mk. VII cartridge.

The charge, consisting of 10 grains of cordite, 20, sliced, is very quick in action.

A strawboard wad is fixed in position in the shoulder of the neck of the case and the case is closed by crimping. Cases manufactured for other types of cartridges are sometimes used for blank and in some cases two wads are used.

Marking.—The base is stamped similarly to the service cartridge, except that L V replaces the VII. Other marking will often be found, however, owing to the use of the cases made for other types of ammunition.

§8.07. ·303-inch drill ammunition.

The ·303-inch drill cartridge is used for the training of troops. It is designated " Cartridge, S.A., Drill, ·303-in., D, Mk. VI."

The case is made of white metal, and, to make it more distinct from the Mk. VII ball cartridge, it has three longitudinal grooves which are painted red.

The bullet is an ordinary Mk. VII bullet, secured into the case in the usual manner, and supported inside by a wooden plug, to prevent it being driven into the case by constant use. There is no cap, and the recess forming the cap chamber is left empty.

Marking.—(*a*) Contractor's initials or recognized trade mark.
(*b*) The letter and numeral " D VI."

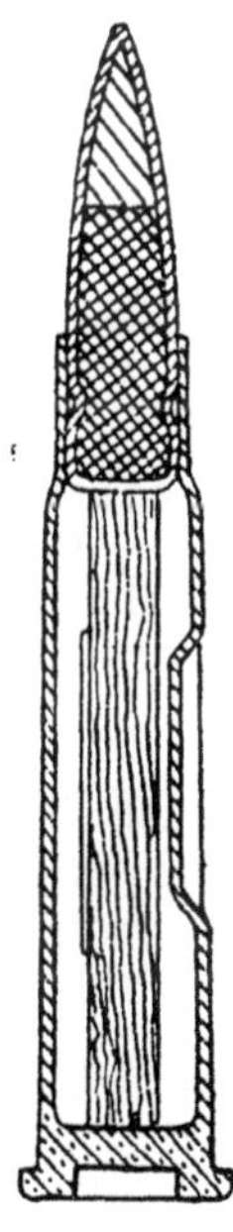

FIG. 8.07.—Cartridge, S.A. Drill, ·303-in. D, Mk. VI.

§8.08. ·303-inch dummy ammunition.

Cartridge, S.A., Dummy, ·303-in., U, Mk. V. For use of inspectors and armourers.

This cartridge is similar to the drill cartridge, except that the white metal case has no longitudinal grooves and the bullet is of gilding metal (90 per cent. copper, 10 per cent. zinc).

The weight and balance of the cartridge is the same as that of the Mk. VII ball cartridge.

Marking.—The base is stamped similarly to the service cartridge, except that " U V " replaces the " VII."

§8.09. Packing of ·303-inch ammunition.

Mk. VII and VIIz ball cartridges are packed as follows :—

(1) 5 rounds in a charger, 10 chargers in a bandolier.
(2) 5 rounds in a charger, 4 chargers in a case charger.
(3) 48 or 50 rounds in a carton.
(4) 250 rounds in a stripless belt.

1,000 rounds in bandoliers or cases charger.	In Box A.S.A. H.1.
600 rounds in cases charger or bandoliers, or 650 rounds in 50-round cartons.	In Box A.S.A. H.2. (This box is obsolescent.)
1,248 rounds in 48-round cartons.	In Box A.S.A. H.13.
500 rounds ; 250 rounds in each of two stripless belts.	In Box A.S.A. H.15.
700 rounds in cases charger or bandoliers.	In Box A.S.A. H.16. (This box is obsolescent.)
1,000 rounds in cases charger or bandoliers.	In Box A.S.A. H.19.
1,300 rounds in 50-round cartons.	In Box A.S.A. H.20.
700 rounds in cases charger or bandoliers ; or 600 rounds in bandoliers ; or 640 rounds in cases charger.	In Box A.S.A. H.21.
900 rounds, 700 rounds or 600 rounds in 50-round cartons.	In Box A.S.A. H.22.

Labels.—All boxes containing Mk. VII or VIIz ammunition bear labels printed in green on white paper. The distinguishing labels bear as a characteristic symbol a green grid, overprinted in black with VII or VIIz as applicable.

Armour-piercing cartridges.

48 or 50 rounds in a carton.	
1,248 rounds in 48-round cartons.	In Box A.S.A. H.13.
1,300 rounds in 50-round cartons.	In Box A.S.A. H.20.

Tracer cartridges.

Similarly to A.P. cartridges.

NOTE.—Before 1926, armour-piercing and tracer ammunition was frequently packed in Boxes A.S.A. H.1.

Labels.—All boxes containing A.P. or tracer ammunition bear labels printed in green on white paper. The distinguishing labels bear as a characteristic symbol, for A.P. a green disc overprinted W I or W Iz in black, and for tracer a green triangle overprinted in black with G I or G Iz.

Special ammunition for the Royal Air Force.

48 rounds in a carton.

1,248 rounds in 48-round cartons. In Box A.S.A. H.13.

NOTE.—Before 1926 special ammunition for the R.A.F. was frequently packed in 48-round cartons, 1,248 rounds in Box A.S.A. H.1 or 768 rounds in Box A.S.A. H.7.

Labels.—Boxes containing this ammunition bear a special label, printed in red on white paper, with the words " Red Label—Synchronised guns R.A.F."

The distinguishing labels are similar to labels of boxes holding ordinary ammunition of the same types, but bear, in addition, the words " Special for R.A.F."

The descriptive labels are printed in red on white paper and also bear the words " Special for R.A.F."

In addition, boxes bear special labels showing the dates for retest of the ammunition.

Blank cartridges.

10 in a bundle.

3,200 or 1,900 rounds in barrels. (Obsolescent.)

1,600 rounds in Box A.S.A. H.13.

Labels.—Packages containing blank ammunition bear labels printed in red on blue paper. The distinguishing labels bear as a characteristic symbol a red grid overprinted in black with LV.

Drill cartridges and dummy cartridges.

10 in a bundle.

1,400 rounds in Box A.S.A. H.13.

NOTE.—Before 1926 drill cartridges were frequently packed in Box A.S.A. H.1 and some were packed in cartons.

Labels.—All boxes containing drill or dummy ammunition bear labels printed in black on cerise paper. The distinguishing labels bear as a characteristic symbol a black grid, overprinted in silhouette, in black, with D VI for drill cartridges and U V for dummy cartridges.

§8.10. Revolver and pistol ammunition.

Revolver and pistol ammunition is intended exclusively for use at short ranges. Its most important characteristic is stopping power.

The minimum striking energy required to stop a man with certainty is about 60 foot-pounds, and this is not sufficient unless the whole of the energy is absorbed in the target.

There are two calibres of revolver ammunition, ·455-inch and ·380-inch, and three types of each, viz., ball, blank and drill. The Cartridge, S.A., Ball, Pistol, Self-loading, is the only example of self-loading pistol ammunition.

§8.11. Cartridge, S.A., ball, revolver, ·455-in. Mk. II.

This is designed for firing in the service type revolver of ·455-inch calibre. The case is of solid drawn brass, provided with a rim, the cap being generally of a similar construction to that of the ·303-inch Mk. VII rifle cartridge, but smaller. It contains ·4 grain of cap composition.

The bullet is of lead-tin alloy (lead 12 parts, tin 1 part), or alternatively of an alloy of 99 per cent. lead, 1 per cent. antimony. It has a cavity in the base. It has three cannelures round the body filled with beeswax.

The bullet is fixed into the case by canneluring the case into the top cannelure of the bullet; the mouth of the case also is coned on to the bullet.

The charge consists of about 5½ grains of cordite 1/·05.

Before 1928 a glazedboard disc was placed as a wad between the charge and the bullet, but such a disc is not now used.

Marking.—The base of the case is stamped with the contractor's initials or recognized trade mark, the numeral II, and since June, 1928, the last two figures of the year of manufacture.

§8.12. Cartridge, S.A., ball, pistol, self-loading, ·455-in., Mk. I.

This cartridge is designed for the Webley & Scott self-loading weapon of ·455-inch calibre. The case is of the semi-rimless type, the cartridge being positioned and extracted by the extractor which fits into the circumferential groove near the base of the cartridge.

The cap is similar to that of the ·455-inch revolver ball cartridge

The charge consists of about 7½ grains of cordite 1/·05. There is no wad. The bullet has a nickel-plated copper envelope and a lead core.

Marking.—The base of the case is marked with the contractor's initials or trade mark, the last two figures of the year of manufacture and the numeral I.

§8.13. Cartridge, S.A., blank, revolver, ·455-in. L, Mk. IIT.

The case is similar to the ball cartridge case.

The cap is of copper and is filled with ·25 grain of detonating composition.

The charge consists of about 8 grains of gunpowder R.F.G.2 or G.12. Over the charge are placed two felt wads and the neck of the case is then crimped over.

Marking.—The base of the case is marked with the contractor's initials or trade mark, the letters and numeral L IIT, and since June, 1928, the last two figures of the year of manufacture.

§8.14. Cartridge, S.A., drill, revolver, ·455-in., D, Mk. I.

The case is of white metal, with three longitudinal grooves painted with red paint at equal distances round it.

There is no cap, and the cap recess is filled up with a red fibre pad, secured by three punch marks in the metal of the case.

The bullet is of lead-antimony alloy (lead 95 per cent., antimony 5 per cent.); this alloy is harder than that used in revolver ball ammunition.

Marking.—The base of the cartridge is marked with the contractor's initials or trade mark and the letter and numeral D I. Before September, 1929, the last two figures of the year of manufacture were also stamped on the base.

NOTE.—The ·455-inch revolver is being superseded in land service by the ·380-inch revolver, but will still remain in use in the Navy, R.A.F. and certain colonies and dependencies.

§8.15. Cartridge, S.A., ball, revolver, ·380-in., Mk. I.

This cartridge is designed for firing in the service type revolver of ·380-inch calibre. The case is similar in construction to that of the ·455-inch cartridge.

The bullet is of an alloy of 95 parts lead, 5 parts tin, or alternatively, of 99 parts lead and 1 part antimony. It has a cavity in the base and there are two cannelures in the body filled with beeswax. The bullet is fixed in the case by a rolled cannelure; the mouth of the case is also coned on to the bullet.

The cap is of brass and is filled with ·2 to ·3 grain of cap composition on which is pressed a lead-tinfoil disc.

The charge is about 4 grains of cordite 1/·05.

Marking.—The base of the case is stamped with the contractor's initials or trade mark, the last two figures of the year of manufacture, the calibre, ·380-inch, and the numeral I. (A quantity of cartridges have been filled with nitrocellulose propellant and these bear the numeral and letter Iz. This filling has been discontinued.)

The annulus of the cap is lacquered dark purple.

Cartridge, S.A., blank, revolver, ·380-in., L, Mk. IT.

The case is similar to the ball cartridge case.

The cap is of brass and is filled with ·2 to ·3 grain of cap composition. It is secured by ringing in.

The charge consists of about 5½ grains of gunpowder R.F.G.2 or G.12. Over the charge are placed two felt wads and the neck of the case is then crimped over.

Marking.—The base of the case is marked with the contractor's initials or trade mark, the last two figures of the year of manufacture, the calibre, ·380-inch and the letters and numeral L IT.

Cartridge, S.A., drill, revolver, ·380-in., D. Mk. I.

The case is of white metal alloy, with three longitudinal grooves painted with red paint at equal distances round it.

There is no cap and the cap recess is filled with a red fibre pad, secured by three indents in the metal of the case.

The bullet is of lead-antimony alloy (lead 95 per cent., antimony 5 per cent).

Marking.—The base of the cartridge is marked with the contractor's initials or trade mark, the calibre, ·380-inch and the letter and numeral D I.

Packing of revolver and pistol ammunition.

·455-inch Revolver Ball ammunition	12 rounds in a carton. 240 rounds in Box A.S.A. H.9.

Labels.—Boxes containing this ammunition bear labels printed in green on white paper. The distinguishing labels bear as a characteristic symbol a green revolver chamber, overprinted in black with II.

·455-inch Self-loading Pistol Ball ammunition.	7 rounds in a carton. 210 rounds in Box A.S.A. H.9.

Labels.—Printed in green on white paper. The distinguishing labels bear as a characteristic symbol a green plimsoll mark, overprinted in black with I.

·455-inch Revolver Blank ammunition.	12 rounds in a carton. 396 round in Box A.S.A. H.9.

Labels.—Printed in red on blue paper. The distinguishing labels bear as a characteristic symbol a red revolver chamber, overprinted in black with L IIT.

·455-inch Revolver Drill ammunition	12 rounds in a bundle. 276 rounds in Box A.S.A. H.9.

Labels.—Printed in black on cerise paper. The distinguishing labels bear as a characteristic symbol a black revolver chamber, overprinted in silhouette, in black with D I.

·380-inch Revolver Ball ammunition	12 rounds in a carton. 240 rounds in Box A.S.A. H.9. 180 rounds in Box A.S.A. H.25.

(The H.25 box will eventually replace the H.9 box for service supplies of ball ammunition.)

Labels.—Printed in green on white paper. The distinguishing labels bear as a characteristic symbol a green oval, overprinted in black with I. The nitrocellulose loaded ammunition in the service bears similar labels overprinted in black with Iz.

·380-inch Revolver Blank ammunition.	18 rounds in a carton. 522 rounds in Box A.S.A. H.9.

Labels.—Printed in red on blue paper. The distinguishing labels bear as a characteristic symbol a red oval, overprinted in black with L IT.

·380-inch Revolver Drill ammunition	12 rounds in a bundle. 384 rounds in Box A.S.A. H.9.

Labels.—Printed in black on cerise paper. The distinguishing labels bear as a characteristic symbol a black oval, overprinted in silhouette, in black with D I.

§8.16. Cartridge, S.A., ball, ·5-in., Mk. IIz.

This cartridge is for use in the ·5-inch Vickers machine gun.

The case is of solid drawn brass and the base is of the rimless type, *i.e.*, instead of a rim, as in the ·303-inch case, it has a groove by which it is extracted from the machine gun. In other respects it is of the same general design as the ·303-inch case.

The cartridge is positioned in the chamber by the shoulder of the case butting against the shoulder of the chamber.

The cap is of brass and is pressed into the cap chamber of the case and ringed in. The joint between the cap and the case is varnished. The cap is filled with 1 to 1·2 grain of cap composition, and is foiled and varnished as in the ·303-inch cartridge.

The charge consists of about 130 grains of nitrocellulose of the type used for the ·303-inch service ball cartridge.

Cordite M.D.T. 7–2 has also been used to a small extent.

When filled with cordite a glazedboard disc or cup is placed as a wad between the charge and the bullet. The cordite filled cartridge is described as Mk. II.

The bullet is pointed and weighs 580 grains. The core is in two parts, the front pointed part of aluminium and the rear part of an alloy of lead 98 parts, antimony 2 parts. The envelope is of steel, coated with cupro-nickel. The bullet has two cannelures, the lower of which is filled with beeswax.

The bullet is secured into the case by indenting the case in three places into the lower cannelure and by coning the mouth of the case into the upper cannelure.

The weight of the complete cartridge is about 1,275 grains.

Marking.—The base of the case is stamped with—

(*a*) The contractor's initials or trade mark.
(*b*) The last two figures of the year of manufacture.
(*c*) The numeral IIz (II in the case of cordite filled cartridges).

The annulus of the cap is varnished with dark purple varnish.

Fig. 8·16.

CARTRIDGE, S.A. BALL, ·5 IN., Mk. II Z/CA/.

SCALE = 1/1.

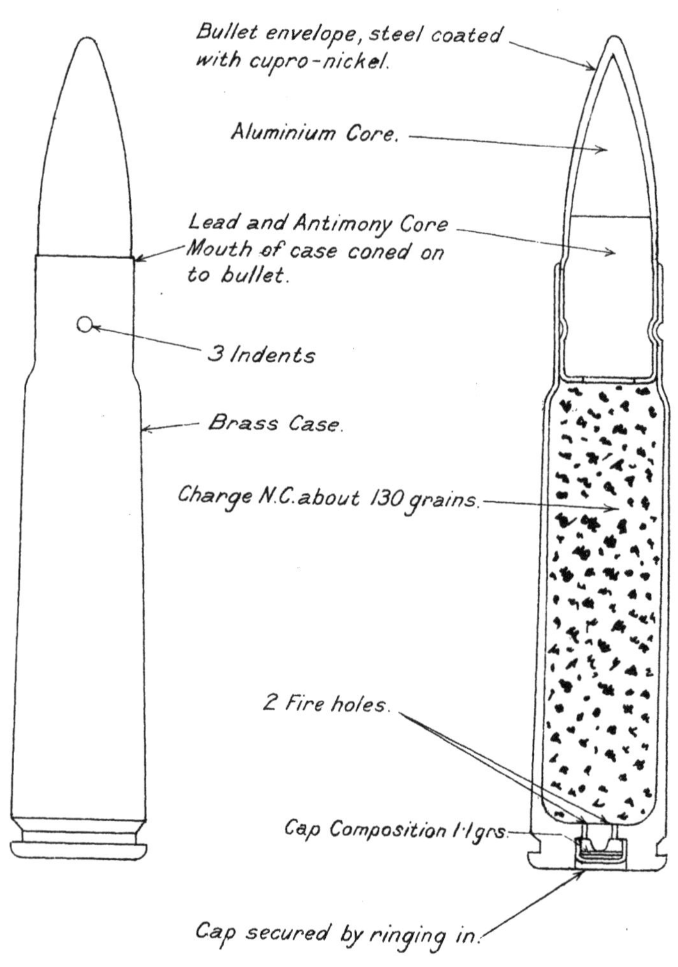

958

Cartridge, S.A., armour-piercing, ·5-in., W, Mk. Iz.

The case, cap and charge are identical with those for the ball cartridge.

The bullet consists of a steel envelope coated with cupro-nickel and a hard steel core. Between the core and the envelope is a sleeve of lead-antimony alloy. The bullet weighs 580 grains.

The bullet is secured in the case in the same manner as the ball cartridge.

Marking.—The base of the case is stamped with the contractor's initials or trade mark, the last two figures of the year of manufacture and the letters and numeral W Iz. The annulus of the cap is varnished with green varnish.

Tracer.—There is no service design of ·5-inch tracer cartridge, but trials are in hand for the settlement of a design.

Packing of ·5-inch ammunition.

·5-inch Ball ammunition	10 rounds in a carton. 370 rounds in Box A.S.A. H.13. 380 rounds in Box A.S.A. H.20.

Labels.—Printed in green on white paper. The distinguishing labels bear as a characteristic symbol a green St. Andrew's Cross, overprinted in black with IIz (II in the case of cordite loaded ammunition).

·5-inch Armour Piercing ammunition	10 rounds in a carton. 370 rounds in Box A.S.A. H.13. 380 rounds in Box A.S.A. H.20.

Labels.—Printed in green on white paper. The distinguishing labels bear as a characteristic symbol a green shield, overprinted in black with Iz.

§8.17. Miniature rifle ammunition.

The service miniature rifle ammunition is the—

Cartridge, Rim-fire, ·22-in., Mk. I—with long case.

There is no detailed service design of this cartridge, and the specification allows for small variations in construction, but lays down a definite standard of accuracy which must be attained at proof.

The term " rim-fire " signifies a cartridge without a cap, the flange or rim of the case being hollow and filled with cap composition.

The cartridge is usually of solid-drawn copper-zinc alloy, but the use of cupro-nickel is permitted.

The bullet is made of an alloy of lead. Its cannelures are usually lubricated with beeswax. The bullet is secured into the case by coning, indenting or crimping.

The charge may consist of cordite, rim neonite or other nitro-cellulose powder. No wad is provided.

The present service type of cartridge is known as "Non-rusting."

Packing.—These cartridges are packed in trade cardboard cartons, each holding 100 rounds. Ten of these cartons are packed in a tin box with a tear-off lid, and ten such tin boxes are in turn enclosed in a Box, Cartridge, Rim-fire, ·22-in H.11, which thus holds 10,000 rounds.

Labels.—Printed in green on white paper. The characteristic symbols on the distinguishing labels for the various types of ammunition are as follows —

(*a*) Cordite ... A green target overprinted in black with I.

(*b*) Smokeless ... A green target overprinted in black with I^{S}.

(*c*) Semi-smokeless ... A green target overprinted in black with I^{SS}.

(*d*) Non-rusting ... A green target overprinted in black with I^{NR}.

Cartridges for horse and cattle killers.

The cartridge used for the service horse killer is the—

Cartridge, S.A., Ball, ·310-in., Mk. I^{Z}.

The case is of brass.

The cap is of copper and filled with 0·4 grain of cap composition.

The bullet is of lead alloy, flat or rounded at the nose and slightly hollowed at the base.

The charge consists of about 5½ grains of Schultze powder, a nitrocellulose propellant.

Marking.—The base of the case is marked with the contractor's initials or trade mark, the last two figures of the year of manufacture and the numeral Iz.

Packing.—20 rounds are packed in a trade cardboard carton and 18 cartons, or 360 rounds, in Box A.S.A. H.9.

Labels.—Printed in green on white paper. The distinguishing labels bear as a characteristic symbol a green skull and crossbones, overprinted in black with I^{Z}.

The cartridge used for the service cattle killer is the—

Cartridge, S.A., Ball, Revolver, ·455-in., Mk. II.

§8.17(*a*). Note regarding packing and labelling of small arm ammunition.

The boxes used for packing Cartridges, S.A., Ball, ·303-in., Mk. VII, contained in chargers and in bandoliers or cases charger are now invariably stained brown.

The boxes used for the packing of S.A. ammunition of any nature in cartons, bundles or stripless belts, are stained green.

A certain amount of carton packed S.A. ammunition in brown stained boxes is still in the service. No more will be so packed.

In order to distinguish by colour between carton-packed and stripless belt-packed S.A. ammunition, the future policy is for boxes containing carton-packed ammunition to be stained a yellowish-green and boxes containing stripless belt-packed ammunition a bluish green. This has not yet taken effect, and there are quantities of carton-packed ammunition in the service in boxes stained bluish-green.

Speaking generally, the information regarding labels given in this chapter relates to labelling since 1928 inclusive. For ammunition packed before 1928, a good deal of the labelling is different, but in all cases the labels give sufficient information to enable the contents of the packages to be readily identified.

§8.18. Inspection and proof of small arm ammunition.

The suitability of small arm ammunition for service purposes is determined by—

(1) Weighing and gauging and visual examination of each individual round.

(2) A firing proof of every delivery to ascertain the ballistic performance of the cartridges and their general shooting qualities.

Under (1) the cartridges are weighed and gauged in automatic machines, which throw out any cartridge which is too light or too heavy, too long or too short, or too big to fit the rifle or machine gun.

Cartridges accepted by the machines are examined visually to see that they are free from damage or defects.

Some rounds from every delivery are broken down, the force required to extract the bullet is measured and the components are weighed, gauged and examined in detail.

Under (2) the cartridges are fired from rifles and machine guns to ensure that the ammunition is fit for use in all the types of weapons for which it may be required. The firing trials are arranged to test the cartridges for accuracy, velocity, pressure, hangfires and set-up of bullets. Firing is also carried out in all types of weapons, new and worn, or out of adjustment within certain allowable limits, to test the ammunition for functioning, freedom from casualties due to defects in case, cap or bullet, and ease of extraction of fired cases.

The number of rounds from each delivery fired in proof is sufficient to ascertain the suitability of the delivery in all these respects. If a delivery fails to reach the standard laid down in the specification it is rejected, or at the discretion of the inspector a further and more comprehensive proof is carried out, to ascertain whether or not the defect which occurred at first proof is characteristic of the delivery.

§8.19. Defects in small arm ammunition.

The following are some of the defects or failures which may be met with in inspection or proof of small arm ammunition.

Inspection defects.—Cartridges long or short, heavy or light, high or low diameter ; thin or defective heads ; scaly, scored or split cases ; defective or loose bullets ; defective caps ; no caps ; incorrect marking ; caps not ringed in or incorrectly varnished ; eccentric cap chambers.

Proof defects.—Cap : Missfires, failure of the cap to ignite the charge, hangfires, pierced caps, escape of gas around the cap, cap out of case.

Case : Burst cases, stretched metal, separated cases complete or partial, splits at the neck, cases hard to extract from chamber after firing.

Bullet : Stripping of the envelope and break up on firing, failure to take the rifling, excessive metallic fouling. Lack of symmetry in the bullet leads to inaccuracy.

Chapter IX

1-INCH AIMING RIFLE AMMUNITION

§9.01. Introduction.

1-inch aiming rifle ammunition is the instructional ammunition used with aiming rifles fitted to guns in coast defence. In use it corresponds to miniature rifle ammunition.

There are both electric and percussion cartridges.

There are two Marks of electric cartridge designated—

Cartridge, aiming rifle, 1-in., electric, Mk. I and Mk. II.

The two Marks differ in the primer, the Mk. I has a primer with a paper cylinder, the Mk. II has a primer with a brass cylinder and a small air space between the gunpowder charge and the closing cup. Refilled cartridges fitted with the new type primers are designated Mk. I*.

The Mk. II cartridge consists of a case with primer, propellant charge and bullet.

The case is of solid drawn brass, with a screw-threaded hole in the base to receive the primer, and a cannelure provided to form a stop for the bullet. It is lacquered internally, with the exception of the threads of the primer hole and that portion which supports the bullet.

The primer is screwed into the case. Its contact pin is supported and insulated from the body of the primer by two ebonite plugs.

A Z.13 iridio-platinum wire bridge (resistance 0·9 to 1·1 ohm) is soldered with pure tin, one end to the point of the contact pin, the other end to the metal body.

The primer body is filled with guncotton dust, which surrounds the bridge and is closed by a perforated glazedboard disc, to which is affixed a paper disc (not perforated), the mouth being burred over on to the disc.

The primer charge is contained in the brass cylinder and consists of rifled pistol powder or gunpowder G.20. The cylinder is closed by a glazedboard cup.

The propellant charge consists of about 160 grains of cordite 3.

The bullet is of lead-antimony alloy (lead 98 per cent., antimony 2 per cent.). Two cannelures are formed round it and filled with beeswax for lubricating purposes. The base of the bullet is fitted with a copper cup.

Marking.—The base of the case is marked with the contractor's initials or trade mark and the numeral II.

Cartridge, aiming rifle, 1-in., percussion, Mk. III.

These cartridges differ only from the electric in having a percussion primer.

Packing of aiming rifle cartridges.

Both electric and percussion cartridges are packed—

6 rounds in a bundle.
114 rounds in Box A.S.A. H.13.
12 rounds in a bundle.
108 rounds in Box A.S.A. H.20.

Labels.—Printed in green on white paper. The distinguishing labels for the electric cartridge bear as a characteristic symbol a green heart, overprinted in black with I, I* or II; and those for the percussion cartridge bear a green diamond, overprinted in black with II.

Fig. 9·01.

CARTRIDGE, AIMING RIFLE, I IN., ELECTRIC, Mk II.

SCALE = 1/1.

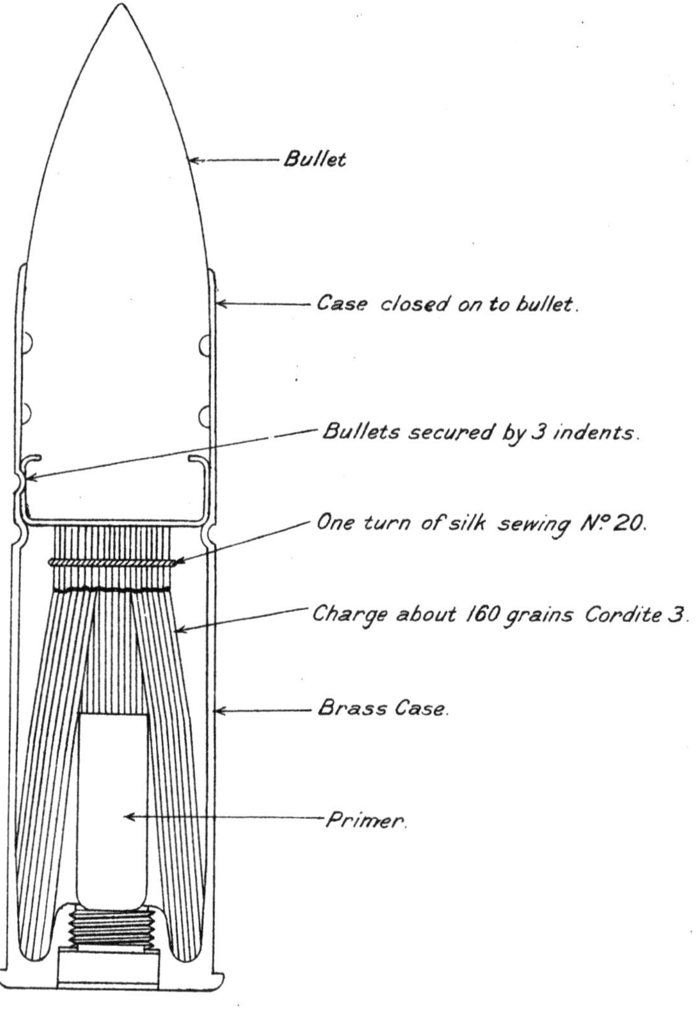

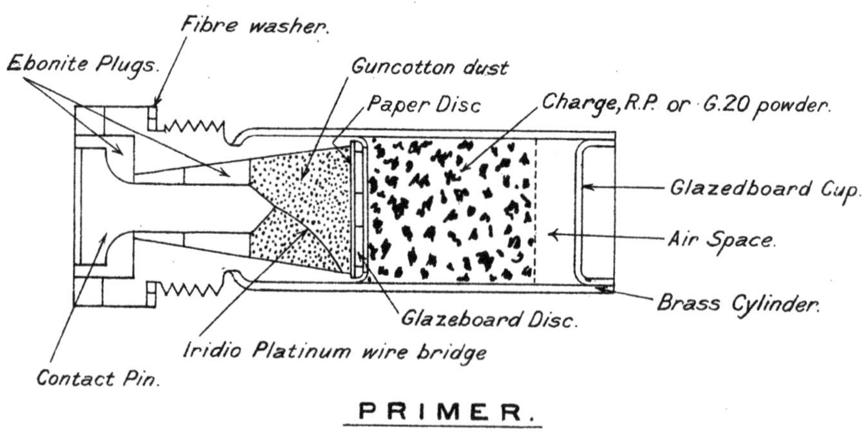

PRIMER.

CHAPTER X

GRENADES

§10.01. Introduction.

A grenade is an anti-personnel bomb, thrown by hand or fired from the rifle.

For many years previous to 1914 little attention had been given to its development. The only grenade in the service was the No. 1 R.L. hand grenade, a percussion bomb mounted on a stick, by which it was held, and fitted with tape streamers to control its flight and ensure nose impact, the only condition under which the direct-action mechanism would operate.

It was discarded, being clumsy, unsuitable for trench warfare, and by no means safe except in the hands of specially trained men.

War development.—A large number of types of grenade were introduced during the Great War. These may be divided into :—

(*a*) Hand grenades, which were of the time-fuze type.

(*b*) Rifle grenades, which were mainly of the percussion graze-fuze type.

These two classes of bomb remained separate throughout the war, with one exception.

The Mill's bomb, developed as a hand grenade, the No. 5, was modified to take a rifle rod as No. 23, thus enabling it also to be fired, and later was further modified, as No. 36, to be capable of being fired from the rifle discharger $2\frac{1}{2}$-inch, introduced in 1917.

§10.02. Hand grenades.

The essential features of all war types were :—

(1) A cast-iron body or tin canister, filled with high explosive and detonated by

(2) A copper detonator, into which was crimped a piece of safety fuze, the other end of which was ignited by

(3) A lighting arrangement.

The fuze was cut to a length suitable for the maximum time of flight, *i.e.*, the longest possible throw.

Primary ignition.—The main developments were in the methods of arranging the lighting of the safety fuze.

In the earlier types, the safety fuze was ignited before being thrown. This was effected by the use of friction lighters of various types.

The main objection to this is that safety fuze is not absolutely reliable and the possibility of premature action in the thrower's hand is always present.

The Mill's type marked a great advance in this respect. The ignition depends on the release of the external lever, which does does not take place till the bomb has left the hand.

This system, unlike the earlier ones, lent itself to rifle firing, since the release of the lever could be arranged to be operated by the discharge of the rifle.

The objectionable features due to employment of safety fuze, however, still remained.

§10.03. Rifle grenades.

General principle of action.—As already mentioned, these developed separately from hand grenades, and depended on " percussion action."

The most common form of firing mechanism was of the " graze " type.

A heavy striker pellet was seated in the rear end of the grenade, and held there by two small retaining bolts, the inner ends of which engaged in the pellet, and the outer ends passed through the wall of the grenade. They were held in position by an external brass collar, which was a close fit round the grenade body and was locked in transport by a safety-pin ; the pin was removed before firing.

The firing of the rifle caused the collar to set back, unmasking the retaining bolts, which fell out, leaving the striker pellet free to move forward on to the detonator in the grenade on impact.

Such a mechanism depends on the grenade hitting head first, and is therefore useless for the purposes of a hand grenade.

Methods of firing.

Rifle rod.—In the first instance rifle grenades were all of the " rodded " type, *i.e.*, mounted on steel rods which fitted into the rifle barrel. These rods also served to steady the grenade in flight and ensure the necessary head-on impact on striking.

There are two main objections to rifle rods :—

(*a*) They are inconvenient to carry on the soldier.
(*b*) They damage the rifle barrel.

The damage to the rifle barrel is due to wave action, which causes a bulge in the barrel at the extreme depth of the tail rod, and immediately ruins the accuracy of the rifle for ordinary shooting.

Rodded signal grenades are obsolescent.

Discharger, 2½-inch.—The idea of using a detachable cup on the end of the barrel (first put into practice with the No. 36 Mill's grenade) gave the double advantage of—

(1) affording a separate combustion chamber for the expansion of gases ;
(2) providing a separate short length of barrel which could be easily renewed if erosion or any other form of damage set in.

By means of the discharger, damage to the rifle barrel was reduced to a minimum, but the No. 36 grenade was too heavy, and the excessive recoil was found to break up the stock of the rifle.

In these early types again, it was sought to give adjustment of range by opening or closing a small port on the side of the combustion chamber.

This method was not at first successful, as part of the charge was found to be unconsumed owing to loss of pressure when the port was open. The principle has been retained, however, and in the latest design of discharger cup, the charge is completely consumed with the port fully open. There is no doubt that the port is the safest and most accurate way of varying the range.

A description of the latest type of discharger will be found at the end of this chapter.

Mill's type of grenade.—The original Mill's type of grenade (No. 5) was a hand grenade, and was designed for ease in holding and throwing. It was corrugated so as to facilitate break up into what were considered useful man-killing fragments, and was filled with a variety of high explosives. Many of these were adapted from commercial explosives to augment war supplies. The most important were ammonal, amatol, and alumatol.

It was gradually developed and improved in details of design. The No. 23 was supplied with a rod, which could be screwed in if the grenade was to be fired from a rifle. The No. 36 was fitted with a screw-on base plate and designed to be used with a discharger, the base plate acting as a rough gas-check, positioner, and guide band to the projectile on firing.

The principle of construction in all the early Marks was similar to the No. 36, which is described below.

§10.04. Grenade ·303-in. rifle, No. 36M. Mk. I.

The body is oval, made of cast iron, and corrugated to facilitate break up into equalized fragments. A screwed plug hole is formed in the side for filling in the high explosive.

The body has a large circular hole formed in the base, which is screw-threaded to receive a centre piece carrying the striker and fuzing arrangements, which are thus separated from the main filling of the bomb.

The lower part of the same screw-thread carries the base plug, which is removed for insertion of the detonator.

The *base plug* has a $\frac{1}{4}$-inch screw-threaded hole to receive the *gas-check plate* when used with a rifle.

This is a steel plate, which is a sliding fit in the discharger and rests on an annular shoulder formed in the metal of the cup.

The aluminium or tinned brass *centre piece* which is screwed into the base of the bomb, consists of two chambers.

In the upper end of the central, and larger, chamber is seated the striker with its compressed spring. When the grenade is prepared for firing, the detonator is inserted from the base so that the cap end of it is seated at the bottom end of this chamber below the striker, the other end formed by the detonator sliding into the side chamber. This eccentricity of the detonator is a disadvantage, as the bulk of the H.E. filling is partially isolated from it by the air gap which the striker chamber interposes, and uneven fragmentation is thereby caused.

The detonator consists of a ·22 rim-fire cartridge, seated in a zinc alloy cap chamber, and is attached to one end of the safety fuze, the other end of which is crimped into a No. 6 detonator. The fuze is bent into a U-shape to suit the centre piece into which the set fits.

The *striker* is of steel, flanged at the bottom to seat the spring; its firing face is shaped for rim firing, and notched to allow the escape of gas through the flange.

The shaft of the striker passes out through the body of the grenade, and is notched at the top to receive the striker lever.

The *striker lever* is a curved steel lever, pivoted on a fulcrum formed on the body of the grenade. The curved lever fits closely to the body, and is retained in position by a pin passing over it and through the fulcrum bracket. The short counter lever fits into the notch on the striker and holds it up against the force of the compressed spring.

Filling.—Trotyl, baratol 20/80, or 40/60, but trotyl is obsolete for future filling.

Amatol, ammonal, alumatol, etc., are obsolete.

Action.—In preparation for action the base plug is unscrewed and the detonator No. 36M grenade inserted.

Preparatory to immediate use, the safety pin is withdrawn and the lever held down by hand during the act of throwing. When fired from a discharger the curved lever rests against the side of the cup, and is held there till the grenade is projected.

On release, the lever pivots on the fulcrum, the striker descends under the force of the spring, throwing the lever clear and firing the cap. The cap ignites the safety fuze, which burns for six seconds before firing the detonator which detonates the filling.

The No. 36M, Mk. I, differs from the No. 36 in having the screw threads of the base plug and centre piece, and the portion of the striker head where it emerges from the top of the grenade, covered with a special waterproofing composition for use in tropical climates.

The No. 36M has entirely replaced the No. 36, which is now obsolete.

Packing.—Box G.5, Mk. III or Box G.36, Mk. I.
Containing—
12 Grenades.
12 Detonators in tinned-plate cylinder.
14 Rifle Grenade Cartridges in tinned-plate box.
12 Gas-check plates.
1 Key.

§10.05. Signal grenades.

The oldest types of signal grenades are rodded. These were all declared obsolete for future manufacture in 1920 and few will be found in the service.

They are made of tinned plate cases and ignited with an igniter consisting of an Eley cartridge head pressed on to a length of safety fuze. A striker, suspended by a shearing wire, and locked, till fired, by a safety pin held over the cartridge in a tubular striker chamber. The igniter screws on to a spigot on top of the grenade. Ignition from above entails the packing of the components in the reverse order to that used in the grenades (subsequently described) fired from a discharger. The parachute is at the bottom of the case and the base cover, which carries the rod, is a sliding fit. The grenade opens at this end when the bursting charge throws out the contents.

The modern signal grenade is fired from a 2½-inch discharger. These are enumerated below with their rodded prototypes.

Rodded	2½-inch Cup	Designation	Contents	Marking
No. 31, Mks. II and III	No. 42, Mk. I	Day Signal	1 smoke candle, Red, Blue, Yellow or Purple	A serpentine line coloured according to candle.
No. 32, Mks. II and III	No. 43, Mk. I	Night Signal	3 stars, Red, Green or Yellow	3 coloured discs according to the colour and sequence of the stars.
No. 38, Mk. II	No. 45, Mk. I	Night Signal	1 changing colour star "Red to Green to Red" and "White to Red to White."	*3 rectangles according to colour sequence of enclosed stars.
No. 51, Mks. I and II	No. 52, Mk. I	Day or Night Signal	3 White illuminating stars	3 white discs and "Day or Night."
	No. 48, Mk. I	Day or Night locality signal	4 flash signals suspended on a string 2 ft. apart and igniting at ½-sec. intervals	No distinguishing marks. Painted drab and a descriptive label attached.

* Coloured bands round the body in the case of the No. 38.

§10.06. Signal grenades 2½-inch, Discharger type.

The general construction of all 2½-inch types of signal grenades is the same.

The body is a rolled paper cylinder closed at the base by a wooden plug perforated by two holes. The base is covered by a brass base cap, which is strengthened by a small steel disc in centre. The cap is perforated by two holes coinciding with the holes in wooden plug.

The front end of the body is closed by a tinned-plate cap.

A *supporting collar* of tin-plate fits over the edge of the discharger cup and prevents the rain getting in when the loaded rifle is ready for use.

The perforations in the wooden plug are filled with lengths of safety fuze to ignite the opening charge, the base ends of the safety fuze are exposed in order that they may be ignited by the cartridge.

The *opening charge,* consisting of a layer of 1-drm. F.G. or G.12 powder, is placed on top of the wooden plug. Over the opening charge is a perforated felt wad, on which rests the star or candle. On the top of the star or candle is a perforated cardboard disc to support the parachute and cord which lies in the front end of the grenade.

The *parachute* of japanned paper is fixed by asbestos cord to the star or candle. The end of the cord is brought through the cardboard disc and is attached to the star or candle.

Action.—On firing, the safety fuze is ignited and burns for about ½ second to allow the grenade to get clear, it then ignites the opening charge which throws out the star or candle, which is suspended by the parachute. At the same time the opening charge ignites the priming (usually quick-match) of the star or candle, the ends of the quick-match being brought down through the hole in the felt wad and into the opening charge.

Packing.—Box numbered according to contents :—

24 grenades.
28 rifle grenade ballistite cartridges.

§10.07. Grenade No. 54, Mk. I (obsolete from 20.11.35).

The body is a symmetrical iron casting, with walls about 3/16 inch thick. Circumferential guides are cast on the exterior and are machined to be a close fit in the 2-inch discharger cup.

The casting is screw-threaded to receive the mechanism chamber at its upper end, and a base plug at its lower end. A side filling hole, closed with a screwed plug, is provided through the wall.

The *holder or mechanism chamber* is a die-casting which is screwed into the top of the body.

At its lower end is secured a copper tube, which, when the chamber is assembled to the body, is flanged over at its lower end and sweated to the seating provided in the base plug hole.

Fig 10·06(a).

GRENADE, SIGNAL, (DAY) Nº 42. MARK I.

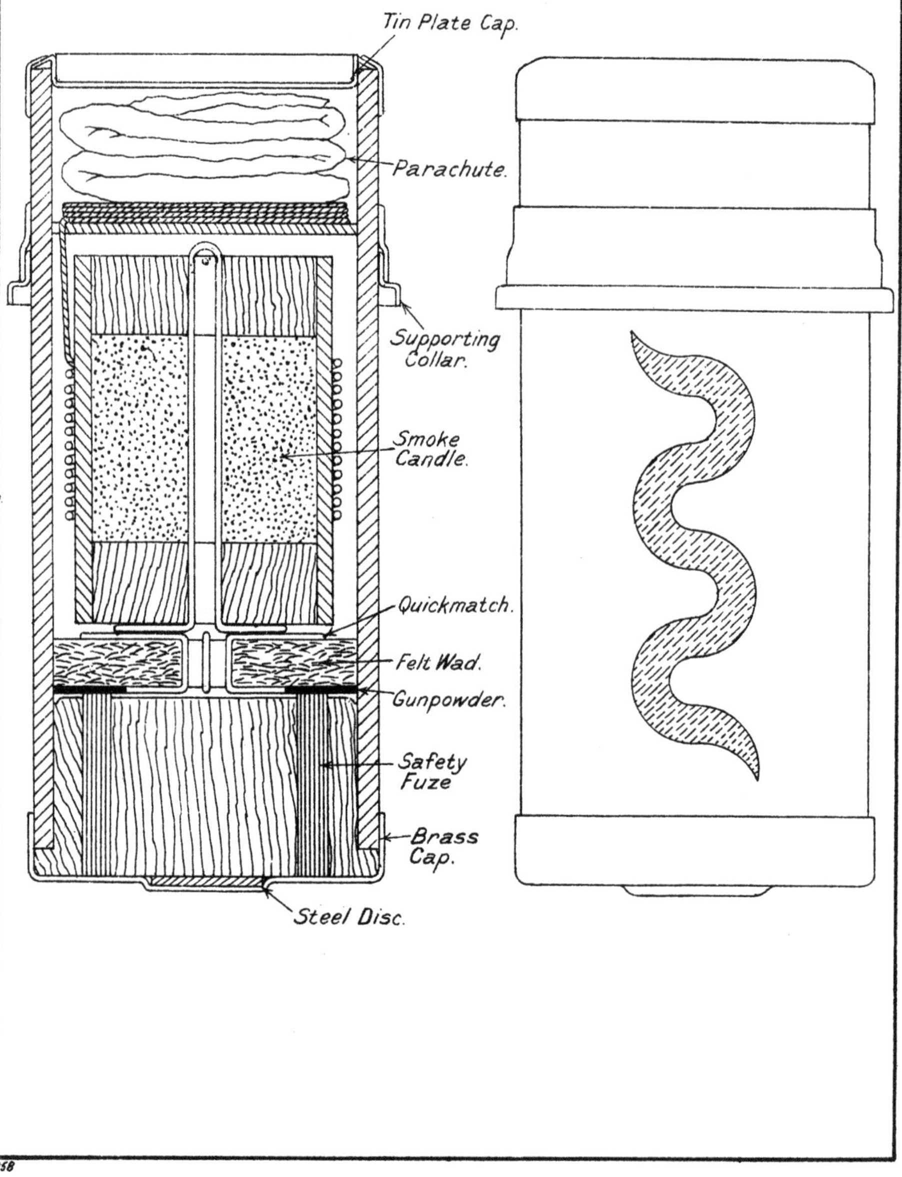

958

Fig. 10·06 (b).

GRENADE, SIGNAL, (NIGHT) Nº 45 MARK I.

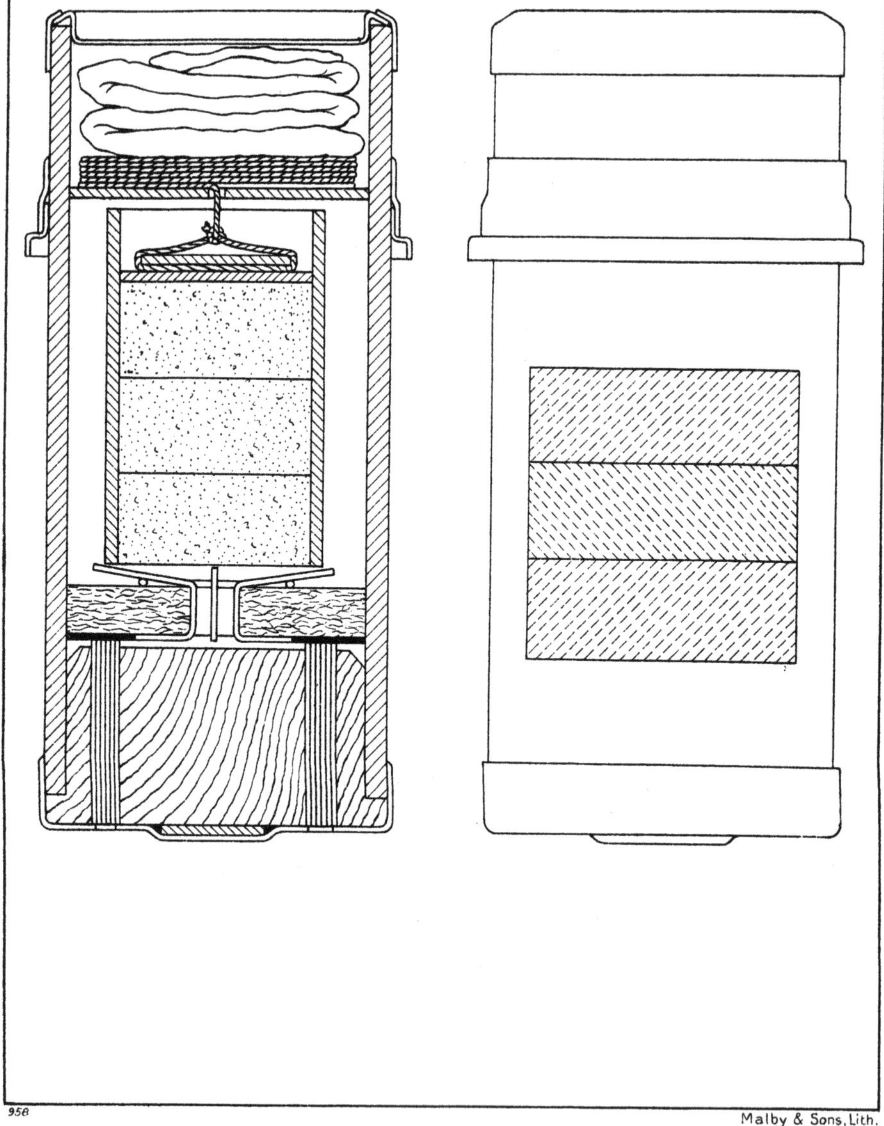

958

Malby & Sons, Lith.

The tube forms the seating for the detonator.

The side of the chamber is drilled to take a safety bolt. A gallery is formed round the head, which is closed by a screwed cap.

The top and bottom of the chamber are formed to provide coned seatings to the firing pellets.

The *firing mechanism* consists of a die-cast *ball* resting on the cup-shaped head of a tinned steel *striking pellet,* the lower end of which consists of a needle stem with a split point. The needle stem slides in the brass *cap pellet,* which is drilled to receive it. A seating is formed at its base end to take a percussion cap, with a hole drilled through the centre of it, which is filled with 1·2 grains of detonating composition " A " mixture. The surface is covered with shellac varnish.

The cap and needle pellets are held apart by a light creep spring.

The *creep spring* serves to keep the ball on top of the striker pellet and the rounded base of the cap pellet against the coned top and bottom seatings, respectively, in the chamber, in which these pellets are otherwise free to move.

Filling.—Baratol.

Safety arrangements.—*(a)* A steel *safety bolt* tapered at the end enters freely through a hole drilled through the stem of the striker pellet.

At its outer end it is pivoted to a metal tag, to which is attached a cotton webbing tape, which is wrapped about 2½ times round the outside of the chamber. A similar metal tag is fixed to the free end of the tape and is weighted with lead.

(b) A *safety cap,* which is locked on over the mechanism chamber as follows :—

Indents are punched in the safety cap, which can only be put on over the mechanism chamber past the cut-away portions round the head of the chamber. The mouth of the cap is pressed on to the rubber washer on top of the grenade body, and the cap is then twisted right-handed about one-tenth of a turn.

This causes the indents to pass under the projecting head of the chamber, so locking the cap firmly, and sealing the head of the grenade against moisture.

Arrangements to prevent tampering.—After assembly as above, a strip of tape is shellacked across the junction of the safety cap and grenade body. The grenade should retain this tape in position until immediately before it is required for throwing or firing.

Loading arrangements.—The base plug is unscrewed, to allow of the insertion of the detonator, and is then replaced. It is fitted with a small rubber pad, on which the base end of the detonator rests. A feather is formed under the plug to enable the plug to be inserted or removed when necessary, with the help of the slot in the bayonet pommel.

Action of grenade. In transport.—The pellets are held apart by the safety bolt.

Preparation for use.—Insert detonator as described above. No safety arrangement is affected by this operation.

Preparation for throwing or firing.—Grasp the bomb in the throwing hand. Grip the safety cap in the other hand. Twist the body of the grenade one-tenth of a turn anti-clockwise and then withdraw it from the cap.

The grenade is now ready to be thrown or placed base first in the discharger ready for firing.

If in doing so the grenade is dropped, the tape will not have unwound sufficiently to pull out the safety pin. The grenade therefore remains locked, the tape can be re-wound, and the grenade recovered.

Grenade in flight.—The natural gyrations of the grenade in flight cause the tape to unwind off the head rapidly and thus withdraw the safety pin.

The firing pellets are now armed, only the weak creep spring holding them apart.

Action on impact.—If the grenade falls on its—

(*a*) Head.—The cap pellet overcomes the weak creep spring by its inertia and draws its cap on to the needle of the striker pellet. The flash from the cap passes direct into the detonator immediately below the chamber.

(*b*) Base.—The inertia of the ball and striker pellet drives the needle of the latter into the cap.

(*c*) Side.—The inertia of the ball, working on the front coned end of the chamber and of the cap pellet on the rear coned end, forces the two pellets together as they slide down these surfaces, thus bringing the cap and needle together.

Transport.—Grenades travel 16 in a box, complete with a separate cylinder containing 16 grenade detonators and a tin box containing 20 rifle grenade cartridges.

The charge of the latter is 30 grains of ballistite.

Grenade, No. 54, *Mk. II*, differs from the Mk. I in having—

(*a*) an improved detonator,
(*b*) a disc on the safety cap,
(*c*) a more efficient seal between junction of cap and body,
(*d*) a needle with a single point.

Considerations affecting present grenade policy.

Although it is fully realized that the No. 36 type of grenade has certain defects, which were the reasons for the production of the No. 54 grenade, the present policy is to continue to use the No. 36, of which large stocks still exist, until such time as experiments in other directions give a somewhat clearer indication as to whether a dual purpose, *i.e.*, rifle and hand grenade, is really required.

Fig. 10·07.

GRENADE, PERCUSSION, 2-INCH, No. 54 Mk I.

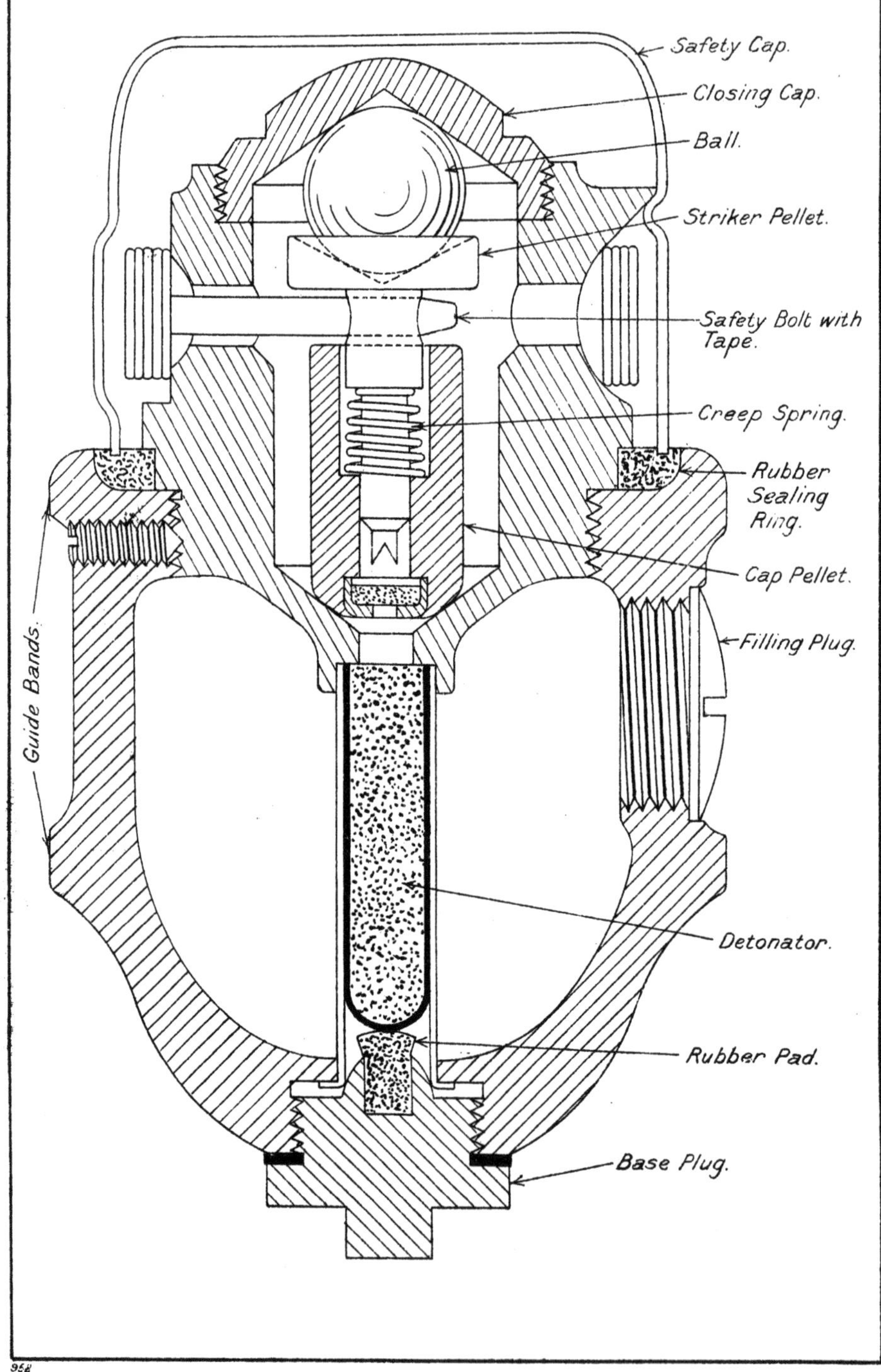

958

The intention, therefore, is to defer any production either of existing types of grenade or a new design until experience with the 3-inch mortar and possibly a small platoon mortar have demonstrated the necessity or otherwise of a rifle grenade.

The No. 54 grenade has so far been open to criticism as a rifle grenade owing to the possibility of prematures due to inverted loading in the discharger cup though the probability of loading in that way even by semi-trained personnel is rather remote.

§10.08. Grenade, 2·5-in. No. 63. Smoke, Mk. I.

The grenade consists of a thin drawn steel cylindrical body closed at one end. The body is filled with smoke composition, primed at the open end with priming composition. In the middle of the latter is a delay pellet, the object of which is to delay smoke emission for some 5 seconds while the grenade is in flight. The

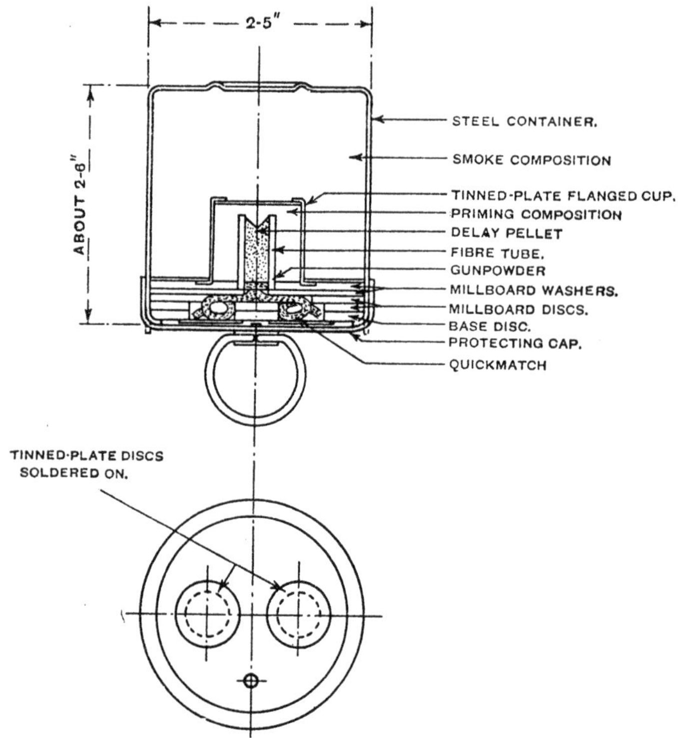

Fig. 10.08. Grenade 2·5-inch, ·303-inch Rifle Smoke.

smoke composition, priming composition and delay pellet are consolidated under considerable pressure.

The top of the grenade is marked "TOP," and has a circular corrugation.

Action.—The grenade is prepared for firing by pulling off the protective cap ; a loop is provided to enable this to be done. The grenade is then inserted into the discharger cup *base* first. When the rifle is fired, the flash from the propellant penetrates the tinned-plate closing discs and ignites the quickmatch. This in turn ignites the delay pellet, the slow burning of which delays the ignition of the priming composition until the latter part of the grenade's flight. The priming composition then ignites the smoke composition.

§10.09. Rifle grenade cartridges.

As previously mentioned, the rifle grenades in the service may be divided into two classes, viz., those fitted with rods, and those without.

Rodded grenades are arranged for firing direct from the rifle; the grenade is held by the rod, which is inserted into the barrel.

The other type of grenade fits into a discharger attached to the muzzle of the rifle.

Owing to the fact that the rodded grenade fills up a portion of the bore of the rifle, the space available for the expansion of the

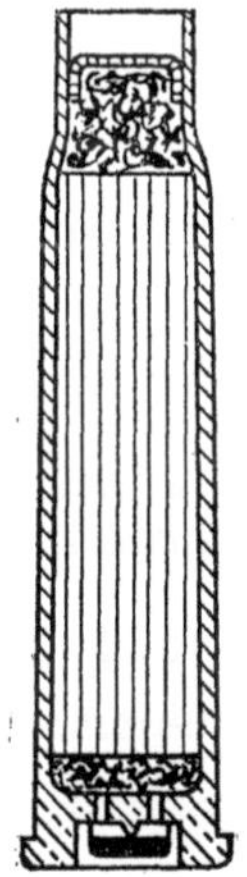

Cartridge, S.A., Rifle Grenade, ·303-in., Cordite, H.Mk. II. (For all rodded rifle grenades)

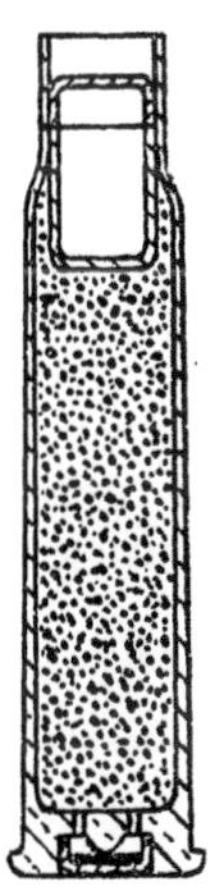

Cartridge S.A., Rifle Grenade, ·303-in., Ballistite, H. Mk. Iz. (For rifles fitted with grenade discharger)

FIG. 10.09.

gases when the charge is fired is considerably less than when the grenade is fired from a discharger. For this reason the same loading would not be satisfactory with both types of grenade, and there are, consequently, two types of rifle grenade cartridges.

The cartridge for the rodded grenade has an ordinary ·303-inch cartridge case and cap, the propellant used being cordite. The charge consists of 43 grains of cordite M.D. 4¼, and, to ensure complete ignition, guncotton tufts are placed both at the top and the bottom of the charge, these being necessary owing to the increased space, and consequently lower initial pressure, behind the grenade rod compared with that behind the ordinary bullet. The mouth of the case is closed by a glazedboard cup.

Marking.—The cases of these cartridges are blackened all over.

N.B.—This cartridge must not be used with a discharger, as the propellant is not sufficiently rapid to ensure complete ignition.

The cartridge for use with a discharger is generally similar in construction but differs in the form of charge used.

The charge consists of 30 grains of ballistite.

No guncotton tufts are employed, and the charge is held in place by a cupped wad of rocket paper and the case is closed by a glazedboard cup.

The cartridge must not be used with a rodded grenade, as the ignition of the ballistite is much too rapid, and dangerous pressures would be produced.

Marking.—The front half of the case only is blackened.

Packing.—Both types of cartridges are issued to the service packed in the same box as the grenades with which they are to be used ; the requisite number being placed in a small tin box enclosed in the main package.

Characteristic symbols.—The symbols for these cartridges when packed in bulk are :—

(*a*) Rifle grenade cordite cartridge—a green club on a white ground, overprinted " H II " in black.

(*b*) Rifle grenade ballistite cartridge—a green spade on a white ground overprinted " H I^{Z}" in black.

§10.10. Discharger, grenade, rifle, 2-in., No. 1, Mk. I.

The new discharger was adopted at the same time as the percussion H.E. grenade, No. 54.

It consists of the following parts :—

1. A *bracket* which forms the means of attachment to the nose-cap of the rifle.

It is intended to be a semi-permanent fitting for rifles used in grenade firing.

It is attached by means of two steel side lugs which engage in the rifle nose-cap recesses, and are tightened by screwing up the cylindrical nuts on the top of the bracket through which the lugs work, using the slot in the bayonet pommel on the feathers of the nut.

The boss on the bracket is cut away to give room for the bayonet ring when the bayonet is fixed. It is drilled centrally and threaded internally to take the stem of the barrel.

The projecting nose of the barrel of the rifle enters the bottom of the hole in the boss, thus centering the bracket which is clamped to the nose-cap.

2. The *discharger barrel* is an elongated steel cup with parallel walls. The screwed stem can be screwed into the bracket until it makes contact with the barrel mouth of the rifle. It can be fitted and removed in a few seconds by hand.

The barrel is designed to give accurate ranging with the No. 54 grenade and the 30-grain ballistite cartridge (range about 350 yards).

Internally above the stem is the combustion chamber. This has been determined as the minimum extra air space necessary completely to convert the 30-grain ballistite cartridge to gas, even though a port, fully open, is used.

The grenade rests on a ledge provided at the top of the combustion chamber.

The parallel portion of the barrel forward of this is made of such a length as to permit of the maximum velocity which the S.M.L.E. rifle has been found capable of standing continuously (1,200 rounds a rifle) with a grenade of the weight of the No. 54.

A toughened steel adapter is fitted to the bottom of the stem of the barrel. This is removable, and replaceable when eroded.

These adapters should be capable of standing at least 2,000 grenade rounds before becoming unserviceable.

§10.11. Markings on grenades (other than signal).

(*a*) *Stampings.*

The following stampings will be found on grenades :—

(i) Serial number and numeral.
(ii) Manufacturer's initials or recognized trade mark and date of completion of empty grenades.
(iii) Manufacturer's and inspection stampings, which are not used for identification purposes in the service.

(*b*) *Paintings, etc.*

Grenades are painted, lacquered or rust-proofed as follows :—

(i) Yellow.—H.E. Grenade, No. 54.
(ii) Lacquered or rust-proofed.—H.E. Grenade, No. 36M.
(iii) Green.—All smoke grenades. Grenades, No. 37, manufactured before November, 1925, were painted black.
(iv) Waterproof drab.—Instructional Grenade, No. 41.
(v) White, tinned or coated zinc.—Weighted practice grenades.

(*c*) *Markings and stencillings.*

The coloured rings and bands are $\frac{1}{2}$-inch wide on all types and natures of grenades and have the following significance :—

(i) A red band on smoke grenades, indicating that the grenade is filled.

(ii) A ring of red crosses on H.E. grenades, indicating that the filling is suitable for use in hot climates.

(iii) A green band, on H.E. grenades, indicating a trotyl or baratol filling.

In the case of grenades filled with baratol the proportion of the mixture, *e.g.*, 20/80 or 60/40, is stencilled in black on the green band.

(*d*) *Painting and staining of boxes.*

Wood boxes are painted or stained as follows :—

(i) Vandyke brown (stain).—Boxes containing grenades other than smoke.

(ii) Green.—All boxes containing smoke grenades.

Chapter XI

DEMOLITION AND BLASTING EXPLOSIVES, AND STORES CONNECTED THEREWITH

§11.01. Introduction.

Blasting explosives may be divided into three classes according to the nature of the work they are required to perform, viz :—

(i) Demolitions in the field, *e.g.*, destroying bridges, railway lines, buildings, etc.
(ii) Underground mining.
(iii) Blasting rock, etc.

The types of explosives used for these purposes have somewhat different qualities, of which the most important are :—

(i) For demolition in the field, the chief requirements are great violence, ease of handling and convenient portability. The work is generally done in the open air, so poisonous gases as a product of explosion do not matter much.

(ii) For underground mining, great power (*i.e.* lifting effect) is of more importance than a localized shattering action, and an explosive that does not produce carbon monoxide on detonation is preferable.

(iii) For quarrying and such-like work, trade explosives may be employed. They are generally plastic, for convenience in loading in small holes or cracks.

The explosives normally used in the service under the above three headings are :—

(i) Guncotton.
(ii) Ammonal.
(iii) Nitroglycerine explosives.

§11.02. Guncotton.

Guncotton is cellulose nitrate (nitrocellulose) containing about 13 per cent. of nitrogen. This approximates to the maximum amount of nitrogen which can be introduced in normal manufacture, and in this highly nitrated form nitrocellulose is only soluble in acetone. It is insoluble in ether-alcohol, and hence goes by the name of "insoluble nitrocellulose" to distinguish it from the "soluble" variety, containing about 12·2 per cent. of nitrogen, which *is* soluble in ether-alcohol. Guncotton is used in the manufacture of Cordite, Mk. I., M.D., M.C., W., soluble nitrocellulose being used for Cordite, R.D.B., S.C., and for N.C.T.

For demolition purposes, guncotton is used in the forms of pressed blocks in two forms, viz :—

(*a*) Wet slabs, containing about 17 per cent. of water.

(*b*) Dry cylindrical primers, having a hole in the middle to take a detonator.

Properties.—Guncotton, when properly made, is a white powdery substance still bearing evidence of the fibrous nature of the cellulose from which it is made. It is odourless and tasteless. Usually a small amount of alkali (chalk) is added during manufacture, so as to ensure the absence of free acids. It is insoluble in water, but can be dissolved by acetone. It contains less than 12 per cent. of constituents which are soluble in ether-alcohol at ordinary temperatures.

(*a*) *Power and violence.*—Wet guncotton, when properly detonated, is only slightly less powerful than dry guncotton, but both are inferior to lyddite, trotyl or 40/60 amatol, and their rates of detonation also are less.

(*b*) *Insensitivity.*—In the dry state, guncotton is very sensitive to friction or percussion, and should be handled with considerable caution. It is also liable to become electrified in dry weather when subjected to friction. It is readily ignited by a flame and burns with great rapidity. Under certain conditions its combustion may develop into detonation.

Wet guncotton, containing 17 per cent. of water or more, is only slightly sensitive to mechanical shock. It smoulders but does not flame when a light is applied to it. It is difficult to bring to detonation, and in order to achieve this, the ordinary fulminate detonator must be augmented by a primer of a more easily detonated material such as dry guncotton or composition exploding.

(*c*) *Freedom from sympathetic detonation.*—Wet guncotton is not very susceptible to sympathetic detonation ; the dry explosive has a greater tendency to detonate in this manner.

(*d*) *Stability.*—Guncotton, even when carefully purified, undergoes a slow process of decomposition at atmospheric temperatures, with the formation of nitric and other acids, and with increase in the solubility in ether-alcohol owing to the formation of lower nitrates. The acids formed would increase the rate of decomposition, but in service guncotton they are rendered innocuous by the presence in the guncotton of chalk (calcium carbonate) which converts them into harmless calcium salts. Owing to the slow rate of the decomposition, a small proportion of calcium carbonate will prevent the formation of appreciable amounts of free acids for a long period.

Wet guncotton has a greater stability than dry, owing to the relatively large quantity of water (17 per cent.) which it contains.

The acid products of decomposition are diluted by this excess of water, and are consequently less harmful.

(*e*) *Effects of moisture and temperature.*—Dry guncotton is not injuriously affected by damp, with the exception that it may become less sensitive. To avoid this, dry guncotton primers are waterproofed by gelatinizing their surface.

Wet guncotton is entirely unaffected by damp. In a dry atmosphere it may become dry, and consequently dangerous. For this reason the tins in which the slabs are packed are fitted with plugs, which may be removed so that water may be added. The correct amount of water needed is given by the difference between the weight of the tin and its contents at the time of examination and the weight marked on the tin. One ounce of carbolic acid (phenol) per gallon is added to the water for rewetting, for the purpose of preventing the growth of micro-organisms (fungi).

(*f*) *Products of combustion.*—Guncotton gives off a large quantity of carbon monoxide on explosion. This restricts its use in confined spaces such as underground mines.

Service uses.—Guncotton is chiefly used for incorporation with nitroglycerine and a stabilizer in the manufacture of Cordites Mk. I, M.D., and W. Other uses are :—

Wet ... Slabs for mines and demolition work.

Dry ... As cylindrical primers for detonating slabs.

As guncotton yarn: in the priming of electric tubes and detonators.

As guncotton dust: in priming composition of electric tubes and detonators.

Soluble varieties of nitrocellulose are used in the manufacture of Cordite, R.D.B. and S.C., nitrocellulose powder and blasting gelatine.

§11.03. Ammonal.

The Service mixture is a grey composite powder consisting of ammonium nitrate 65 per cent., trotyl 15 per cent., aluminium 17 per cent., and charcoal 3 per cent. It is much more powerful than guncotton, but is slower in action and consequently produces a better lifting action in mined charges though it is less shattering in effect. It absorbs moisture readily and deteriorates rapidly with exposure. Being a powder it must, in small charges, be used in a container; it is therefore, not so well adapted for demolition purposes as guncotton.

The fumes are only slightly poisonous, so this explosive is very suitable for underground mining operations.

§11.04. Nitroglycerine explosives.

Glycerine explosives are used for blasting rock, etc. They are very violent, and being fairly sensitive are not so safe in storage and transport as guncotton and ammonal.

Dynamite No. 1.—This is a plastic explosive consisting of about 75 per cent. of nitroglycerine mixed with " Kieselguhr," an inert porous earth, included for the purpose of getting the nitroglycerine into a plastic state suitable for handling. It is extensively used commercially in mines and quarries and was the first substance of its kind, being introduced soon after the discovery of nitroglycerine by Sobrero in 1846.

Dynamite cannot be used after exposure to wet, as the nitroglycerine at once separates from the earth and is then dangerous to handle. Like most nitroglycerine mixtures, it freezes at 40° F. It can be recognized in its frozen state as it is then harder, more brittle than plastic, and slightly lighter in colour.

Complete detonation with frozen dynamite is not possible; it should, therefore, be thawed before using, but this is a dangerous operation, and should be left to experts.

It should not be exposed to the rays of a tropical sun.

The effect of dynamite is more local and shattering than guncotton. It is not so safe to handle as either guncotton or ammonal; but, being plastic, is especially adapted for use in small spaces.

Dynamite No. 2 consists of about 18 per cent. of nitroglycerine mixed with 71 per cent. of potassium nitrate, 10 per cent. charcoal and 1 per cent. paraffin wax. It is a much later product and is in reality a form of gunpowder with nitroglycerine mixed with it. It is less powerful than dynamite, No. 1.

Blasting gelatine is the most violent of all the glycerine explosives. It was introduced by Alfred Nobel in 1875, and was the forerunner of the modern propellant (*i.e.*, it originated the discovery that nitrocellulose could be gelatinized by nitroglycerine). It is made by dissolving finely divided nitrocellulose in nitroglycerine in the proportions of about 6 to 94. The product is a gelatinous mass, the colour of honey, varying in consistency from a tough leather-like substance to a jelly. It is practically unaffected by water.

Gelatine dynamite and *Gelignite* are blasting gelatine with the addition of about 20 per cent. and 40 per cent. respectively of a mixture of potassium nitrate and wood meal. They are somewhat milder in action than blasting gelatine.

§11.05. Methods of firing.

The principal methods of firing these explosives are by :—

(*a*) Safety fuze.
(*b*) Instantaneous fuze.
(*c*) Electricity.
(*d*) Fuze Instantaneous Detonating.

The stores required are described below.

§11.06. Fuze, safety, No. 16, Mk. I.

Safety fuze was invented by William Bickford in 1831.

In its modern form it consists of a train of F.G. or G.12 gunpowder enclosed in jute yarn contained in a tubular wrapping of waterproofing composition which is in turn protected by a covering of waterproofed tape, yellow in appearance.

It burns at the rate of 1 yard in 125 to 165 seconds, and will burn under water. It is easily ignited by a portfire.

To prepare the fuze the gutta-percha must first be removed by an oblique cut and the powder laid bare, both at the end in contact with the charge and at that which is to be ignited.

Fuze, safety, No. 11, *Mark I,* is similar but is coloured black, and the rate of burning is from 70 to 110 seconds a yard.

§11.07. Quick-match, 4-thread and 6-thread.

This is made of cotton wick boiled in a solution of sulphurless mealed powder and gum, and afterwards dusted over with sulphurless mealed powder before it is quite dry. Unenclosed it burns at the rate of 30 ± 5 seconds a yard, enclosed it is practically instantaneous. It is used principally for priming, and in the manufacture of instantaneous fuze.

§11.08. Fuze, instantaneous, Mk. IV.

Consists of two or more strands of quick-match with a waterproof tape covering, painted red. It burns at the rate of 30 yards a second. This fuze is used in the manufacture of certain aircraft and trench warfare detonators, etc.

Fuze, instantaneous, Mark V (*commercial pattern*).

This fuze has a powder train with a waterproof covering and is provided with a fixed exterior snaking of cotton yarn, or other suitable medium painted orange colour. It burns at the rate of 30 yards a second. *Use,* for training purposes.

§11.09. Fuze, instantaneous detonating (" Cordeau Detonant ").

(*a*) Fuze instantaneous detonating, picric acid is contained in a tin tube.

(*b*) Fuze instantaneous detonating, trotyl in lead or tin tube.

It is used principally—

(1) To fire a number of charges simultaneously when firing by safety fuze, or in conjunction with electric firing to obviate complicated connections and circuits.

(2) To avoid the use of excessive lengths of safety fuze in certain demolitions, *e.g.*, mined charges, when firing by fuze.

Fuze, Instantaneous Detonating, must be detonated; its action is practically instantaneous. It will not detonate if ignited.

The explosive in " Fuze, Instantaneous Detonating," deteriorates if exposed to air; before using, therefore, about 6 inches of open end should be cut off. The ends, if they are likely to be left in charges for any length of time, should be protected.

Mk. III (filled trotyl) is superior to earlier Marks.

§11.10. Detonator, No. 8, Mk. VII.

For use with instantaneous and safety fuze.

(*a*) The Detonator for use with safety fuze consists of a solid drawn copper tube, partly filled with 30·8 grains of fulminate composition, the remainder of the tube being left empty to receive the safety fuze.

The detonator is painted red, and is packed in tinned-plate cylinders, painted red.

In trade use the equivalent of the No. 8 detonator is known as " No. 8 Commercial Cap."

Caution.—It is important to remember that it is most dangerous to handle detonators roughly.

When wet, fulminate of mercury is harmless, but when dried it regains its original sensitivity and violence.

Any detonator that has missed fire or is injured in any way should at once be destroyed.

Before inserting a detonator into a guncotton primer, force the rectifier up to the full extent to which the detonator is to enter, and then withdraw the rectifier by twisting it.

§11.11. Detonator, No. 27, Mk. I.

This detonator is identical with the Trade Standard No. 8 aluminium tube (concave base) detonator.

It consists of an aluminium tube with a concave base, and is charged with corned C.E. On top of the C.E. is pressed A.S.A.* composition. The surface of this composition is formed with a flash cavity to facilitate ignition.

The outside of the detonator is painted red.

Use.—For initiating detonation in Fuze, Instantaneous Detonating. A union, fuze and detonator, No. 2, is used to connect up.

§11.12. Detonator, electric, No. 9, Mk. IV.

The detonator consists of the following principal parts :—

Ebonite head, brass socket with tinned-plate tube, brass cup, two electric leads and tinned copper pole pieces.

---

* A.S.A. composition consists of Lead Azide, Lead Styphnate and Aluminium.

The tube is charged with 30 grains of fulminate of mercury, it is soldered to the brass socket and the bottom is closed by a plug of shellac putty. The priming composition of guncotton dust and mealed powder surrounds a platinum silver wire bridge, and is contained in a cup, the hole being closed by a vegetable paper disc.

The connecting wires pass through the ebonite head and are soldered to the pole pieces. The ends of the connecting wires are bared, and are of unequal length to minimize the risk of a short circuit when connecting up.

A current of 1 ampere must fire the detonator, and the resistance of the bridge is to be between 1·5 to 1·8 ohms.

Marking.—The head and socket are painted *yellow* and the tube portion *red.*

Packing—The detonators are packed 25 in a tinned-plate cylinder which is painted half *yellow* and half *red.* A "rectifier guncotton primer" is also included in the tinned-plate cylinder.

Detonator, Electric, No. 13, *Mark III, is obsolescent.* It differs from the No. 9 in a few details.

Marking.—Upper half is painted *white* and the lower half *red.*

§11.13. Fuze, electric, No. 14, Mk. IV.

This fuze is used for exploding charges of gunpowder, etc.

In construction it is similar to the No. 9 detonator except that the magazine is charged with G.20 gunpowder, instead of fulminate of mercury.

Marking.—The fuze is painted *white* all over, and packed 25 in a tinned-plate cylinder which is also painted *white.*

Chapter XII

PYROTECHNIC STORES

§12.01. Introduction.

Under this heading are included rockets, lights, portfires, signal and illuminating cartridges, signal grenades, flares, smoke producers, etc.

Signal grenades have already been dealt with in Chapter X on grenades.

§12.02. Rockets.

Rockets are employed in the land service for signalling.

A rocket consists of a cylinder, closed at the head, and having a vent at the rear end. In this cylinder is a quick burning composition with a conical recess up the centre.

The ignition of the composition causes a pressure of gas in the rocket, and this gas, escaping out of the vent, presses against the air and propels the rocket in whatever direction the head may be pointing.

It is necessary to provide some means of keeping a rocket travelling in the direction in which it started, for, if the rocket were a simple cylinder, it would tend to turn over and over, because the centre of gravity is continuously altering as the composition burns away. Therefore the rocket is kept straight by the attachment to the rear end of the cylinder of a long stick, or a short stick with a tail of rope.

Instructional labels are pasted on the bodies of all rockets.

In order to reduce the area over which the destructive effect of the accidental ignition of a store of rockets would extend, most types of rockets are fitted with a beechwood plug in the vent, which must be removed when the rocket is required for use. A rocket so fitted will burst on ignition instead of being projected.

Rockets are issued usually in tinned-plate cylinders, the cylinders being provided with a stripping band in the case of rockets which require to be hermetically sealed for preservation purposes, or are fitted with lids having bayonet joints with the body, where hermetic sealing is not essential.

For rockets with enlarged heads, the cylinders are made to the form of the rocket.

The cylinders are issued packed in wooden boxes or metal-lined cases.

The types of rocket that may be met with are :—

(1) " Rocket, Signal," containing a number of stars.

(2) " Rocket, Light, Parachute," containing one star provided with a parachute for suspending it after ejection.

They are obsolete for future manufacture, and, when stocks are used up, will be replaced by signal grenades.

§12.03. Rockets, signal.

These are designated Service, Red, Blue or Green, according to the colour of the stars.

Each nature is made in two weights of 1 lb. and ½ lb. The coloured rockets are similar in construction to the service rocket except for the colour and number of the stars, and also their heads are larger and are fitted with a round cap.

The colour and number of the stars are :—

Colour	1 lb.	½-lb.
Service (White) ...	28 stars	20 stars.
Red, Blue, Green	49 stars in 7 tiers ...	30 stars in 6 tiers.

The general construction of a signal rocket may be seen from the diagram.

The body of the rocket is painted drab colour, and the head red, blue, or green according to the colour of the stars.

Method of firing.—Attach the rocket to the stick, as described by the label on the body, and remove the screw plug from the vent. Place the stick in sand, or more conveniently, support the rocket in the required position (almost vertically) by passing the stick through a couple of iron loops on a post in such a manner that no resistance is offered to the stick going up with the rocket.

To fire the rocket, ignite the priming in the vent by means of a portfire.

The stick used is 5 feet long and tapered to the end ; in confined spaces a short stick about 1 foot 6 inches long with a 5-foot rope tail is more convenient.

§12.04. Rocket, light, parachute, 1-lb., Mk. II.

This rocket is similar to the Rocket, Signal, but has a single red or white star with a parachute to suspend the star after ejection. The parachute is in the head of the rocket and is separated from the star by a cardboard disc.

The rocket is provided with fuze which ends in contact with a friction board glued to the side of the body. A match for striking the friction board and igniting the rocket is carried in a cardboard cylinder attached with adhesive tape to the side of the rocket above the socket for the stick.

The vent is closed with a screwed wood plug which must be removed before firing.

Fig. 12·03.

ROCKET, SIGNAL, 1 LB., SERVICE, MARK III.

SCALE – 1/2.

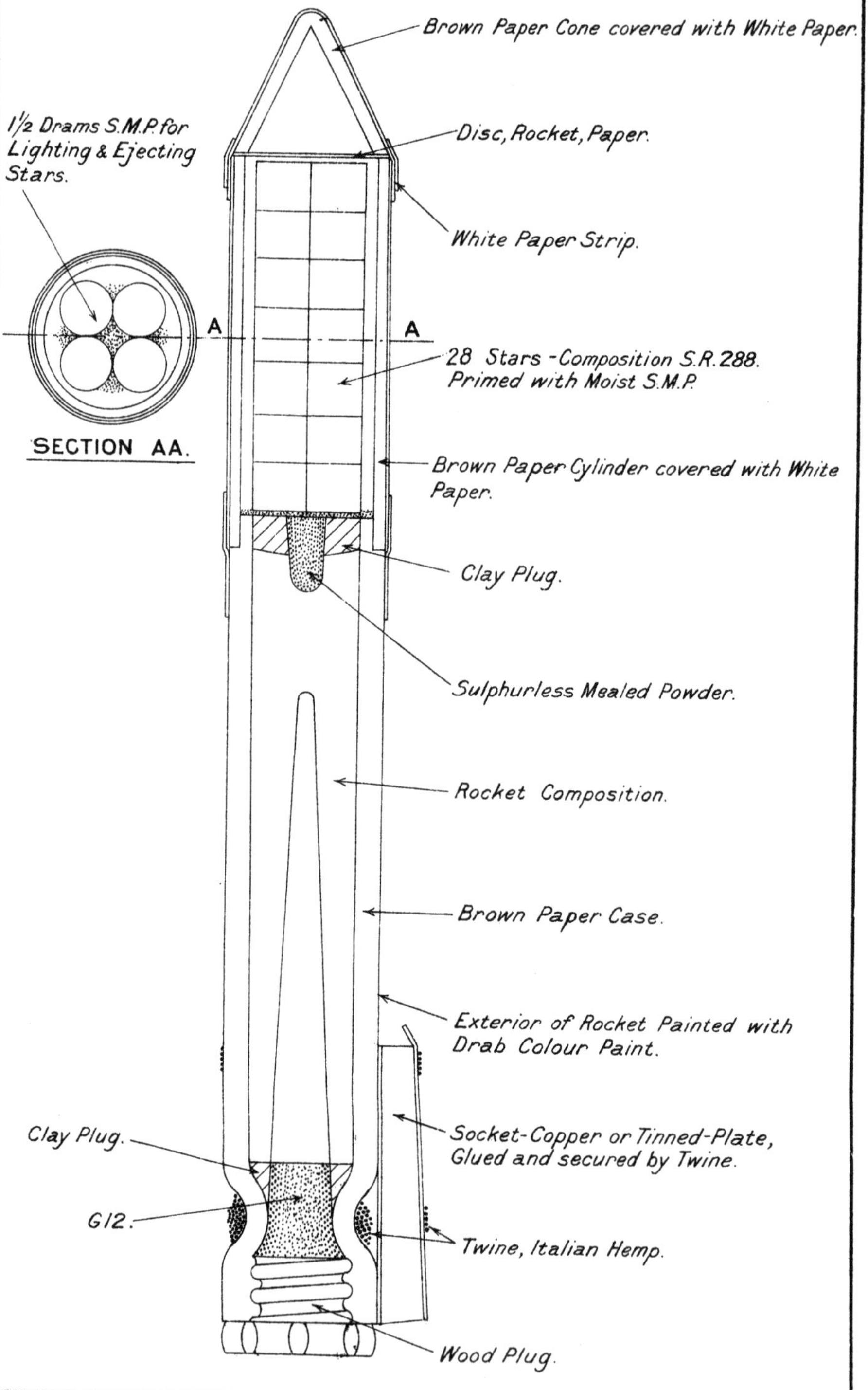

Brown Paper Cone covered with White Paper.

Disc, Rocket, Paper.

1/2 Drams S.M.P. for Lighting & Ejecting Stars.

White Paper Strip.

A A

SECTION AA.

28 Stars - Composition S.R. 288. Primed with Moist S.M.P.

Brown Paper Cylinder covered with White Paper.

Clay Plug.

Sulphurless Mealed Powder.

Rocket Composition.

Brown Paper Case.

Exterior of Rocket Painted with Drab Colour Paint.

Socket-Copper or Tinned-Plate, Glued and secured by Twine.

Clay Plug.

G12.

Twine, Italian Hemp.

Wood Plug.

958

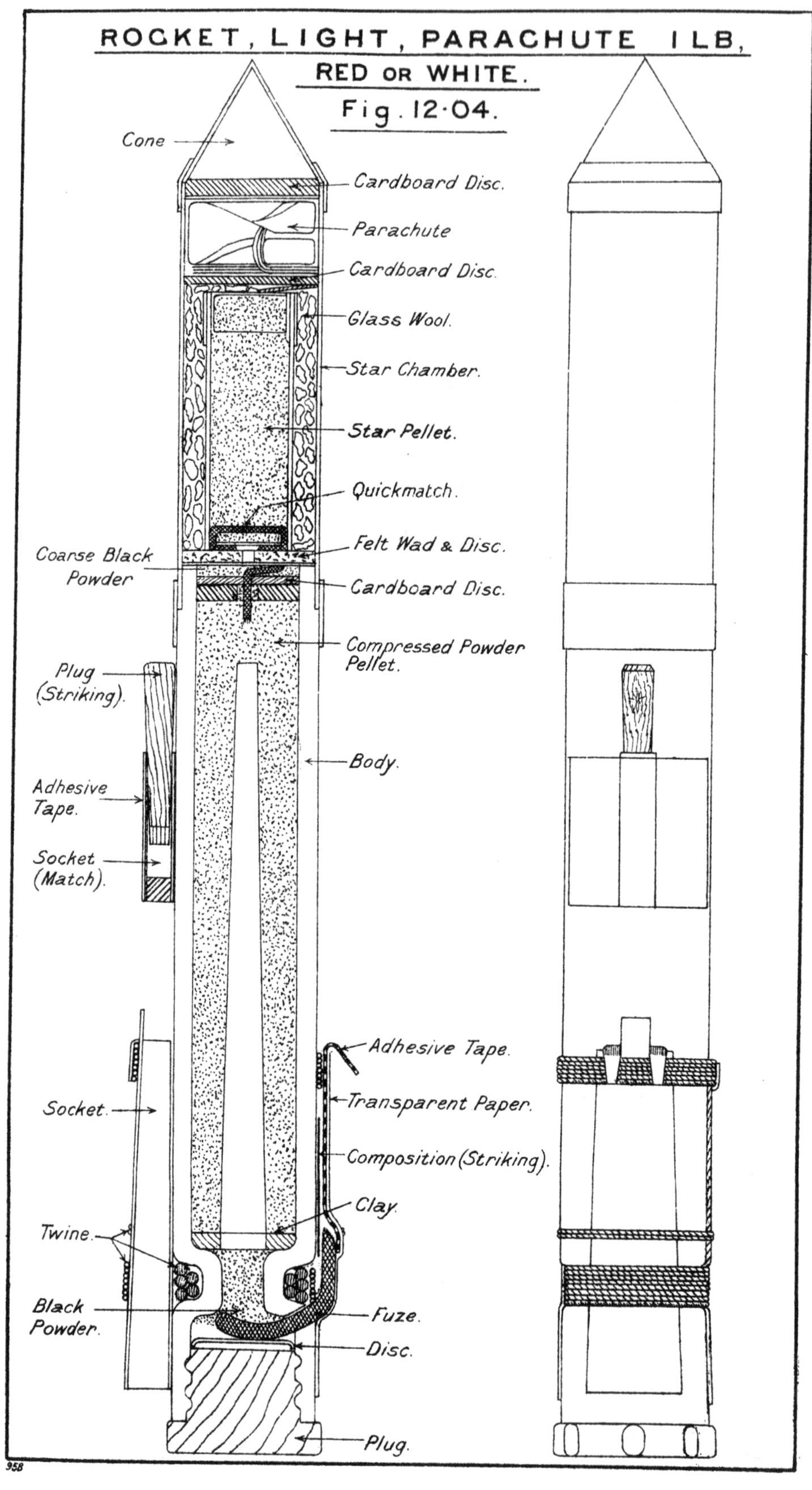
ROCKET, LIGHT, PARACHUTE 1 LB,
RED OR WHITE.
Fig. 12·04.
Cone
Cardboard Disc.
Parachute
Cardboard Disc.
Glass Wool.
Star Chamber.
Star Pellet.
Quickmatch.
Felt Wad & Disc.
Coarse Black Powder
Cardboard Disc.
Compressed Powder Pellet.
Plug (Striking).
Body.
Adhesive Tape.
Socket (Match).
Adhesive Tape.
Transparent Paper.
Socket.
Composition (Striking).
Clay.
Twine.
Black Powder.
Fuze.
Disc.
Plug.
958

Fig 12·05.

LIGHT, LONG, { BLUE MK III. / GREEN MK III. / RED MK III.

SCALE ⅓.

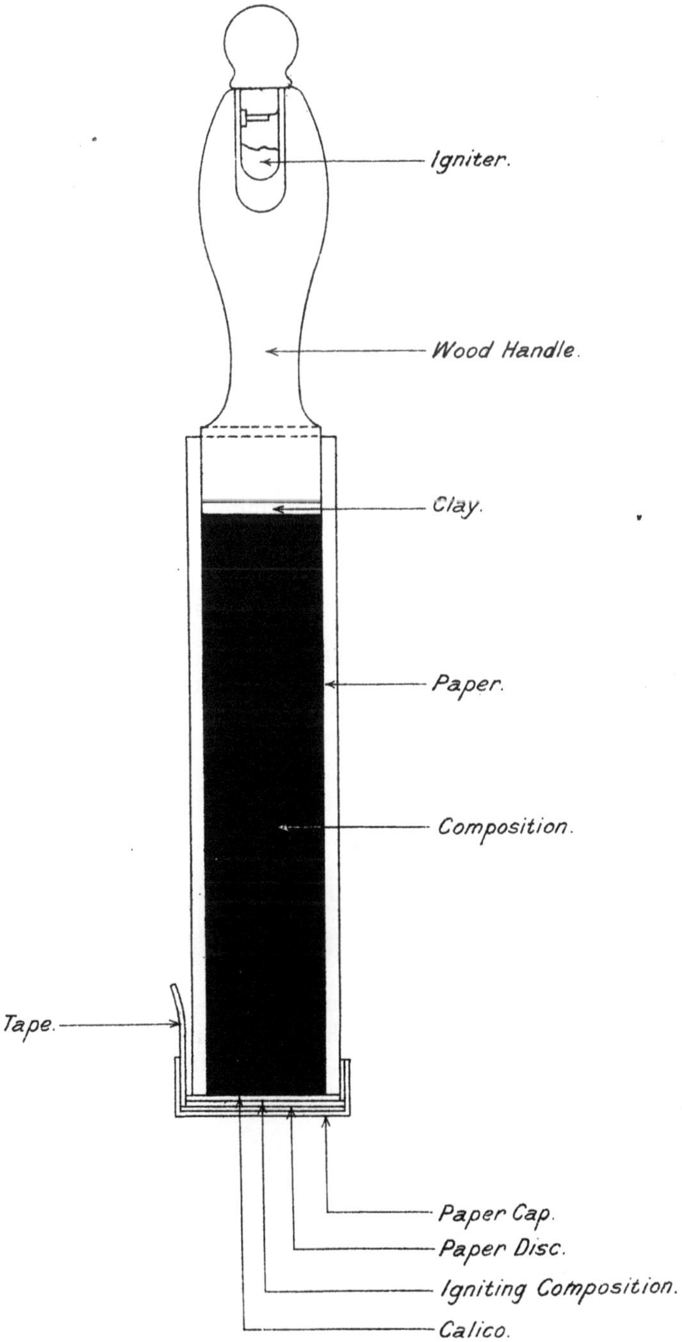

Igniter.

Wood Handle.

Clay.

Paper.

Composition.

Tape.

Paper Cap.

Paper Disc.

Igniting Composition.

Calico.

§12.05. Lights.

Lights are obsolete for future manufacture and when existing stocks are used up will be replaced by ground flares. There are two lights at present in the service.

(1) *Lights, Long, Blue, Green and Red, Mk. III.*

The light consists of a rolled paper case into which is pressed a column of light-giving composition. The top of the composition is covered with a thin calico disc smeared with ignition composition covered and protected by a paper disc and a coloured paper cap. A piece of tear-off tape is fastened under the disc and the end is carried up the side of the light and protrudes free beyond the end of the paper cap.

At the bottom of the composition is a plug of clay. A beechwood handle, having a recess at the lower end for a wood plug, is inserted in the bottom of the case till it bears against the clay plug and secured with shellac. In the recess at the end of the handle fits a beechwood plug covered at its top end with striking composition.

To ignite the light.—The paper disc and cap are first pulled off by means of the tape. The igniter plug is then withdrawn from the handle, and, with the light pointing away from the body, is drawn across the priming composition in the head. On no account is the igniting composition to be struck with the primed end of the plug. A label giving full instructions for lighting is pasted on the body of the light.

Each light is packed in a tinned-plate cylinder and 24 thus enclosed in a box. The red and blue light burns about 2¼ minutes, the green light about 1 minute 40 seconds.

They are painted externally according to the colour of the light.

(2) *Light, Short, G.S., Mk. II.*

This is a white light similar to (1) but is shorter and burns about 1½ to 2 minutes. It is painted drab.

§12.06. Portfires.

Portfires are used for incendiary purposes. They are generally lit by slow-match or any other handy means, and are not extinguished by water. To put them out, the burning end should be cut off. A stick is supplied, fitted with a steel socket and thumb-screw.

Portfires "friction" and "self-igniting" may be met with. These are only half the length of the common portfire with corresponding shorter burning, and are fitted with a friction lighting device. They are trade supplies.

§12.07. Slow-match, Mk. I.

This is made of pure hemp slightly twisted and boiled either in a solution of wood ashes (potassium carbonate) in water or in a solution 8-oz. saltpetre to 1 gallon of water.

It burns at the rate of 1 yard in 8 hours.

Use.—For lighting portfires, etc.

§12.08. Signal cartridges.

These are used for signalling purposes. They were originally issued in two sizes, viz., 1-inch and 1½-inch ; in future, 1-inch only will be used in the land service and 1½-inch in the air service.

The 1-inch cartridges are fired from a pistol of 4-inch barrel almost vertically into the air ; they have a range of about 125 yards and the stars burn for about 7 seconds.

The construction of a typical cartridge may be seen from the diagram.

The signal stars are either green or red. A cartridge with yellow star is used by certain colonies, etc. To distinguish the cartridges they are marked near the mouth of the case with a coloured band according to the colour of the star, and secondly, a circular label of the same colour of the star or, alternatively, a white label with the word " Red," " Green " or " Yellow " printed on it is attached to the closing disc.

§12.09. Illuminating cartridges.

These are also fired from pistols of 4-inch barrel, at an angle of 45 degrees to obtain a horizontal range of at least 200 yards.

The star travels for about 50 yards after firing before the light is visible, and should burn brilliantly for at least 8 seconds.

The cartridge is marked by a white circular label on the closing disc with the words " Illuminating Dark Ignition."

Nomenclature and packing.

(i) Cartridges, signal, Green, 1-in., Mk. IVT.
(ii) Cartridges, illuminating, 1-in., J., Mk. IIT.

Packing.—Box A.S.A. H.13, containing 150 cartridges.

Symbols.—Signal cartridges : (Typical).

A blue hexagon ended dumb-bell with the word " GREEN " on the ends in white and overprinted " IVT " in black.

Illuminating cartridges : (Typical).

A blue triangular ended dumb-bell on white ground overprinted " J IIT " in black.

§12.10. Ground flares.

These are used for signalling and illuminating purposes, and will supersede Lights, Long and Short.

Fig. 12·08

CARTRIDGE, SIGNAL, 1 INCH.
TYPICAL.

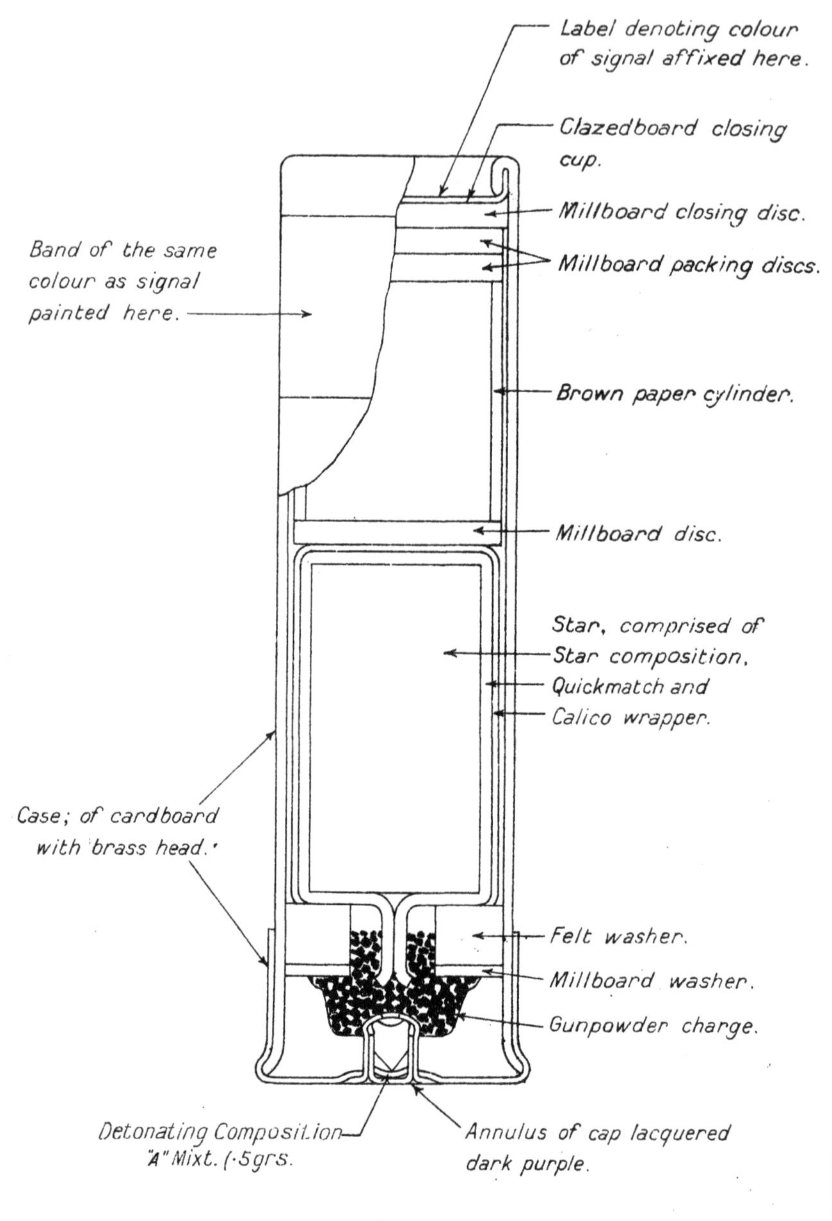

Malby & Sons, Lith.

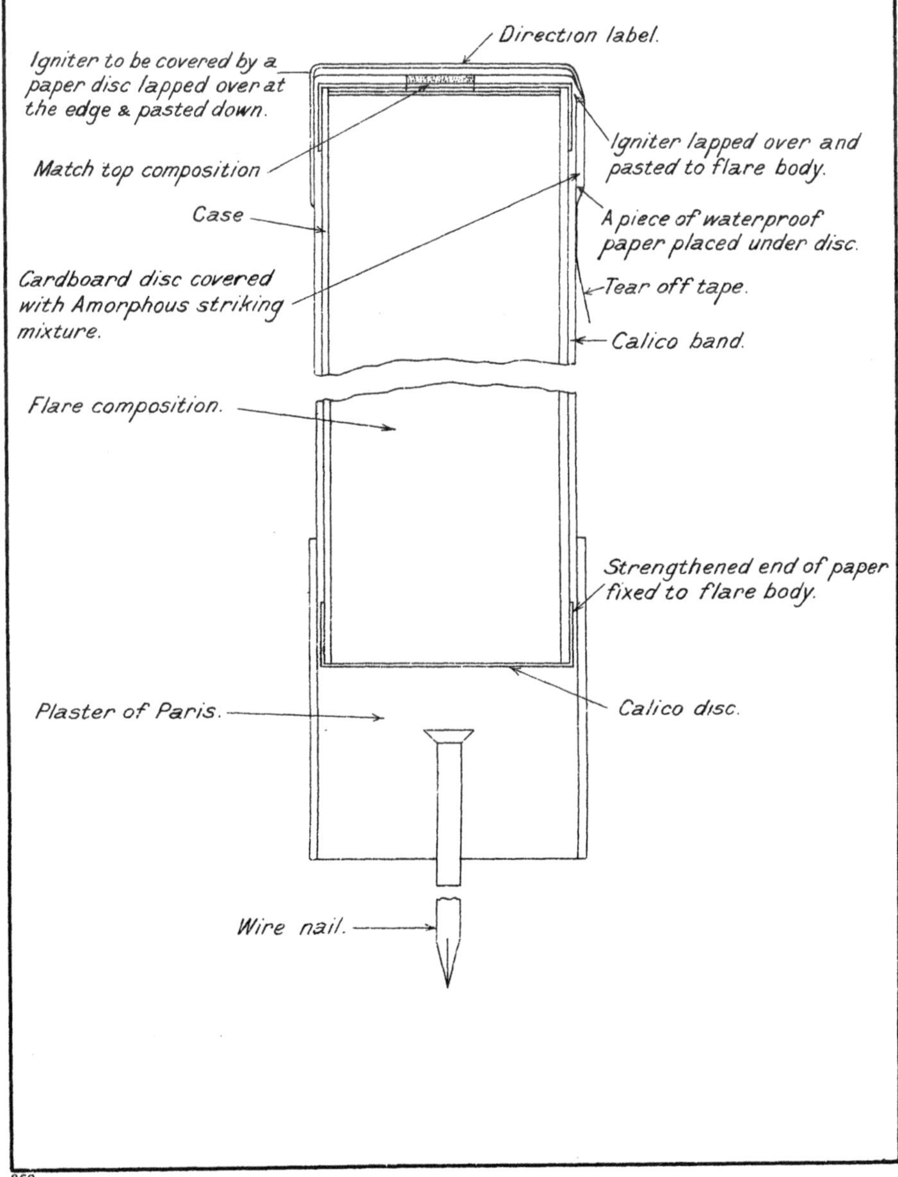

Fig. 12·10.
FLARE, GROUND, 1/2 HOUR, RED, MARK I.
SCALE = 2/3.
Direction label.
Igniter to be covered by a paper disc lapped over at the edge & pasted down.
Match top composition
Case
Cardboard disc covered with Amorphous striking mixture.
Igniter lapped over and pasted to flare body.
A piece of waterproof paper placed under disc.
Tear off tape.
Calico band.
Flare composition.
Strengthened end of paper fixed to flare body.
Plaster of Paris.
Calico disc.
Wire nail.

The various flares in the land service are :—

(1) Flare, Ground, ½-hour, Red.
(2) Flares, Ground.

3 inches by 2 inches diameter, Red, Yellow, Green, White.
1¾ inches by 2 inches diameter, Red, Yellow, Green, White.
1½ inches by 1½ inches diameter, Red, Yellow, Green White.

The diagram of (1) gives the general construction of a flare.

To ignite the flare, the tear-off tape is pulled sharply upwards, thus removing the top cardboard disc and baring the igniter, which is then ignited by rubbing the amorphous phosphorus disc sharply across the match composition at the top.

The flares mentioned in (2) differ from (1) in not being fitted with the socket containing plaster of paris and the nail, and also in dimensions. Their times of burning vary from 2½ minutes to 40 seconds according to size and colour (yellow longest, red, green, and white shortest).

A paper label, of the same colour as the composition in the flare, gives the designation of the flare and instructions for use ; it is pasted on the top of each flare.

§12.11. Smoke producers.

There are two smoke producers which fall under the heading of pyrotechnics.

(*a*) *Candle, Smoke, Ground, Mk. II, Type S1.*

The candle consists of a cylindrical case of tin-plate with a spring-on lid. The case is filled with smoke composition and a friction ignition arrangement contained in a tin thimble protrudes through a hole in the lid. On a tear-off tape are fastened a tin cap which lies over and protects the friction composition in the thimble and a cardboard disc covered with striking composition. The whole is covered with a waterproof paper cover from under which the end of the tear-off tape protrudes.

To ignite the candle the tear-off tape is pulled sharply upwards, thus baring the igniter, which is fired by rubbing the striking disc firmly across the blob of friction composition. The smoke should last from 3 to 4 minutes.

This is obsolescent and is being replaced by—

(*b*) *Generator, Smoke, No. 5, Mk. I.*

In this generator the cylinder is made air-tight, being closed by a baffle plate. This is raised in the centre to accommodate the igniter. The raised portion is perforated and closed with a soldered tear-off plate (fitted with a D ring to assist removal). Two striker sticks are attached to the top of the baffle plate by adhesive tape and the whole is closed with a lid fastened by adhesive tape.

To fire the generator, remove the lid, put a finger through the D link and tear-off the igniter cover and strike the exposed igniter with the prepared end of one of the sticks. The smoke should last from $5\frac{3}{4}$ to 8 minutes.

§12.12. Signal, vertical light ray.

Yellow to Red, Mk. I.
Yellow to Green, Mk. I.

This store has been introduced for use by artillery survey personnel in locating survey stations. The signal consists of a tinned-plate cylinder closed at the top with a millboard and tinned-plate disc and at the bottom (ignition end) with a book-muslin disc, a perforated tin-zinc alloy disc and a red shalloon disc.

The signal is primed at the bottom and filled with yellow composition topped with red or green composition. The unit is then pressed in a floating mould to about half its original size.

A label " Yellow to Red " or " Yellow to Green " as applicable is fixed to the top.

The signal is fired vertically from the 2-inch discharger. The flash from the propellant burns through the shalloon disc and ignites the priming through the holes in the perforated disc.

The signal rises to a height of 400 to 500 feet leaving a trail of smoke and yellow flame which changes at the vertex to green or red. The yellow light lasts about 3 seconds and the second colour about 2 seconds.

§12.13. Thunderflash, Mk. I.

This consists of a cylindrical brown paper body closed at the base with a wood plug and containing a 1·8 inch length of safety fuze followed by a $\frac{3}{8}$-inch length of quick-match and surrounded by 10 grains of loose composition. The body near the top is choked to the fuze with two turns of twine and the space above the choke filled with priming and covered with a paper cap, secured with a paper band, over a tear-off tape.

A label giving the designation of the store and instructions to " Throw away immediately the friction is struck " was attached to the body of the first supplies, which were made by Brock's and provided with an igniter. In later supplies the label reads " Throw away immediately the priming is ignited."

This store, which is obsolete for future manufacture, is used for simulating gunfire for the training of horses or at displays.

Fig. 12·11 (a).

CANDLE, SMOKE, GROUND.

MARK II.
TYPE "S1."

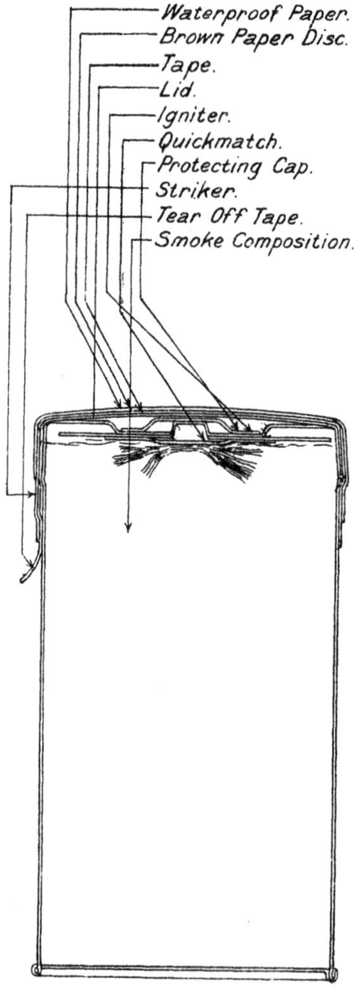

958.

Fig. 12·11. (b).

GENERATOR, SMOKE, No. 5 Mk. I.

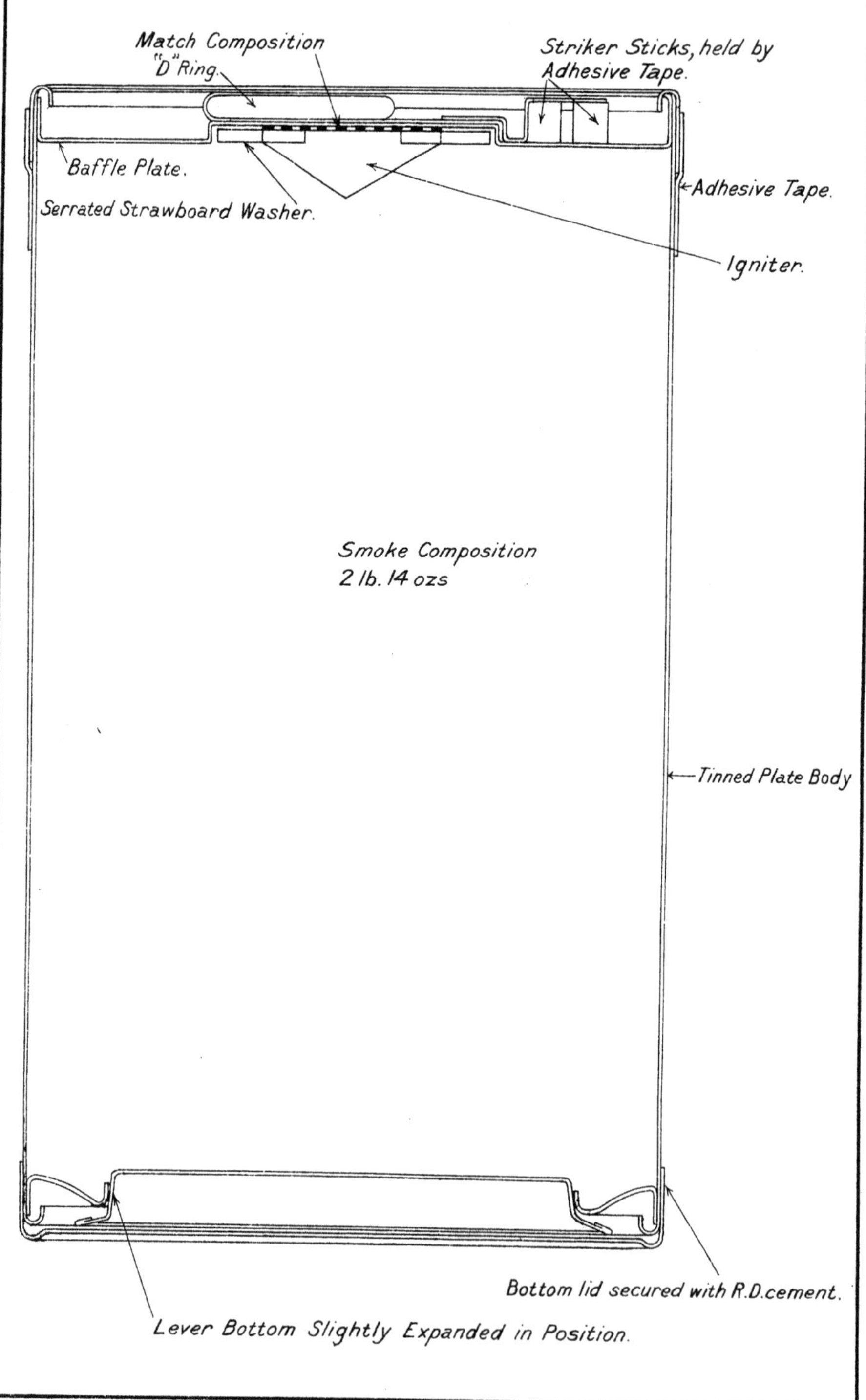

958.

Fig 12·12.

SIGNAL VERTICAL LIGHT RAY.

YELLOW TO GREEN.
YELLOW TO RED.

DIAGRAM OF ASSEMBLY (BEFORE PRESSING.)

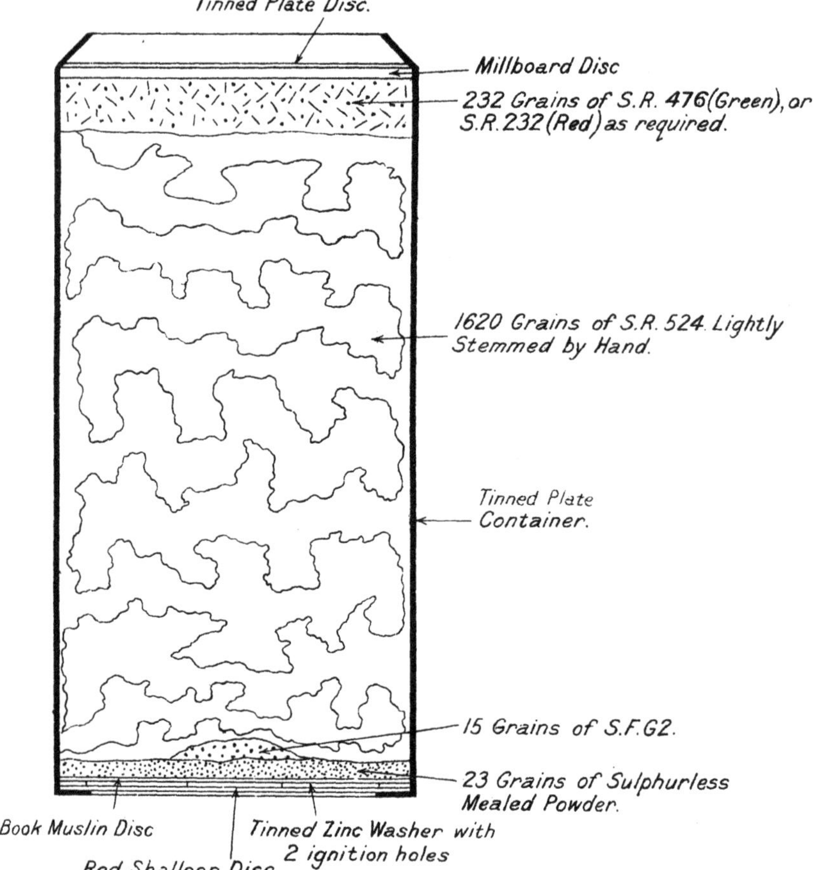

After assembly the Signal is compressed under a load of 18 Tons. Final length about 1·65 Inches.

958.

Appendix I

ARMOUR AND ITS ATTACK

Armour.—Iron protection for ships was first suggested about a hundred years ago. Since that time, armour has developed in successive stages governed by the improvements in the manufacture and treatment of steel. Continuous rivalry has existed between guns and armour, and neither side has kept the advantage for very long; an improvement in the attack producing greater defensive power, and this in turn leading to a more effective projectile.

The armour plates of the present day may be classified as :—

(1) C. (cemented) armour with a hard face.
(2) N.C. (non-cemented) armour homogeneous in structure.

In addition to these, non-cemented plates of special quality are used for special purposes where thin armour is required, such as turret roofs, armoured decks, and gun shields in the land service.

Functions of armour.—The functions of armour are broadly :—

(1) In the Navy and coast defences to keep out shell or the fragments of shell burst on impact on the armour.
(2) In the field, to keep out shrapnel and rifle bullets or fragments of H.E. shell.

As regards ships, the problem of designing the armoured protection is governed to a great extent by the weight that can be allotted for the purpose in conjunction with that required for engines, armament, etc. The armour is therefore not the same thickness all over the ship. Thick armour is placed over the "vitals," *i.e.*, engines, magazines, etc., and main armament, and thin armour over the secondary armament and in the decks.

Fortress guns are given protection calculated to keep out the shell of guns of approximately equal calibre.

The armour for tanks is a matter still in the experimental stage. It is a problem similar to that of the light cruiser and the destroyer, where protection must be sacrificed to mobility.

Light guns in the field are protected by a thin shield calculated to keep out rifle and shrapnel bullets. Further protection is prohibited by considerations of mobility.

Attack of armour.—The object in the attack of ships is to perforate the armour and burst the shell inside the ship, so as to wreck the internal structure and machinery and disable the personnel or, when perforation is not obtainable, to penetrate or break the armour.

For heavy armour, the armour-piercing shell with its hard and massive point and small burster is the only projectile likely to be successful, whereas semi A.P., common pointed, or common (H.E.) shell would be used in other cases according to the strength of the armour and the power and calibre of the gun.

Penetration of armour.—In the destruction of armour the striking energy of the shell is converted into destructive effect on the target. The estimation of this destructive effect is complicated by the many ways in which the striking energy is dissipated. The high velocity and great violence exerted produce effects which it is very difficult to measure. Energy is dissipated in heat, and work is lost by the actual motion imparted to the fragments. Owing to the uncertainty regarding the relative importance of these and other effects, there is a limit to the degree of accuracy with which results may be calculated.

Probably the most favourable conditions are those in which the projectile perforates, leaving as clean a hole as possible, while remaining itself unbroken but with little spare force. For this reason, the thickness of wrought iron that projectiles fired from different guns could perforate, has been found to provide a convenient method of comparing their powers.

Two methods have been recognized by which armour may be destroyed :—

(*a*) " Penetration " meaning that the shell has made a hole in the armour, whether it has burst in the plate or in front of it.

(*b*) " Perforation "—meaning that the shell has got through the plate in a fit condition for bursting on the far side.

" Perforation " is the criterion by which shell are judged to-day.

Appendix II

FORMULÆ FOR PERFORATION OF WROUGHT IRON

The formulæ that have been used for determining the perforation of wrought iron depend on certain assumptions which are only very partially true. These are :—

(1) The act of perforation is assumed to be that due to punching, and therefore depends on the circumference of the hole made, or on that of the projectile.

(2) The resistance of the plate is proportional to some function of its thickness.

The first, and probably the simplest formula based on these assumptions, was suggested by Fairburn, who gave :—

$$\frac{Wv^2}{2g} = \pi d t^2 K$$

Where W = the weight of the shot in lb.
v = the striking velocity in f/s.
g = the force of gravity.
d = the calibre of the shot in inches.
t = the thickness of the plate completely perforated in inches.
K = a certain constant, whose value depends on the quality of the plate, etc.

To obtain the answer in foot/tons, 2,240 must be embodied in the denominator on the left-hand side of the equation.

This formula was used for a long time in the Department of the Director of Artillery in the empirical form :—

$$\frac{Wv^2}{2g} = \pi d t^{1.6} \times 2{\cdot}53.$$

This formula may be used to obtain the total energy " E " of the shot, in order to ascertain the penetrating effect produced on steel or cast iron that cannot be perforated.

Another standard of comparison is the energy " e " per inch circumference. The value of this depends on the same assumption, that the plate is perforated or sheared at the circumference of the projectile, and therefore resists the passage of the shot in proportion to the circumference of the hole that the shot makes.

Thus the formula above might be used to give three results :—

(1) The total energy " E," representing the total available blow.
(2) The energy " e " per inch circumference of the shot, or penetrating figure.
(3) The actual thickness " t " of wrought-iron which can be perforated under the given conditions.

Inglis, Maitland, De Marre, Tressider, Booker and others have produced formulæ which can all be reduced to the same general form, with different constants and in some cases with the indices altered to attempt to bring them into line with experimental results.

Tressider's formula was based on the assumption that perforation varies with the power (as opposed to the energy, as in other theories) absorbed in punching.

Hence he gave :—

$$t = K \times \frac{Wv^3}{d}$$

where K by experiment = $\log^{-1} \bar{9}{\cdot}159$.

Appendix III

FORMULÆ FOR PERFORATION OF HARD-FACED ARMOUR

No reliable formula for the penetration of steel or compound armour has been found, but penetration must depend on the strength of the material. The tenacity of wrought-iron is more or less uniform while that of hard-faced plates is not so. The best that has been done so far is to apply a modification of the factor based on the latest experiments, but any change in the nature of the shell or plate renders them quite unreliable.

Although wrought-iron has long ceased to be used as armour, it is convenient to use it as a standard for comparing the values of different plates and projectiles.

The " Factor of Merit " of a plate is the ratio between the thickness of wrought-iron and the thickness of the plate under consideration, both of which can be just penetrated by a given projectile at the same striking velocity.

The " Factor of Perforation " is the ratio between the thickness of wrought-iron which the round under consideration can just penetrate and the actual thickness of the plate fired at, irrespective of whether the plate is penetrated under the conditions or not.

The " De Marre Coefficient " is the ratio of the striking velocity of a projectile against a plate to the velocity that would be required by the same projectile to penetrate a similar thickness of ordinary steel, as estimated by the De Marre formulæ.

These coefficients are often used to compare the efficiency of different shells under conditions of striking velocity and angle of impact, which are not too greatly dissimilar for a fair comparison by the De Marre formulæ.

The shell that succeeds in penetrating a plate at a lower De Marre coefficient is the more efficient shell.

The De Marre coefficient can be read off a diagram.

The De Marre formulæ are :—

for mild steel $$t^{1.43} = \frac{W(v\cos\theta)^2}{d^{1.53} \times \log^{-1} 6{\cdot}0186}$$

for C armour $$t^{1.43} = \frac{W(v\cos\theta)^2}{d^{1.43} \times \log^{-1} 6{\cdot}218}$$

for N.C. armour multiply the number of inches of C armour by 1·2 to find the penetration.

θ = the angle of penetration to the normal of the plate.
t = the thickness of the plate in inches.
W = the weight of the shell in lb.
v = the striking velocity in f/s.
d = the diameter of the shell in inches.

The results given by these formulæ are good only for the nature of shell with which the variables were determined ; also they deal with penetration not with perforation, but the De Marre coefficients derived from them are used to compare the severity of the conditions of trial of shell that have perforated.

Tressider's formula given above may be used to calculate perforation of armour in conjunction with a suitable figure of merit for the armour in question, but the figure of merit must be obtained from actual trials, and varies with the nature and thickness of plate as well as with the nature, velocity, and diameter of the projectile.

A later formula given by Tressider for bare perforation velocity is given by

$$v_p = \left[\frac{T^2 d \times \log^{-1} 8{\cdot}841}{W}\right]^{\frac{1}{3}}$$

where T is the thickness of C plate expressed in its equivalent of wrought-iron, and corrected for angle of impact, thus :—

$$T = \frac{t \times K}{\cos \frac{\theta}{2}}$$

where t is the actual thickness of C plate.

Table of Equivalents (K).

	In.
Mild steel per inch of thickness	1·25
High tensile ,, ,, ...	1·3
C armour ,, ,, 4 in.	1·8
,, ,, 6 ,, ...	1·75
,, ,, 8 ,, ...	1·7
,, ,, 10 ,, ...	1·65
,, ,, 12 ,, ...	1·60
,, ,, 14 ,, ...	1·54
,, ,, 15 ,, ...	1·52

The relative efficiency of C armour, therefore, diminishes as the thickness increases. This is due to the fact that the heat treatment, tempering, and other manufacturing processes can be better controlled when dealing with a smaller mass of metal. The same remark applies to the hardening of A.P. shell : the larger the shell the less its relative efficiency.

Tressider's formula for remaining velocity behind the plate after perforation :—

$$v_r = \left[\frac{v^3 - v_p{}^3}{v(1 + Q)}\right]^{\frac{1}{2}}$$

$$\text{where } Q = \frac{td^2 \log^{-1} 1{\cdot}2526}{W}$$

In this case t is the thickness, no multiplication by figure of merit being necessary. The coefficient of obliquity is discarded as having an insignificant effect on the result.

This formula is useful in calculating the length of delay to be expected from a delay fuze for which the time is known between the impact of the shell and the burst.

Length of delay $= v \times t$ seconds.

It is also useful to know the velocity of a shell after perforating ship's armour of any thickness and at any angle of attack.

These formulæ have been used to give an approximate idea of what the perforation of a shell will be under any given conditions, but it must be remembered that both individual shells and individual plates vary in quality. Consequently the factors used and the equivalents in wrought iron of C armour which are found for one set of conditions may not be correct for another. It is indeed a little surprising that the formulæ have given as good results as they have when the assumptions on which they are based and the factors omitted are taken into consideration.

For instance, the first assumption implies the punching out of an approximately cylindrical plug from the plate. Actually the plug, if formed, is not usually cylindrical at all, and the hole made in the plate is only cylindrical for a short distance, the greater part of it more nearly approximating to a kind of paraboloid. In many cases no plug is formed, but the plate opens at the back in a star-shaped opening and " petals " of steel bend away to allow the shell to pass.

Again the formulæ omit all reference to the shape of head (c.r.h.) though this has been found by experiment to have a definite effect on the perforation velocity. Other factors which may affect the result are the ballistic length of the head and the shape of the armour-piercing cap.

In the present state of our knowledge of the subject, the only certain results are those obtained by experiments, by actual firing, and from these alone can fairly reliable deductions be made.

Appendix IV

MEANS OF FIRING THE PROPELLANT CHARGE

The earliest contrivances for firing cannon were red-hot spikes or bars which were thrust into the vent, thus igniting the charge. A pair of bellows for heating the priming irons formed part of the equipment of the artillery in the fourteenth and fifteenth centuries.

Owing to the gradual increase in the size of cannon, this method was doubtless found to be both inconvenient and dangerous, and before the middle of the fifteenth century, it was practically superseded by the plan of priming the vent with loose powder. It has not been established how this priming was ignited in these early times, but the probability is that a heated iron or match of some description was used. The latter supposition is considered to be the more correct, although no mention is made of a match or linstock for holding the same until a century later.

No further improvement was effected till the seventeenth century, when gunners of that period had recourse to the occasional employment of quickmatch, which, when placed in the vent, acted as a weak tube. Quickmatch used for this purpose was known as " porte-feu," a name we still retain in " port-fire." This may be considered as the primal origin of " the tube," a fact borne out by the French word " étoupille," a diminutive of " étoupe," meaning " thread or match prepared in a particular way."

The next great advance in the priming of guns was made in the first half of the eighteenth century. This step was the introduction of the " quickmatch tube," the original " tube proper " found in the science of artillery. This tube, introduced somewhere about 1765, consisted of quickmatch surrounded by a tube of tin primed on top.

A piece of tin tubing, dimensioned to fit down closely into the vent, had a piece of quickmatch threaded through it. The top, covered originally with a paper cap filled with mealed powder moistened with spirits of wine, was eventually closed by a flannel cover soaked in a solution of saltpetre in alcohol.

This great stride radically affected the action of artillery, by introducing three main advantages :—

(*a*) Increased rapidity of fire.
(*b*) Protection of the vent from erosion.
(*c*) Greater safety to personnel.

These " quickmatch tubes " soon proved, however, to be too weak, and it was discovered that a tube containing composition with a central cavity was much more efficacious as a substitute, since it produced a much stronger flash on ignition. In 1768 we hear of such tubes being introduced into France ; the composition being mealed powder moistened with spirits of wine.

Various materials were used for the bodies of such tubes. Originally the most usual was tin, which soon fell into disrepute as it was considered to act injuriously on the composition it contained, and to be very susceptible to corrosion. The French employed reeds and continued to do so for many years. Copper was tried and used, which naturally was preferable to tin on account of its non-corrodibility. Paper appears to have been employed by the Dutch, and such metals and alloys as pewter, zinc, thin iron and brass have been used on various occasions with more or less success.

Quill was finally introduced in 1778.

Copper, quill and, for emergency, paper were the materials selected for our service.

Until this time there was no mechanical means of firing the tube, the ignition of which had to be effected in every case by means of a match or port-fire; a store which was introduced as an article of equipment about 1700. Although the flint-lock had been known since the beginning of the sixteenth century, no serious attempt seems to have been made to adapt it to firing cannon until 1778, when Sir Charles Douglas urged the use of flint-locks and quill tubes for firing guns in the Royal Navy. There was, however, considered to be a grave danger connected with using metal tubes (which, with the exception of those made of paper and reeds, were the only ones known in our Service) in wooden ships, and the Admiralty therefore would not entertain Sir Charles' idea. He thus bought musket locks at his own expense, and caused them to be let into pieces of wood wired on to the guns of his own ship. He also purchased a stock of goose quills and necessary ingredients with which he had tubes made up under his directions.

On Sir Charles's appointment to Captain of the Fleet, Captain Gardner succeeded him in command of H.M.S. *Duke*, and in the Battle of Gibraltar Bay, in April, 1782, the quick and efficient firing of that ship was very conspicuous, and led to a great tactical success. This settled the question of the efficacy of flint-locks, but the war at sea also being settled, locks were not officially introduced into the Navy till 1790.

These original flint-locks were found to be disadvantageous owing to the single flint, and an improved lock containing two flints firing a tube with a better method of priming was introduced into the Navy in 1818. The use of this type of lock, adopted for the land service in 1820, never became general, and tubes for the artillery still continued to be ignited by port-fires.

At this period each type of ordnance had a different tube, the length of each tube being that of the vent of the gun with which it was used. In some cases tubes were made pointed and of a sufficient length to pierce the cartridge, thus doing away with the separate operation of "pricking." But this latter arrangement was found to lead the gunner into difficulties; for the explosion of the charge, tending to distort the end of the tube, made withdrawal difficult and led in some cases to the spiking of the pieces.

Tubes of differing lengths were also found to be inconvenient; but for a long time it was considered necessary that the end of the tube should touch the charge in order to ensure proper combustion. When, however, the truth was realized, tubes were made to a uniform length; but this change was not introduced till somewhere about the year 1820.

About 1824, a new "match tube," very similar in design to the tubes just described, was proposed by Lieut. Fynmore, R.N., and adopted for flint-locks. This tube was the recognized equipment for all ordnance until superseded by one of the percussion principle several years later.

The next great improvement was the introduction of the percussion or detonating principle as applied to tubes. It rendered firing more rapid and accurate and effected a great economy in material and space by the abolition of port-fires, priming powder, etc., in the battery.

The first tubes of this description were invented by Mr. March of the Royal Arsenal Surgery, and consisted of a quill body with a side quill filled with detonating composition. The body was 2½ inches long and the side quill 1 inch long. The side quill having a priming of 0·2 inch of mealed powder, continuous with the powder in the main tube, was otherwise filled with a mixture of equal parts of chlorate of potash and sulphide of antimony. The tubes were varnished over completely with red sealing-wax dissolved in spirits of wine.

These tubes, known as "Rectangular percussion quill tubes," were approved for the Navy in 1831, and for the Army on 21st November, 1845; the Royal Artillery being supplied with them on 20th May, 1846.

With this tube the lock was fixed to the vent field out of the way of the explosion of the vent, an undoubted advantage, but the tube as originally designed was slow in its action. Improvements were suggested to remedy this defect, and eventually the idea of tying the detonating quill across the top of the main tube was evolved, and some of these " crutch-tied " tubes found their way into the Service. Finally, the " cross-headed detonating tube " recommended by the late Colonel Dansey, R.A., was approved on 9th September, 1846, by the Master-General of the Ordnance.

This improved tube consisted of a main quill about 2½ inches long, bored near the top to receive at right angles a cross-head of a pierced pigeon's quill or " snipe," the two being made fast by a twist of fine silk. The cross-head was filled with a mixture of potassium chlorate, sulphide of antimony and ground glass. The portion of the main quill above the " snipe " was filled with L.G. powder, and the open ends sealed with shellac putty. The body of the tube was varnished with black, and the head and snipe with red varnish.

The tubes were ignited by percussion, the blow being delivered by a hammer fixed on to the gun for that purpose and thrown over on to the tube by the pull of a lanyard.

This tube was, therefore, approved in 1846 for use in sea and land service with the exception of field pieces, with which common tubes and port-fires still continued to be employed.

Percussion caps to be fitted to tubes were proposed in 1844 by Major Jacob and experiments with them proved very successful.

The next advance was the introduction of the friction tube. A tube of this type was laid before the Royal Arsenal officials in 1841 by Lieut. Siemens, of the Hanoverian Army, but, owing to defects, was rejected. Some quill friction tubes were ordered to be made for experiment on 21st May, 1841, designed by Colonel Dansey. They were not, however, introduced ; and it was not until the year 1851 that Mr. Tozer of the Royal Laboratory succeeded in perfecting the copper friction tube which was adopted for the land service for all types of artillery on 24th June, 1853. It was not recommended for the Navy on the ground that any metal tube would be objectionable and highly dangerous between decks. However, on the recommendation of Colonel Boxer, a quill friction tube was introduced for naval service on 16th July, 1856.

The friction tube marked a further milestone on the road to knowledge, and its advantage over the detonating and common tube with port-fire was so obvious that it was finally adopted for universal service, the detonating tube being declared obsolete on 6th September, 1866.

With very slignt alterations the friction tube has remained the same to-day as it was 60 years ago.

The copper friction tube consists of a body of sheet copper about 3 inches long driven with mealed powder and pierced with a central hole. The top is closed with shellac putty and varnished paper, and the bottom by a disc of varnished paper. The nib-piece contains a copper friction bar roughened on both sides and smeared with a detonating composition of chlorate of potash, sulphur and sulphide of antimony. The nib-piece is pressed down on to the sides of the friction bar, and the projecting part of the latter has an eyepiece into which the hook of the lanyard fits.

The quill friction tube is of a very similar design, the body being made of a goose's quill. In the head is a little detonating composition through which a roughened friction bar, fitted with an eyelet, passes. A binding of fine copper wire serves to support the top when the tube is placed in the vent ; and in some types a loop attached to the head, passing over a friction tube pin in the gun, supports the tube against the pull of the lanyard on firing.

The Navy originally objected to friction tubes on the ground that the copper friction bars, which would litter the decks after firing, tended to cut the men's feet. It is interesting to note that the same objection was also raised against Major Jacob's percussion caps. The Navy, however, became reconciled when quill was substituted for copper in their case, although the objection to the friction bars must still have remained valid.

In 1852 and 1855 curious accidents occurred at Woolwich connected with the proof of guns, which revealed the danger lurking in the method of firing adopted in the butts at that period. In order to allow the gun detachment time to retire safely to cover during the actual firing of the gun under proof, a piece of port-fire, intended to burn for about three minutes, was placed over the priming in the vent and attached to the gun by means of clay. This procedure was clumsy and in the end proved extremely dangerous, and thus Mr. McKinlay, the proof master at that period, turned his attention to the firing of guns by galvanic agency. In 1856 he submitted his " galvanic tube " which was approved on 9th February, 1856.

This tube may be said to have been the first " electric tube " adopted in the British Service.

This does not, however, point to the first application of the principles of electricity for firing guns or initiating explosions. The first employment was in 1751 when Franklin carried out experiments on the ignition of gunpowder by this means. Priestley in 1767 followed in the same direction ; the electricity employed being " frictional " in both cases. In 1831 Moses Shaw of New York applied frictional electricity to mining purposes and succeeded in detaching large masses of rock with the aid of gunpowder and fulminate of silver. In 1842, 1843 and 1845 experiments were carried out by other investigators, who succeeded in igniting explosives over considerable distances by means of frictional electricity.

In 1853 the induction coil, invented by Ruhmkorff, turned the scientific mind of the day in the direction of carrying out blasting operations by means of the high-tension current.

The voltaic or galvanic cell as a means of exploding mines seems to have been first entertained by Mr. Hare about the year 1831. The idea became realized in 1838 when Mr. Roberts devised his electric fuze or cartridge, but the system does not seem to have been applied to the firing of ordnance till about 1853 in which year we hear of a gun mounted at Dover being fired by voltaic electricity from Calais.

Nothing definite was approved for the service, however, until the introduction of Mr. McKinlay's galvanic tube before mentioned.

This tube consisted of a quill body 2¾ inches long, driven and pierced in the usual manner. On the top of the body was secured by means of shellac varnish an almost hemispherical boxwood head ¾ inch in diameter. The head had a central hole in which the body fitted. Two holes on either side accommodated two copper bushes, to the ends of which was soldered a bridge of fine steel wire in the cup on the underside of the head. The cup was filled with L.G. powder. Wires from the cell, fitting into the two copper bushes, supplied the current which heated the bridge to redness, and fired the tube.

These tubes were used for proof and experimental work from 1856 to 1862, when they were, to a great extent, superseded by Abel's electric tubes. Galvanic tubes were formally pronounced obsolete in 1866.

Meanwhile experiments in high-tension electricity had been carried on in regard to mining and similar operations. In 1851 Messrs. Statham and Brunton invented " Statham's fuze," an apparatus depending for its action upon a spark passing through priming material, instead of an incandescent wire. This fuze, originally intended to be fired by low tension, was eventually adapted by M. Ruhmkorff and Colonel Verdu, of the Spanish Army, for high-tension currents.

Mr. Henley, at the Royal Arsenal, Woolwich, continued experimenting in the same direction. Even with his powerful instrument, however, he found that the ignition of gunpowder by this method was very uncertain, and that a more sensitive priming composition must be discovered. Finally, after long and careful researches by Messrs. Abel and Wheatstone, a priming composition, made up of copper sub-sulphide, copper sub-phosphide, and chlorate of potash, proved quite suitable ; and Sir Frederick Abel introduced his electric tubes, the first issue of which took place in 1862.

These tubes were officially introduced as " Tubes, electric, high-tension, Abel's Pattern I " in 57/24/4124, dated 27th January, 1866, in List of Changes, paragraph 1201.

At this point, both in regard to " tubes, friction," and " tubes, electric," the stage covered by List of Changes, Volume I, dated 1860, has been reached, and further details of tubes introduced since that date can be obtained by reference to that publication, if desired.

In 1860, therefore, there were seven types of tubes in the service, common tubes of quill or copper, friction tubes of quill or copper, the Dutch paper tube, the percussion tube, and the galvanic tube.

These have all been described with the exception of the Dutch paper tube. This was a common tube lighted by a port-fire, the case being constructed of rolled paper instead of copper or quill. Its nomenclature arose from the fact that the Dutch were the first to use paper for this purpose in the eighteenth century. At this period such tubes were not an article of Royal Laboratory manufacture, and were only made for a special service or during an emergency.

Abel's high-tension tube consists of a quill body of the usual length, driven and pierced in the orthodox manner. The head of beechwood received the body of the tube and carried two insulated copper wires attached to copper bushes. The bare ends of the fine wires are buried in the special priming composition before mentioned, the other ends being fastened to the copper bushes. Being designed for high-tension currents, the insulation was of fine gutta-percha and the spark gap in the priming composition was 1/16 inch across.

These tubes were used in the land service for proving guns and in the Navy for firing simultaneous broadsides. They were, at the time of their introduction, considered an important advance in design, and great advantages were claimed for this " magneto-electric " tube over its earlier " galvanic " prototype. They remained as a service store until finally declared obsolete on 29th January, 1892.

In 1878 the low-tension electric tube was introduced, the pendulum having again swung in the direction of direct-current agency, where it has remained up to the present, a fact no doubt accounted for by the improved methods of cell making and transmission of electrical energy in general. The tube, in its essentials, consisted of an ebonite head, two insulated copper wires, a platinum bridge, a priming of guncotton dust and mealed powder, and a quill body containing mealed powder. It was originally introduced for firing radial-vented R.M.L. guns, but latterly its use has been restricted to the destruction of unserviceable cordite : the latest mark became obsolete for land service in 1919.

The last great step in the evolution of the tube was the invention and introduction of a type that sealed the vent, called generically a " vent-sealing " tube. The adoption of this tube allowed greater accuracy in the practical application of ballistics, as a sealed vent allowed the amount of available powder gases in rear of the projectile to be calculated to a much greater nicety. This step was taken in the early 'eighties, when " Tubes V.S., mechanical and electric 6-inch, 80-pr. B.L." were introduced on 17th April, 1882.

V.S. tubes conferred three striking advantages :—

(1) Shooting became more accurate.
(2) The life of the vent was prolonged owing to the cessation of erosion due to the hot escaping powder gases.
(3) The tubes being retained in the guns after firing, the danger to gun crews from flying pieces of metal, so accentuated in the case of axial-vented guns, was reduced to a minimum.

Sealing was obtained by the pressure and heat of the explosion expanding the brass walls of the tube tightly into the tube chamber, thus preventing the escape of gas past the tube. A lock retained the fired tube, which was subsequently extracted by some suitable mechanism.

A few years later, Tubes, V.S., Electric V. and M., and Tubes, V.S., Friction V. and M. were adopted for service. They were somewhat similar in shape to the later P-tubes, and were made of brass. Both electric tubes were on low-tension principle and had two wires proceeding from the head; the friction tubes had a central friction bar or wire, which remained in the tube and acted as a seal against internal gas escape after firing. The V-pattern was used with the heavier R.M.L. guns, and the M. pattern with the older pattern B.L. guns, from 4-inch up to 12-inch respectively.

The next tube to make its appearance was the P tube, which was approved for use in 1884. This tube again marked a definite advance, as precautions were taken in its design to render it gas-proof both internally and externally. Outside escape of gas was prevented by the usual method of the expanding metal case, but internally sealing was effected by means of cones, which blew back and jammed on the shock of discharge.

This tube was designed in two types, the one electric with wires leading from the head, and the other percussion; the anvil in the latter's head firing a cap on being struck by the hammer or striker of the lock.

In 1894 the T-type of vent-sealing tube came into existence in the form of the "tube friction T." These were secured in the breech by a bayonet joint, the tube having a friction bar secured in the head. The bar was pulled by a lanyard at right angles to the vent, internal sealing being secured by means of a copper ball and coned seating in the interior of the tube. Electrical tubes of this shape having two wires proceeding axially from the head followed, and were in turn succeeded by "tubes friction push T" and "tubes percussion T"; the latter not being introduced until the middle of the Great War.

The last definite type of tube to be evolved was an electric P tube, without wires, called "Tube, V.S., electric wireless P." The original mark was introduced on 25th June, 1895, and was clearly an advantage in many ways over the wired tubes of former years. It increased the rate of fire, and, with the corresponding introduction of special locks, added to the general efficiency of guns on fixed mountings.

Variations have, of course, been introduced into the original P design, and certain tubes of a larger capacity have been made, but it portrays the high-water mark of electric tube construction, just as the "Tube, V.S., Percussion" stands out pre-eminently in percussion tube design.

These tubes are now almost universally used in coast defence, both in B.L. and in Q.F. guns, although an electric primer made to screw into the base of the cartridge case and similar in action to a wireless electric tube was developed as early as 1889.

In Q.F. field guns percussion primers are fixed in the brass cartridge cases, thus doing away with the necessity of a separate tube.

T tubes are still used for some guns and howitzers with bayonet joint vents, and other howitzers fitted with special locks employ blank S.A. cartridges known as "Tubes, percussion, S.A. Cartridge" for initiating the combustion of their charge.

Such innovations, however, are due almost entirely to the high price and difficulty of manufacture experienced during the Great War, and cannot be regarded as any specific progress in the realm of design, the highest attainment up to the present being the latest mark of the percussion and wireless type of P tube.

The terms "sealing" "wireless" and the letter "P" are now omitted from nomenclature of tubes.

APPENDIX V

PROPELLANTS

Gunpowder

The origin of gunpowder is lost in the mists of antiquity and the name of its inventor, together with that of the country in which he first saw light, have alike eluded the researches of the modern student.

This powder was certainly known to Berthold Schwartz and also to Roger Bacon, whose recipe, written in the thirteenth century, has been handed down to us. In any case it was the first propellant used with cannon and remained as such in splendid isolation for over five hundred years.

Gunpowder consists of an intimate mixture of saltpetre, sulphur and charcoal, the proportions of each having varied from time to time. In early centuries the three substances were mixed together in quite an arbitrary fashion, the amount of each constituent being entirely at the discretion of the individual maker concerned ; and it was not until comparatively recent times that any standardized specification for powder was adopted.

Roger Bacon's ideal mixture was 41·2 parts saltpetre, 29·4 parts charcoal and 29·4 parts sulphur. Another recipe written about 80 years later postulates 66·7 parts saltpetre, 22·2 parts charcoal, and 11·1 parts sulphur. During the centuries that followed, further changes in composition took place until finally in 1781 the present proportions of 75, 15, 10 were fixed.

The main drawback, originally experienced in gunpowder, was its great liability to absorb moisture owing to the hygroscopic nature of saltpetre. Powder, therefore, was often found to be moist and quite useless when required in an emergency ; and such a contingency was not one to be lightly dismissed in those days as it rendered an army's artillery incapable of inflicting the slightest damage on an enemy's troops.

Consequently, many references to unserviceable powder and to the means taken to prevent such a state of affairs occur in old records. In the English store accounts for the years 1372–1374 there are entries of payments for faggots for drying purposes, while in 1459 the Scotch were endeavouring to preserve their stocks by storing them in waxed canvas bags. Owing to the general humidity of the atmosphere at sea, the Navy were always the chief sufferers in this respect, and for this reason ships, after the middle of the 17th century, carried nothing but corned powder. This form was considered to be more immune from the effects of moisture than the fine variety formerly used. The trouble still existed, however, for we are told that in an action fought off Grenada in July, 1779, the English shot would not reach the French men-of-war, as the powder in the barrels had coalesced into large lumps in the midst of which were visible segregations of saltpetre.

The gunpowder of the Middle Ages, originally prepared in a very fine state of division known as serpentine, suffered from the following disadvantages :

(1) It was extremely sensitive to moisture.
(2) It had a tendency to separate out into its components during transit.
(3) It left a large residue after firing.
(4) It required very careful ramming.
(5) It burnt very slowly and over-ramming caused combustion without explosion.
(6) It always gave rise to quantities of fine explosive dust.

Several methods were introduced during the 15th and 16th centuries to overcome these defects, some of which have already been mentioned. For a long time the ingredients were carried separately and the gunpowder mixed locally as required, in order to prevent stratification and minimize the risk of premature explosions due to its production of dangerous dust; also the adoption of paper or linen cartridges was advocated to prevent to some extent the effects of excessive fouling.

All such expedients were only regarded as temporary make-shifts and corned powder gradually superseded serpentine. Corned powder was known in the 15th century and was employed for small arms in England before 1560. It had, however, two qualities operating against its universal use at this period; its cost was excessive and its power too great for the early models of cannon.

It was the old story of one department of human knowledge outstripping another. This grained form of powder did not, therefore, become general for artillery till the latter end of the 16th century, when the engineer had caught up with the chemist and designed a gun strong enough to withstand the higher pressures involved.

The following advantages were claimed for corned powder:—

(1) It was much less susceptible to damp, especially when glazed.
(2) It deposited less residue after firing.
(3) It did not resolve itself into different strata during transport.
(4) It required less careful ramming.
(5) Owing to the size of its grains and consequently greater surface and air-spacing, it was consumed so rapidly that there was little or no escape of gas through the vent. Consequently, weight for weight, it had 33⅓ per cent. more power.
(6) It produced far less dust.

The early troubles in connection with gunpowder centred themselves around saltpetre. Until the middle of the 19th century the only available method of procuration was by extraction from naturally occurring deposits. In Europe few localities existed in which this salt could accumulate to such an extent as to render its manufacture and purification profitable, except such places as underground cellars, caves and stables. The Orient, therefore, was the main source of supply of this commodity and its salesmen charged accordingly.

European Governments, soon realizing the rapacity of Eastern merchants, sought to conserve all possible sources of supply within their own dominions. In France, saltpetre commissioners were appointed in 1540 and confirmed in their appointment by edict in 1572. These officers had the right of entry into any stable, cellar, pigeon-loft, cattle-shed, or sheep-pen in order to gather their provender, and could undertake operations in any private house where efflorescence on the walls was suspected.

In England the same procedure for obtaining this valuable product existed, and in 1558 Queen Elizabeth granted a monopoly for gathering and working saltpetre to Richard Hills and George and John Evelyn for a period of 11 years. This monopoly included the whole of the South of England and the Midlands, except the City of London and 2 miles outside it; and in 1596 it was extended to include London and Westminster.

At the beginning of the 17th century, the East India Company commenced importing saltpetre into England. The Company erected powder mills in Surrey, and when its charter was renewed in 1693 it agreed to supply 500 tons of saltpetre to the Ordnance every year. From that date, therefore, the supply of saltpetre became more regular and the manufacture of English gunpowder was placed on a firmer footing.

The utility of gunpowder was quickly realized soon after the introduction of cannon, and traders were quick to grasp the lucrative nature of its commerce and manufacture. There was a powder mill at Augsburg in 1340, one at Spandau in 1344, and another at Liegnitz in 1348. At this early date it was often made in private dwelling-houses too.

Although in 1338 gunpowder is mentioned among the stores at the Tower, and in 1461 a "powder-house" was installed in that fortress, its manufacture did not take place in England until the reign of Queen Elizabeth. Before her accession supplies were imported from Europe, but owing to the threatening attitude of Spain at this time, the Government issued patents for the manufacture of gunpowder as a monopoly and the Evelyn brothers before mentioned appear to have been the first to have produced powder on any considerable scale.

Mills, therefore, began to spring up about the countryside. In 1555 there was one near Rotherhithe, and in 1561 others were found at Long Ditton, Leigh Place near Godstone, Faversham, and Waltham Abbey, the latter finally becoming a Royal Ordnance Factory for explosives.

The weight of the charge in proportion to that of the projectile has always varied from time to time. In fact the assessment of powder for gun charges appears to have been almost as arbitrary as the composition of the propellant itself. During the middle of the 15th century the charge was usually fixed at one-ninth the weight of the stone shot, but towards 1500 it was increased from that figure to one-fifth and even to one quarter.

Different natures of ordnance also had dissimilar charges for the same weight of shot, and the variability of the composition of gunpowder at this date makes it impossible to lay down any hard-and-fast rule. Even two centuries later, when artillery became a more precise science, we are still confronted with the same multifarious set of figures ranging from one-half up to one-thirty-second.

Finally, a complete set of firing charges for all classes of projectiles from all natures and calibres was laid down on 24th March, 1863; the weights selected following roughly the somewhat simple rule of thumb given below :—

(1) *For shot from guns.*
The mean service charge: about one-quarter of the weight of the projectile.

(2) *For shell from guns.*
The mean service charge: about one-sixth to one-twelfth the weight of the heaviest projectile.

(3) *For carronades.*
The mean service charge about one-twelfth the weight of the projectile.

(4) *For mortars.*
The charge varied with the range.

Up to the Crimean War there was little or no segregation of various classes of powder in this country, but the following classification was laid down on 9th December, 1858.

Class I.—New powder.
Class II.—Powder returned from the Royal Navy and found to be quite serviceable after proof and examination.
Class III.—Returned powder, which, although broken and dusty, is dry and serviceable as bursting charges.
Class IV.—Returned powder which can only be used for shell filling after restoring.
Class V.—Unserviceable powder only fit for extraction.

Previous to 1865 the following principal powders were used in the British service, the changes in the nomenclature shown being approved on 9th January, 1865 :—

(*a*) Mealed powder.
(*b*) Mealed pit powder.
(*c*) "Common F.G." afterwards renamed "F.G."

(*d*) " Medium rifle " powder, afterwards renamed " Shell F.G."
(*e*) " Started F.G.," afterwards renamed " Exercising F.G."
(*f*) " E.R." (Enfield Rifle) or " J2," afterwards renamed " R.F.G."
(*g*) " Common L.G.," afterwards renamed " L.G."
(*h*) " Common L.G. classes 3 and 4," afterwards renamed " Shell L.G."
(*i*) " Started L.G.," afterwards renamed " Exercising L.G."
(*j*) " A_4," afterwards renamed " R.L.G."

Charges for guns previous to the introduction of rifled ordnance were always made of large grain black powder. Rifled guns, however, fired a far heavier shell, about three times the weight of that of a smooth-bore of the same calibre, with the result that the pressures developed during firing become excessive and a slower burning powder became imperative. R.L.G. powder was therefore introduced on 11th April, 1860, for Armstrong guns under the name A_4. This powder had a higher density and a larger grain; and as it was considered an improvement on L.G. powder the latter was made obsolete for future manufacture on 21st February, 1866. The existing stocks were still to be used up with S.B. guns, but after these were exhausted R.L.G. was to become the service charge for these guns also.

The Armstrong guns fired a charge of one-quarter the weight of their shell, but, as the size of rifled guns increased, the problem of obtaining slower burning powders had to be seriously faced owing to the pressures developed by increased charges in forcing the shell to conform to the twist of the rifling. The main factors influencing the rate of burning are density, hardness, shape and size of grain, amount of glaze and moisture; and, not unnaturally, the easiest factor to adjust—namely, the size of the grain—was the first to be tried.

The R.L.G. powder answered quite well for guns of small calibre, but when R.M.L. guns of 7-inch and upwards were introduced it was found advisable to adopt a still slower powder. Experiments therefore were carried out with this object in view, and the gunpowder committee recommended the temporary adoption of pellet powder on 21st August, 1866. This powder was provisionally approved on 18th May, 1867, but never attained the dignity of a service store as further experiments by the Committee on Explosives, set up on 8th May, 1869, under the presidency of Colonel C. W. Younghusband, R.A., F.R.S., led to the approval of Pebble or " P " powder for all gun charges over 40 lb. and all battering charges on 22nd October, 1870. Theoretically there was little difference between pellet and pebble powder, but the latter appeared to be more uniform in its quality and gave equal velocities for lower pressures. " P " powder had the high density of 1·765 and was much larger in the grain than R.L.G.

On 1st July, 1870, the service powders were re-designated for the second time. Minor alterations have also taken place in regard to the classification list of 1858, and various powders differing in grain and density for special services have been introduced. R.F.G.^{2} was approved in 1873 for Martini-Henry rifles, and P^2, approved in 1876, superseded pebble powder for the largest R.M.L. guns such as the 12-inch as it was larger in grain than its predecessor.

It now became realized that the rate of burning could not be controlled indefinitely by increasing the size of the grain, and that the regularity of burning depended on other considerations besides size and density. Therefore, experiments led to the adoption of moulded or prism powder—so called from the shape the grain finally took after several experiments—for the heaviest R.M.L. guns and for B.L. guns of a somewhat smaller calibre. Prism^1 black, introduced in 1881, was moulded into regular hexagonal prisms of about 1 inch in height and approximately 1·4 inches in length of side. It was used for cartridges in the 12·5 and 17·72-inch R.M.L. guns, and some of the B.L. variety up to 8-inch. Prism^2, introduced in the same year, was made on an even larger scale, but it was eventually ordered to be used up at practice.

Black powders were soon superseded by brown or "cocoa" powder in B.L. guns. It earned its sobriquet from its appearance and was first made in Germany. This powder, known as "Prism brown" was introduced in 1884. It was made from slack burnt charcoal and succeeded very well as a slow-burning powder, but, being less sensitive to ignition than black powder, it required a small primer of the latter to start its combustion when formed into a cartridge. Besides the different type of charcoal employed, its composition varied from the standard as follows :—

Saltpetre	79
Charcoal	18
Sulphur	3

Finally E.X.E. was introduced in 1887 to supersede P^2, prism¹ black and prism² in all R.M.L. guns and all 6-inch B.L. guns except Mark V. It is of the same size as prism¹ black, but has an indented ring round the centre perforation on one face as a distinguishing mark. It is of a dark slate colour, due to the admixture of two charcoals in its composition.

E.X.E. and prism brown, therefore, remained in use as the propellants used for R.M.L. and B.L. guns until the introduction of cordite.

Smokeless Powders.

Guncotton is the basis of nearly all smokeless powders. This substance, which is obtained by nitrating cotton, was discovered by Schönbein, Professor of Chemistry at the University of Basle in 1846, during experiments he was making upon highly oxidized bodies. Pélouze had forestalled him to a certain extent by making an explosive in 1838 by the action of nitric acid on cotton, but he missed the important step of adding sulphuric acid as a dehydrator.

Schönbein kept his patent a secret and tried to dispose of it to various Governments. In the same year guncotton was also discovered independently by Professor Böttger of Frankfort-on-Maine, and the two scientists agreed to share whatever profits the invention might bring them. In 1846 Schönbein came to England, carried out successful demonstrations, and taking out a patent, made arrangements for a British firm to proceed with manufacture. In France also the preparation was carried out. In 1847, however, explosions took place in both countries, leading to such loss of life that manufacture was entirely suspended and not attempted again for another 16 years.

Guncotton was found in practice to have much too high a rate of burning to permit of its use as a propellant, and many explosions occurred in European countries as a result of trying to adapt it to such uses. In 1852 Baron Von Lenk, who had been interested in this subject for some years, did a good deal of pioneer work in this direction in Austria. Explosions, however, always resulted from his efforts and the Austrian authorities finally refused to countenance the idea.

Other countries then followed where others had led, and in 1863 Sir Frederick Abel began extensive researches in England and ultimately developed methods of manufacture whereby all the impurities in the nitrocellulose were removed, the preparation rendered far less dangerous, and the finished guncotton made safe and reliable.

This substance was used solely as a disruptive at this time, as its properties in this respect had been recognized since its inception. All attempts to use it as a propellant failed owing to its porosity, as under the high pressure developed in the bore of a gun the interstices were penetrated by the hot gases, with the result that the whole charge was ignited practically instantaneously and the gun destroyed.

In 1846 Sobrero, Professor of Chemistry at Turin, discovered nitroglycerine. Owing no doubt to its highly dangerous nature, it was never applied practically except in small quantities for medicinal purposes. In 1859–1861, however, Nobel experimented with it and regardless of accidents placed its manufacture on such a world-wide basis that by 1873 fifteen factories had been built for this purpose in Europe and America alone.

The first successful smokeless powder was made by Major Schultz of the Prussian Artillery in 1865. It consisted of a species of nitro-lignose impregnated with saltpetre or barium nitrate. Another highly successful smokeless propellant was E.C. powder manufactured by the Explosives Company at Stowmarket. This was introduced in 1882 and consisted of a mixture of nitro-cotton and nitrates of potassium or barium, in the form of grains hardened by being partially gelatinized in ether-alcohol.

These powders were, and still are, highly successful as charges for sporting gun cartridges, but their action was too rapid for rifled weapons. In 1885 the introduction of small calibre magazine rifles made the problem of securing a suitable smokeless propellant urgent. It became realized that to obtain the degree of slowness required the structure of the original cellulose had to be entirely destroyed by gelatinizing it. Vieille in France produced the first good smokeless rifle powder in 1886, and designated it Poudre B after General Boulanger. Ballistite, invented by Nobel, followed in 1888, and other powders began to appear on the Continent.

In England experiments were being actively pushed forward by Sir Frederick Abel, Sir James Dewar and Dr. Kellner, and in 1889 patents were taken out in the name of Abel and Dewar on behalf of the Government for cordite. It was introduced into the Service in 1891.

Cordite, Mark I, had the following composition :—

	Per cent.
Nitro-glycerine	58
Guncotton...	37
Mineral jelly	5

In addition to its smokeless nature and non-fouling properties cordite was very much more powerful weight for weight than gunpowder, owing to its greater heat content. At the same time, owing to its homogeneous character, its rate of burning was much slower and could be more easily controlled. The result was that, although a lower maximum pressure could be used than in the case of gunpowder, a high pressure could be maintained over a much longer period of time.

This had a beneficient effect on gun design. The weight of metal at the breech could be suitably reduced, and a longer and lighter weapon produced.

Cordite Mark I, however, had a very high temperature of combustion owing to its large proportion of nitro-glycerine. This caused excessive erosion and consequent wear of gun. Cordite M.D. or modified was therefore introduced in 1901 to remedy this state of affairs. It was a slower burning propellant and gave out less heat on combustion. Its composition was :—

	Per cent.
Nitro-glycerine	30
Guncotton...	65
Mineral jelly	5

A heavier charge of M.D. than of Mark I was required to produce the same ballistic results, but the life of the gun under the more moderate temperature conditions of discharge was greatly prolonged.

Owing to its method of manufacture, cordite can be made in almost any desired shape or size with very little extra trouble, the orifice of the final die deciding the ultimate form of the "cord." In this manner the most convenient shapes suitable for different types of cartridges can be studied with ease and produced if required. Every diverse form of section, giving rise to its own special form-factor, has a marked effect upon the internal ballistics of a gun, and the question is too deep to be touched upon here. Suffice it to say that round, tubular, oval, and chopped cords, strips, flakes, sea-weed and other forms have been tried with success at different times for special purposes and cordite is more polymorphous than any other smokeless powder.

During the Great War, other propellants, such as N.C.T., N.C.Z., R.D.B., etc., were introduced as emergency powders owing to the shortage of acetone, but such substitutes had not the keeping qualities of cordite. Cordite had stood the test for 30 years. It has survived two wars, has been stored in the worst climates, has been subjected to drastic treatment in various ways, and has never failed yet. For this reason it must be considered one of the most successful smokeless powders ever produced, and it still remains the standard propellant in our services.

Cartridges.

Cartridges were first of all introduced to diminish the pernicious effect of fouling. The original serpentine, especially if moist, left such an amount of residue in the chamber and first few inches of the bore after firing off a few rounds, that re-loading became a matter of some difficulty.

Cartridges made of linen or paper bags are therefore mentioned in 1560, but the probability is that they did not come into even partial use till the early part of the 17th century when they were employed for rapid firing. From this time onwards their use became more and more general.

Many materials have been tried in this connection; paper of all kinds, "double-paper," "cured-paper," and "paper-royal," together with parchment, bladders, canvas, linen, merino, wildbore, and bombazette. The two substances most commonly employed, however, until the introduction of flannel, were paper and parchment.

All cotton and linen stuffs had the great disadvantage of never being completely consumed on discharge, a fact which necessitated the use of a wad-hook and rendered the service of the guns dangerous and slow. Parchment, also, had its drawback; it tended to shrivel up under the heat of combustion and choke the vent.

In 1778 Sir Charles Douglas suggested the use of flannel or serge to obviate these difficulties. His plan, however, was not immediately adopted and paper cartridges still continued to be used. In the meantime the objection to paper was partially overcome by placing flannel bottoms on the cartridges, a device occasionally carried out as in the case of H.M.S. *Iron Duke*, where the ammunition was thus fitted out at Sir Charles Douglas's own expense.

Flannel for the use of field and naval services was finally introduced about the beginning of last century, and by 1828 paper in this respect had almost entirely disappeared.

Only the best flannel or serge was used in cartridge making, and two advantages over paper were claimed for it:—

(1) It was more completely consumed on discharge.
(2) It was stronger.

For the old S.B. cannon, cartridges were made in two shapes—

(1) Cylindrical—for all except Gomer-chambered ordnance.
(2) Conical—for all Gomer-chambered ordnance.

The next fabric to be tried was silk-cloth which was introduced for land service blank cartridges on 20th March, 1868. Its qualities were so good that it has remained as the standard material until to-day. S.B. cartridges at this time were choked and hooped with worsted.

Cartridges for R.B.L. guns were naturally cylindrical and were formed of two pieces of serge, one rectangular for the body and the other circular for the bottom; they also had lubricators and were choked with twine. All cartridges manufactured since 25th August, 1862, were hooped with blue worsted braid.

As guns increased in size it was found that flannel was not strong enough to hold the heavier charges, and, therefore, silk cloth was introduced for all Woolwich guns on 8th March, 1875, the case being hooped with silk braid and choked with silk twist. Red shalloon was afterwards adopted for smaller natures of cartridges, and these two materials together with brass in the case of Q.F. guns, are the ones in use to-day.

Since the introduction of cordite, a primer of powder has always been placed at the base of the cartridge to facilitate ignition, and the rigid outline of the former has no doubt helped to strengthen the cartridge and make its construction easier.

What may be the future development of cartridges and propellants, none can tell; but unless the present fundamental principles are changed, any improvements would appear to be in the nature of differences in degree rather than in kind.

Appendix VI

PROJECTILES

The first cannon in Europe is said to have been designed in 1313 by Berthold Schwartz, a German monk, and therefore the projectile, with reference to the gun apart from the " ballista," could not have existed before the 14th century.

In the beginning, it was only natural that men should conceive of loading their cannon with missiles similar to those they were accustomed to discharge from their cross-bows. We therefore find that the original " projectiles " consisted of bolts and iron darts feathered with brass.

The dart, weighing about 7 oz., was wrapped in leather to ensure a tight fit being made in the bore, as even the gunner of those far-off days had some conception of the disadvantages arising out of excessive windage.

Inaccurate and feeble as these feathered barbs must have been in their effects, their use apparently lingered on for 250 years, for Sir Francis Drake, in a return to the Government dated 30th March, 1588, mentions arrows and muskets amongst the stores of his ship.

Shot

Shot, if we may apply the term to the round crude stores of antiquity, had been used as a weapon of offence for centuries, and it was again only natural that such projectiles should be tried in the primitive gun. The earliest type was fashioned of stone and was under experiment in France in 1346. Stone shot were also employed in Italy in 1364 and formed part of the equipment for the armament of Brest Castle in 1378. From this date they continued to be used almost universally until the 17th century, although balls of iron and lead had begun to supersede them in isolated cases long before this period.

The earliest mention of iron shot in England occurs about 1350, and other records show them to have existed in Europe during the latter part of the 14th century. Bronze, as a material for shot, appears about the same time, and in 1346 this country sent out leaden shot for the defence of Calais.

Although metal projectiles were known in the 14th and 15th centuries their use was by no means universal, as the expense of casting alone far exceeded the cost of stone ; and size for size, owing to its greater density and weight, a shot of this description required a heavier charge of powder to throw it the same distance.

Range, not striking energy, was the criterion of successful shooting in bygone days, and therefore the iron or leaden ball required more propellant to produce the same spectacular effect. The increased weight of powder undoubtedly tended seriously to injure the somewhat feeble and badly constructed cannon of the period, and thus caused stone shot to remain the favoured projectile until such time as the gun became really capable of withstanding greater chamber pressures.

Round shot improved but little in the course of centuries, except perhaps in regard to methods of casting, but a new departure was introduced in 1579 by Stephen Bathory, King of Poland, who formulated the idea of firing red-hot shot.

Balls of clay, heated to redness, and flung from slings had of course been used in Cæsar's campaign in Gaul, but the idea of adopting this principle to gun ammunition was not apparently evolved till the 16th century. The full value of this method of attack was not, however, appreciated till the siege of Gibraltar in 1779–1783, probably owing to the difficulty and danger of loading an incandescent body into a powder-filled gun ; an operation which remained impracticable till the thick wet wad had been devised.

Cast-iron spherical shot were made in 12 different sizes for the British service, and even as late as the middle of the 19th century these were occasionally heated to redness for a specific purpose such as the destruction of shipping or stores ; such a practice, however, ceased entirely upon the introduction of Martin's shell in 1857.

These solid iron shot were used against infantry in massed formation, masonry, and wooden shipping.

On the introduction of ironclads, however, a projectile of greater penetrative power had to be designed and solid steel shot were manufactured, the first issue being made to H.M.S. *Hector* on 6th February, 1865. Only the larger type of gun, namely the 68-pr., 100-pr., and 150-pr. fired these projectiles introduced for the purpose of piercing ship's plates.

As the quality of the steel had to be very carefully watched during manufacture to secure this result, their production was expensive and, not proving an unqualified success, they were declared obsolete for future manufacture on 24th October, 1866. Steel shot were replaced in 1867 by chilled iron spherical shot, and this material was considered more than a satisfactory substitute owing to the success which had attended its use in the manufacture and performance of elongated projectiles.

With guns of position and with those cast in bronze round cast-iron shot were riveted to wooden bottoms, and were issued " prepared " or " unprepared for bottoms " or " riveted " according to their eventual method of use.

Steel and chilled-iron shot were never attached to wooden bottoms but were always issued loose.

Round shot continued to be used as long as the S.B. gun remained a service weapon.

A new development occurred after the Crimean War. It was an event of such far-reaching importance in the science of artillery that henceforward ideas flowed along other channels, and the future course of gunnery and ballistics became fundamentally changed.

This innovation was the introduction of the breech-loading gun and elongated projectile ; the Armstrong breech-loading system being adopted in 1858.

Major Palliser's shot, approved on 21st October, 1867, effected an improvement over the ordinary elongated solid shot. It was adopted for the larger types of R.M.L. ordnance rifled on the Woolwich principle, and remained a good many years in the service. It naturally underwent changes consequent upon the various improvements in gun design, but in itself it remained essentially the same. It finally became obsolescent on 11th October, 1909.

The last of the line of this type of projectile was the A.P. shot, designed for the penetration of armour. It was introduced on 28th June, 1887, being considered more menacing than shell against the ship of that day. The latest mark used in land service equipments disappeared at the end of 1921.

With its passing, the curtain is rung down on a phase of projectile evolution which had held the stage for over 500 years.

Case Shot

The value of being able to throw a hail of small projectiles at bowmen, pikemen, and musketeers in massed formation was realized in very early times, and the origin of " case " was probably coeval with that of shot.

Originally, this effect was produced in two ways, either by mounting a number of small bombards on one carriage, or by filling an envelope or canister with small missiles and discharging it from a single piece.

The first method may be considered the prototype of the Gatling and the machine gun, whilst the second was the forerunner of modern " case."

At its inception this canister shot consisted of old iron, nails, bolts, flints and gravel, well calculated to injure the person. This medley, known as " langridge," was loaded into the cannon and, no doubt, proved effective at short ranges. Later the idea of placing this " scrap " in a case or covering was conceived, and we are told that " case shot " proper was first used in the defence of Constantinople about the middle of the 15th century.

A projectile known as " hail shot," and marking a further stage along the road to efficiency, was used about a century later. It consisted of a leaden cylindrical box filled with a species of iron " hail," the composition of which may be imagined, and a charge of powder. The fuze, a simple match, was attached to one end of the cylinder in the axis of the shell, and, in loading, this end was placed towards the rear, so that the combustion of the charge could ignite the fuze.

At first sight this projectile appears to belong more to the early history of " shrapnel " than " case " ; and, in fact, the Germans did claim that a countryman of theirs forestalled General Shrapnel's invention. This German shell, invented by Master Gunner Zimmermann in 1573, was somewhat similar to " hail shot," and had an effective range of some five or six hundred yards. Its bursting charge, however, filled half the leaden jacket, thus clearly divorcing it from Shrapnel's principle, in which the bursting charge is a minimum compatible with opening the envelope.

The developments in " case " consisted in standardizing the size and weight of the bullets and in improving the cylinders. All types of ordnance fired these shot, but the light nature was abolished on 25th November, 1830, ordinary case having 41 bullets to the pound being approved for future use in the field.

Before 1860 all case shot were fitted with wooden bottoms. Owing, however, to difficulties of supply, Colonel Boxer proposed on 26th May, 1859, that tin or iron plate should be substituted. This change was approved on 27th March, 1861, for all case shot used in the garrison or naval service, wooden bottoms alone being retained for those fired from bronze guns.

Iron case shot for the 8-inch, 10-inch and 100-pr. were not approved till 25th January, 1866.

In 1866, therefore, case shot was divisible into three main classes :—

(1) Tin case with wooden bottoms (original type) ;
(2) Tin case with iron bottoms (approved 27th March, 1861) ;
(3) Iron case (approved 25th January, 1866) ;

all of which consisted essentially of cylindrical containers filled with sand-shot of varying size and sawdust or shavings. Certain types were fitted with handles.

With S.B. guns the envelope needed only two characteristics :—

(1) Sufficient weakness to break up at the muzzle on firing.
(2) Sufficient strength to resist the strain of transport.

On the introduction of rifled guns, however, another consideration became important ; the envelope must have sufficient strength to withstand the possibility of setting up on discharge. Such a contingency would cause it to take the rifling, and were that to happen, the dispersion of the bullets laterally would be very great and their effective range to the front very small.

In the improved types clay and sand replace sawdust and shavings, and the envelopes themselves are made of tougher material in order to stand up to the increased chamber pressures. Certain patterns, such as the later marks for the 12-pr. and 15-pr., were fitted with a copper band to prevent the escape of gas towards the muzzle.

With the exception of such slight modifications, however, case shot has not materially altered in design during the last 60 years.

The use of case shot has also remained unchanged by the passage of time. It is principally directed against troops in mass, men in narrow defiles, and in the days of sailing ships against rigging. In more recent years R.M.L. case shot of large calibre, containing chilled iron balls averaging 1 lb. each in weight, have been employed for the defence of narrow channels against attack by torpedo-craft; such a use has now, however, become obsolete.

In the field, also, case shot has been given its death-blow by the use of shrapnel fired with fuze set 0, a practice which is handier, more efficient, and more economical in space and transport.

Case shot, therefore, as a current projectile only survives in the smaller natures on fixed mountings, such as are used in the armament of tanks, where a Mark V for the 6-pr. has been sealed as recently as 1919.

Grape Shot

The early history of grape shot is in the main contemporary with that of case, for, like case, it may be considered as an improved form of "langridge." Mention is made as early as the 14th century of "sacks filled with stones" fired from cannon, which, like "canvas cartridges containing small balls," were nothing more or less than the primary forerunner of the quilted grape shot.

Bar and chain shot belong to grape rather than to case shot, and such methods of attack and defence were used soon after the introduction of artillery.

The grape shot of a hundred years ago was of the "quilted" type and consisted of an iron plate, from the centre of which passed up an iron spindle, round which were piled sand shot enclosed in a canvas bag, which was drawn together between the balls or "quilted" by a strong line. It was owing to its rough likeness to a bunch of grapes that such a projectile received its name.

In 1853 a new pattern, proposed by Mr. Caffin, of the Royal Laboratory, was approved on 2nd September. It was not, however, manufactured till 1856.

Caffin's grape shot consisted of four horizontal iron plates, the lowest one wrought, the others cast, through which passed a wrought-iron spindle bolt-headed at the bottom and screw-threaded to receive a nut on top. Between the plates were three tiers of sand shot.

Grape was made in 10 different sizes, from that of a 6-pr. up to a 10-inch. After 1856 all grape, except that of the 10-inch gun and carronades which were enclosed in a cylinder-like case, was of Caffin's design; quilted grape was still found in the service after this date, though it became obsolete for future manufacture.

The advantages claimed for Caffin's pattern were :—

(1) Less perishable.
(2) More portable.
(3) Greater destructive power.
(4) Interchangeability of parts.

Grape was used for the same purpose as case, but its larger and heavier shot made it effective up to 600 yards, a longer range than the latter, and gave it greater powers of destruction against rigging.

It became obsolete for the Navy on 20th February, 1866, and its manufacture was discontinued for the land service after 20th September, 1868, when all quilted and carronade grape-shot were ordered to be broken up on return to Woolwich. An exception was made for local issue when such patterns might still be retained if serviceable. Caffin's pattern for all natures of ordnance and the special type for the 10-inch gun were still, however, to be issued until the stock became exhausted.

Grape shot, therefore, just passed out of existence with the S.B. guns from which it was fired.

Common and A.P. Shell

The idea of a projectile, capable of breaking up and thus enhancing its powers of destruction, had been present in the human mind from very early days, for it was obvious what advantages such a missile would confer upon its possessors. History, therefore, teems with allusions to vessels containing igneous compositions, known as " fire-pots," " thunder of the earth," etc., which broke up during flight.

Before the introduction of gunpowder, however, such missiles must have been " incendiary " rather than " explosive " in their nature. In either case, it appears certain that they were thrown by hand and sling, or discharged from catapults before they were adopted as accessories for cannon.

Shell were undoubtedly fired on occasion and their ideal value was recognized, but shot were still found to be more efficient. The reason is not far to seek. Man's imagination had outrun his practical ability to manufacture, and the want of a sound fuze prevented the supersession of shot by shell in the Middle Ages.

The only method of igniting the early fuze was either by placing the shell in the bore, fuze towards the charge, so that the combustion of the latter fired the former, or by loading the projectile fuze towards the muzzle, when it was lighted by a match or port-fire thrust down the bore.

The result in either case was that the injury intended to be meted out to the enemy usually fell on the unfortunate gunners by the bursting of their weapon. If history could tell us, it would be interesting to learn how often the firing of a shell entailed the destruction of the early gun.

All explosive and incendiary shells were originally termed grenades, the name being derived from the pomegranate owing to the similarity of its seeds to grains of powder. The word " shell " was a later derivation from the German " Schale " meaning the " outside rind or bark."

The use of grenades was first extended to mortars, to which their application was long restricted. These projectiles were either round or oblong in shape, the former retaining the name of grenade whilst the latter received the appellation of " bomb." This word, in common with " bombard " and " bombardier " came from the Greek on account of the great noise experienced in firing the primitive cannon.

The Venetians are reported to have fired mortar shells at Jadra in 1376, and shells fitted with fuzes were used in 1421 at the siege of St. Boniface in Corsica. Such primitive projectiles consisted of two hollow hemispheres of stone or bronze joined together by means of a hoop of iron, a hinge, and keys, the fuze being a sheet-iron tube, filled with priming, riveted to one of the hemispheres.

Other military writers speak of shells being used at various stages during the 16th century, and in 1543 we have evidence of mortars, and shells filled with " wild-fire," being manufactured in England. It was not, however, till the 17th century that their use became general in Europe.

About 1700 shells began to be employed for horizontal fire, being projected from howitzers loaded with a small charge, and in 1779 experiments were carried out which demonstrated that they could be fired from guns with heavy charges. This combination of their special properties with the advantages of increased range was recognized as a valuable asset, and shells were therefore introduced for guns about the end of the 18th century.

The first recorded employment of bombs at sea was by the French on 28th October, 1681, at the bombardment of Algiers.

Materials for shell differed in these early days. Iron was the most popular, but bronze, lead, brass and even glass were tried on occasions.

During the 19th century common shell consisted of hollow spheres of iron cast with fuze holes and filled with gunpowder, and any subsequent alterations were chiefly improvements in the fuze rather than in the shell. About 1860 three types existed in the service—common, naval and mortar—

differing only in minor details, with a thickness of metal about one-sixth of their diameter and a weight equal to two-thirds of that of solid shot of the same calibre.

In order to ensure shells being loaded fuze towards the muzzle they were attached to wooden bottoms. These " sabots " were also intended to reduce jamming during loading, and rebound effect from the sides of the bore on firing. The necessity for such discs was recognized by a committee of artillery officers in 1819, and on 25th November, 1830, all wooden bottoms were henceforth ordered to be made of a uniform thickness of half an inch.

Elongated projectiles became a necessity on the introduction of rifled guns, and shells coated with lead to take the rifling were approved for the Armstrong equipments. This innovation of the rotating projectile, whereby accuracy and range were enormously increased, marked a great advance in the artilleryman's attainments.

Trouble with the breech mechanism and obturation, however, led to the temporary abandonment of the B.L. principle, and a reversion to the muzzle-loading type, known as the R.M.L. gun, took place. Here the lead-coated projectile proved unsuitable, and one with studs to take the rifling was devised in its stead. Studs by themselves, however, were found to be unpractical; not only did they weaken the walls of the shell, but they allowed such an amount of windage as would soon render the gun useless.

Some means, therefore, had to be devised to overcome this rush of gas with its consequent pernicious erosion and loss of accuracy. The earliest adopted was the papier maché wad cupped to take the base of the projectile and placed between the charge and the shell. These were immediately superseded by " Bolton Wads," introduced in January, 1869, which were made of pulp prepared from 75 per cent. of old rags, known as " tammies " or " woollens," and 25 per cent. of old tarred rope. They were formed in a mould and varnished when dry. Proving entirely useless, however, they were withdrawn from the service on 22nd August, 1872.

Meanwhile, a committee had been appointed to investigate the question of adopting means for the prevention of scoring in the bores of heavy R.M.L. guns. It reported in July, 1872, and finally in January, 1877, with the result that the copper gas-check was introduced in August, 1878. These gas-checks were attached to the base of studded projectiles, and in 1879 a " ribbed " or " rotating " gas-check, capable of fitting the grooves and so eliminating the studs, was tried.

These ribbed gas-checks were found to confer advantages with regard to range and accuracy which the studded shell did not possess, and in 1881 the automatic gas-check was adopted, which not only engaged the rifling, but also had the property of being loaded independently.

On firing, the gas-check gripped the base of the shell, became firmly attached, and made the latter conform to the twist of the rifling.

The introduction of the B.L. guns led to the development of the Vavasseur copper driving band, which is essentially the same as the modern band of to-day.

The foregoing details of lead-coated projectiles, gas-checks, and driving bands are naturally applicable to all forms of projectile, with the exception of " case " and " grape."

Common shell were either nose-fuzed, or pointed and fuzed at the base, from the latter, modern armour-piercing shell may be said to have developed.

The first pointed shell for the penetration of armour was introduced by Major Palliser in 1863. It was made of chilled cast-iron with an ogival head of 1½ calibres radius, and possessed the following properties :—

(1) Intense hardness.
(2) Crushing strength.
(3) Brittleness.
(4) Increased density.

It was manufactured by pouring the molten metal into an iron mould, the result being the production of a hard white iron. Later, a sand mould was employed for casting the walls in order to ensure that the head alone became chilled.

These shells were very similar in shape to modern A.P. shell. The head was solid and the walls thick, a comparatively small cavity being left for the bursting charge of powder. Their use was most successful against the wrought-iron armour of the period.

During 1880–1890, steel shell and armour began to make their appearance, and although at first such projectiles were out-matched by hard steel armour, it was recognized that steel had great advantages over iron as a shell body. It developed the force of explosion, and was better able to withstand the increased shocks of discharge occasioned by the growing use of comparatively high velocity B.L. guns.

The first steel shell were cast, a process later followed by forging, a method of treatment which still further toughened the material.

Gunpowder was used as a filling for all explosive shell until January, 1896, when lyddite was adopted. After that date, common shell were either filled powder or H.E. Owing, however, to the danger then considered to exist in filling base-fuzed shell with anything except powder, all H.E. projectiles were nose-fuzed till 1913.

Steel A.P. shell were introduced into the service in 1900. They were cast or forged with hardened heads and pointed tips. They differed from ordinary common pointed shell in having a much greater thickness of metal in the head.

Shortly after the appearance of A.P. shell mild steel caps were introduced. They were fitted over the nose of the shell and were found to support the hardened steel point at the moment of striking ; they thus led to increased penetration, especially with high striking velocities at normal impact. The caps, at first cylindrical, were afterwards streamlined to diminish the air resistance.

After 1913, A.P. shell were given H.E. fillings, and disruptives such as lyddite, trotyl, amatol, etc., gradually took the place of gunpowder.

The latest type of projectile intended for the attack of armour is capped and known as the A.P.C. shell, although the present cap is no longer soft.

Common shell filled H.E. have now been included in field artillery equipments owing to the lessons taught by the Great War.

With the passing of years, the original shape of head has been modified, and shells of 4, 6, and even 10 calibres radius have now been designed.

These long-pointed heads considerably improved the range, but in many cases diminished the accuracy of shooting owing to their increased transverse moments of inertia. False or ballistic caps, on the other hand, were found to increase the range without decreasing the accuracy, but they introduced difficulties with regard to fuzing, etc.

Incendiary Shell

Incendiary missiles were originally employed for two purposes, arson and illumination. For the latter use only star-shell remain, all other methods now in existence being pyrotechnical and thus outside the scope of this appendix.

The practice of hurling burning compositions against an enemy dates from remote ages and is probably contemporaneous with the dawn of history itself. Such incendiary compositions were certainly known to the inhabitants of Asia in very early times, as they are depicted on certain bas-reliefs in the British Museum.

Probably the earliest form to be employed was scalding water poured down from ramparts on to the heads of foremost besiegers. The substitution of oil for water marked a further step in this unpleasant practice which culminated in molten lead, the use of which survived into comparatively recent times.

Mention has already been made of red-hot balls of clay and iron which were used previous to the Christian era.

In later centuries the Greeks appear to have specialized in incendiary compositions which probably consisted of a mixture of sulphur, tar, gum, bitumen, naphtha, resin, or of some similar inflammable compound.

They were employed as early as the 5th century B.C. with great effect, and receptacles full of burning matter was thrown by hand and flung by machines at the siege of Syracuse and Rhodes in 413 and 304 B.C.

This type of mixture had continued in use for over 1,000 years, when suddenly in A.D. 673 the Mediterranean world was startled by the appearance of " Greek " or " sea " fire invented by Kallenikos. This fire was either projected from syphons placed in the prows of ships or thrown in pots and phials, and apparently inspired the warriors of the day with terror. It was first employed with great success at the siege of Constantinople in A.D. 674–6, and as it was used by the troops under Jaroslaus in 1043 and also in the battle of Rhodes in 1103, " sea-fire " must have held the field for nearly five centuries. The secret of its composition was well guarded and has remained a mystery to this day, notwithstanding the shrewd hypotheses as to its ingredients which have been put forward from time to time.

The discovery of gunpowder in the 13th century at once widened the field of research, and increasing numbers of compositions were tried. The greater proportion of these, however, were hurled from machines, as the practical difficulties of firing an igneous projectile from a gun were enormous in those early days.

One of the first incendiary shells was invented by Valturio in 1460. It consisted of two bronze hemispheres fastened together by bands containing a small hole to admit the flame to a quill of incendiary matter placed in the compressed gunpowder charge. Another shell of about the same period was similarly constructed of two hemispheres of iron containing pitch and resin.

From that time onwards there seems to be little progression until the carcass appears on the scene. This projectile was invented by a gunner in the service of Christopher Van Galen, the fighting Prince Bishop of Munster in 1672. The name apparently arose from the binding of the original " fire-balls " with iron hoops covered with canvas and quilting cords, a fact necessitated by the growing charges of the gradually improving cannon. Primarily oblong to contain a maximum of incendiary matter, their flight proved to be so erratic as to compel the adoption of a spherical form. Gradually the iron skeleton and canvas covering gave place to a thick spherical shell, vented in order to allow the egress of flame when the composition became ignited. The thickness of the walls was then reduced in order to increase the internal capacity of the shell. This reduction, however, was carried to excess and many carcasses broke up in the bore. To remedy this defect, the interval between carcass and charge was filled with turf during the siege of Quebec in 1759.

The first spherical iron carcass had no stipulated number of vents, four, five and even one or two being sometimes found. By 1828, however, all carcasses were being made with four vents. The Crimean War, nearly 30 years later, showed this number to be a failure, and experiments were carried out in 1855 to determine the happy medium. A three-vented carcass emerged as the result.

The old-fashioned oblong carcass disappeared about the time of the battle of Waterloo, but its spirit lingered on in the " ground light ball " which survived till the year 1883. The modern 3-vented carcass was not, however, sealed till July 9th, 1860.

Until 1854 a primitive kind of star shell was in vogue for incendiary purposes. It consisted of a carcass, filled with a number of stars of " Valenciennes Composition," *i.e.*, a mixture of saltpetre, sulphur, resin, antimony and linseed oil, which on bursting scattered around its blazing contents. The stars, however, had an unhappy knack of exploding and vitiating the desired result. This type of carcass was officially abolished in 1863.

The modern carcass, which was fired from all natures and calibres of ordnance, from the 12-pr. upwards, except the 100-pr., consisted of a hollow spherical iron shell pierced with three vents. As the thickness of metal was slightly greater than that of the corresponding common shell, they were consequently heavier. They were filled with a composition consisting of saltpetre, sulphur, resin, sulphide of antimony, turpentine, and tallow put in hot, three holes being made in the filling in prolongation of the vents. The holes were driven with fuze composition and matched with quick-match to ensure ignition. The vents were plugged with brown paper and further secured by kit plasters. Before firing, the plasters and plugs had to be removed and the priming exposed.

Carcasses burnt with a violent flame and were difficult to extinguish. Sometimes they exploded. Their great drawback, however, was their rapid deterioration in store, and it was due to this fact that they ceased to be included in equipments some 50 years ago, although they still continued to be made and issued for special purposes as long as the S.B. gun remained in the current armament of the nation.

The next incendiary shell to occupy public attention was Martin's. This shell, filled with molten iron, was proposed by Mr. Martin, a civilian, in March, 1855. Experiments followed in April, 1856, and the projectile was finally approved for service on 29th October, 1857, the 8-inch variety being introduced by order of Lord Panmure. The latest pattern was inaugurated on 10th February, 1860, in which year on 30th May the 10-inch shell was also approved.

These two sizes were the only ones ever made.

The shell consisted of a spherical cast-iron hollow body coated with loam on its interior surface, and filled before loading with molten iron poured through a filling hole. It had a thickening of metal at the base to withstand the shock of discharge and a corresponding thickness at the head with a flattened internal surface to cool the top layer of red-hot metal below its melting-point. By this method the shell, when filled, sealed itself by forming a solid plug in the filling-hole.

The walls were cast thin at the sides in order to facilitate the break-up of the shell on impact, and the consequent dispersal of its molten contents.

The coating of loam and cowhair acted as a non-conducting medium. It prevented the shell body from becoming unduly heated and helped to keep the filling longer in a semi-molten condition.

Martin's shells were used principally against wooden ships, but also in a minor degree against buildings and other combustible targets.

The Ordnance Select Committee recommended the adoption of these shells for four reasons :—

(1) They were easily filled.
(2) They were easier to handle than red-hot shot.
(3) Their use was less likely to burst the gun than red-hot shot.
(4) Their incendiary power was greater than that of red-hot shot.

Martin's shell were declared obsolete in 1869.

Between the disappearance of Martin's shell and the beginning of the 20th century there was a long hiatus during which no shells of an incendiary nature were contemplated. In 1911 one designed by Dr. Hodgkinson was approved, but it remained for the Great War to re-initiate this type of projectile. The latest shell are filled with thermit, carcass composition, and other incendiary mixtures, and are employed for destroying any objective capable of being smitten by fire.

Turning now to the alternative rôle of burning compositions, that of illumination as apart from arson, the " fire-ball " emerges as the principal type of illuminating missile. These were used throughout the centuries, and as it is only comparatively recently that any real distinction can be drawn between the two missiles, the history of illumination is synonymous with that of incendiarism, and the terms "fireball" and "light-ball" were interchangeable.

Balls of tow steeped in incendiary composition, stone and iron shot smeared with composition, and paper shells filled with composition were used indiscriminately, but the plan usually adopted for making a flare was to place the combustible matter into a stout canvas bag, strengthened with wooding and often reinforced with saucer-shaped iron tops and bottoms.

This latter type flourished during the 17th century and was almost identical with the later "light-ball" of the 19th century, with the exception of the iron ribs by which the ends of the latter were connected and the projectile strengthened.

At the beginning of the last century there were five ground light balls in existence, *i.e.*, 13, 10, 8, 5½ and 4⅖ inch, but the largest appears to have become obsolete about the time of the Napoleonic Wars. The others were still service equipment in 1867, although their use was, to a great extent, superseded by the parachute light ball, a superior illuminant in every way.

The ground light ball consisted of a wrought-iron frame partially covered with canvas, filled with a composition of saltpetre, sulphur, resin, and linseed oil put in hot. Holes were then made in it driven with fuze composition and matched in a similar manner to the carcass.

Their time of burning varied from about 9 to 16 minutes and their weight was about ½ to ⅗ that of common shell of the same calibre. They were fired with very reduced charges and were exceedingly difficult to extinguish, water having no effect.

Ground light balls became obsolete on 30th August, 1883.

Parachute light balls were introduced by Colonel Boxer in 1850 and patterns were sealed on 4th September of that year, although none were manufactured for the service until five years later. These lights, fired from a gun, were an extension in principle of the rocket parachute light invented by Sir William Congreve at the beginning of last century.

Experiments with a view to perfecting the design continued, and in 1863 new patterns were submitted. Finally the latest form was approved on 2nd January, 1866, which was again slightly amended five months later on 21st June.

Parachute light balls, made in three sizes, viz., 10-inch, 8-inch and 5½-inch, consisted of two outer and two inner tinned iron hemispheres, the two outer were lightly riveted together, the two upper connected by a chain. The inner upper hemisphere had a depression at the top to admit the bursting charge and fuze.

A quick-match leader conducted the flash from the bursting charge to the fuze composition in the lower inner hemisphere, the inner upper hemisphere containing the parachute tightly folded up. To ensure the latter opening, a cord was passed between its folds and through a hole in the top of the parachute, and was fastened to the upper inner hemisphere, so that when the hemisphere was blown away the cord was pulled through and the parachute expanded. The lower inner hemisphere contained the composition of saltpetre, sulphur and red orpiment. A hole was bored and driven with fuze composition and matched in the ordinary manner; this hemisphere was connected with the parachute by cords and chains.

The bursting charge was issued in the parachute, and, on firing, the outer hemispheres were blown away, liberating the burning light.

About 1870, instructions, similar to those for carcasses, were issued as to the manufacture and use of parachute lights on account of their brief life.

With the introduction of the rifled gun, a decided improvement in the form of the first star shell made its appearance on 29th May, 1873, when a pattern was approved for the 7-pr. R.M.L. gun.

This type consisted of a thin iron shell having a chamber in the base to take the bursting charge, a wrought-iron disc, and 13 stars of paper filled with barium nitrate, chlorate of potash, magnesium powder and boiled oil. The shell was closed with a wooden head having a gun-metal socket to accommodate the fuze.

Star shell, though improved in many ways, remain essentially the same as when they were introduced. The only alteration of note was the use of parachute stars during the Great War.

Smoke Shell

As chemists have only recently made such enormous strides in knowledge, and as chemical industry is a comparatively new branch of commerce, shells of this description were not used before the Great War.

The absence of such shells must not be taken, however, to connote a kindlier disposition on the part of our forefathers, but rather to expose a lack of knowledge and applied craftsmanship. This is evidenced by the fact that smoke balls, " which we have a way of preparing so that during their combustion they cast forth a noisome smoke, and that in such abundance that it is impossible to bear it," were in existence in the 17th century.

The composition used in these smoke-balls was saltpetre, coal, pitch, tar, resin, saw-dust, crude antimony and sulphur, which was generally placed in canvas sacks. Later the employment of paper shells to contain the composition came into use.

This type of projectile lingered on till it finally became obsolete on 30th August, 1883, although it is doubtful whether any had been fired previously for many years.

Latterly, they were constructed in five sizes and consisted of a concentric paper shell about 1/15th of its total diameter in thickness. They were filled with a mixture of L.G. powder, saltpetre, pitch, coal and tallow. The vent was then driven with fuze composition, matched, and covered with kit plaster. During filling, sulphur and coal dust were sprinkled in three times.

They were intended—

(1) To suffocate or expel the enemy when working in casemates, mines, or between decks ;
(2) To conceal operations ;
(3) To serve as signals ;

and in this last connection it is worthy of note that six ships, fitted out in 1852 for an Arctic expedition, carried a supply of these projectiles.

Segment Shell

Segment shell were introduced with R.B.L. guns, the first being that for the 12-pr. on 13th April, 1860. This projectile consisted of a very thin cast-iron cylindro-conoidal shell about 2½ calibres in length lined with cast-iron segments built up in layers, having a cylindrical powder chamber in the centre. The base was closed with a cast-iron disc. A thin coating of lead alloy extended from shoulder to base and flowed in between the segments lining the powder chamber, giving great weight and solidity.

The shell in this manner exhibited great resistance to external pressure, whilst a small bursting charge could open it. It was employed as a kind of intermediary between case and shrapnel, although in some cases it was used simply as shot or common shell.

Segment shell could only be fired from Armstrong guns, but its action was met to some extent by the adoption of " Ring shell " for B.L. and R.M.L. guns in 1891.

This shell was made of cast iron, cast on a core of rings, each ring being arranged to prevent breaking joint with another. The rings were also weakened to break up when the shell burst.

It contained a bursting charge of special powder.

Ring shell became obsolete for the Imperial service on 3rd December, 1920, and are now only used in India.

Shrapnel

A shell which could burst into a number of small fragments, and thus become effective against isolated parties of men, was sorely needed during the siege of Gibraltar, and the void was temporarily filled by adopting the plan of Captain Mercier, 39th Regiment. This officer conceived the idea of firing the 5·5-inch Royal Mortar shells with short fuzes from 24-pr. guns. This proposal was adopted and proved very efficacious throughout the siege.

It was realized to be nothing more than a makeshift, however, and, after the siege, the thoughts of artillerymen turned towards the permanent solution of this problem.

The outcome was the far-reaching invention of Lieut. Henry Shrapnel, R.A., who, in 1784, proposed his " spherical case shot," a former designation of the shrapnel shell.

This projectile originally consisted of a thin iron shell filled with musket or carbine balls, sufficient powder being inserted in the balls to cause the bursting of the shell when ignited by the fuze.

Shrapnel was not intended to supersede case or grape shot, but to act in a similar capacity at longer ranges ; in fact, General Shrapnel, speaking of his own invention, says " the object now accomplished is the rendering of the fire of case shot effectual at all distances within the range of cannon."

The idea embodied in this shell was then entirely novel. Until the beginning of the 19th century all shell depended on their bursting charge for the dispersal of their parts. Such charges, therefore, were a maximum, and all attempts to improve the efficiency of the burst was effected by their increase.

Shrapnel's principle, however, legislated for a minimum charge consisten with opening the envelope, so that the liberated bullets travelled forward with the velocity of, and in the same direction as, their ruptured case. It was this significant fact which differentiated shrapnel from any other projectile, and made its effect so deadly, compared with others, against personnel in the open.

The unsteadiness of the old-time spherical shell during flight, whose fragments were initially propelled by its own bursting charge, often caused the pieces to be blown in any direction except the correct one, thus nullifying in many cases its man-killing properties. Such a contingency could never occur with shrapnel shell, and therein lay its claim to recognition.

That claim did not, of course, remain unchallenged, and both French and German writers tried to demonstrate that General Shrapnel merely improved on existing ideas.

These contentious arguments died in the course of time and Shrapnel's claim was finally recognized. As a mark of respect to its distinguished servant the British Government issued, on 11th June, 1852, an order that " spherical case shot " should in future be known as " shrapnel shell " in honour of its inventor.

Like most epoch-making inventions it was shelved for some years, and not until 1792 was the proposal formally submitted to a committee of artillery officers. Four years afterwards, however, the idea was seriously considered, and Major Shrapnel, as he was then, fired some of his shells containing leaden bullets before the Ordnance Select Committee on 3rd June, 1803.

As a result " spherical case shot " were finally approved for the service, and the first recorded case of their use in action was against the French on 17th August, 1808, at the battle of Rolica.

The valuable properties of the original shell were, however, neutralized to a great extent by its liability to premature action. This naturally led to a serious depreciation of its efficiency, and from extensive series of trials carried out in 1819 and 1852 it was discovered that 17 and 22 per cent., respectively, of the shell fired were failures.

This fault was originally considered due to the setting down of the fuze composition, together with the concussion of air at the moment of firing; it was also thought to arise in some cases from the shell being too weak to withstand the shock of discharge. In 1850, Colonel Boxer introduced his fuze to obviate this and the walls of the shell were thickened, but the trouble still continued to be prevalent.

The main causes were then attributed to be :—

(1) The fuze being driven out by the weight of bullets in the shell on firing allowing the flash to pass from the propellant to the bursting charge.
(2) The friction of the bullets generating sufficient heat to fire the bursting charge.

Two attempts were made to remedy this state of affairs—by coating the interior of the shell with cement to diminish the friction; by fixing the balls with pitch, sulphur, plaster of paris, and other materials, and by reducing the propellant charge. Neither of these remedies proved satisfactory.

Extensive experiments were then carried out to establish finally the cause of these premature explosions, and the trials of 1852, 1853 and 1854 proved beyond doubt that the intermingling of bullets and powder was the contributing factor.

Colonel Boxer then suggested in 1852 the separation of the bullets from the bursting charge by means of a diaphragm. Experiments with shell constructed on this principle were most satisfactory, and on the 1st October, 1853, the Ordnance Select Committee recommended the manufacture of a large number of them with a view to their introduction into the service. Boxer's diaphragm shell were thereupon provisionally approved on the 11th October, 1853.

Meanwhile, in order to deal with the difficulty of the original pattern shrapnel in store, a conversion was approved on 23rd March, 1854, by which the bullets and powder were separated by means of a socket and tin cylinder introduced at the fuze-hole. These shell were called "improved shrapnel shell," and were never manufactured as new stores, all being conversions from existing stocks.

The diaphragm shell underwent one or two minor alterations and, as amended, were provisionally approved on 29th September, 1858. They were not finally approved till 27th September, 1864.

Diaphragm shell were made for 11 different natures of ordnance ranging from the 6-pr. to the 150-pr., the 10-inch being an exception. They were fired from guns, howitzers and carronades.

The shell consisted of a thin cast-iron casing, weakened by four grooves down the sides to make it open out, thickened at the junction of diaphragm and shell to prevent disruption into two pieces, strengthened at the fuze-hole to support the socket, and increased at the base in all natures above the 12-pr. to withstand the shock of discharge.

A wrought-iron diaphragm divided the shell into two unequal parts, the smaller forming the powder chamber and the larger holding the bullets of lead and antimony packed in coal-dust. A socket, fitting down into the shell through the diaphragm, acted both as a funnel for introducing the bullets, and as a receptacle for the fuze. The charge of pistol or F.G. powder was inserted through a loading hole closed by a plug.

The shells were issued either :—

(1) Empty, loose, prepared for bottoms for India only.
(2) Empty, riveted.
(3) Filled.

In every case all shrapnel were issued with bullets, the filling only referring to the charge. Their weight when filled approximated to about seven-eighths that of a solid iron shot of the same calibre.

Six advantages were claimed for the Boxer diaphragm type over that originally proposed by General Shrapnel. These were :—

(1) Premature action completely overcome.
(2) Full service charges could be used, thereby increasing the range and its effects.
(3) Reduction of the dispersive effect of the bursting charge on the bullets.
(4) Adhesive action of the bullets overcome.
(5) Keeping qualities increased.
(6) Greater safety in transport when filled.

The improved shrapnel, which was recognizable by its projecting socket, became obsolete a few years after its introduction to the service and disappeared.

The introduction of the rifled gun may be said to have ushered in the era of modern shrapnel so far as its internal arrangements are concerned. The R.B.L. shell consisted of a cast-iron body weakened internally by longitudinal grooves lightly attached to a head of elm covered with a wrought-iron skin. In the base of the body was formed a chamber which accommodated a tin cup holding the charge. Over the cup rested a diaphragm pierced with a central hole, through which passed a tube connecting the fuze socket and magazine. On the diaphragm rested the weight of the lead and antimony bullets set in resin. A primer was screwed into the top of the central tube to facilitate the passage of the flash from fuze to bursting charge.

The shell of to-day differs but little from its Armstrong predecessor, although certain variations founded on experience have been introduced. For instance, the weakening grooves have disappeared, cast-iron has been superseded by cast and forged steel, and heads are now easily attached by rivets and twisting pins.

Some natures were provided with a tin cage or cylinder in which the bullets reposed, others dispensed with wood in the head. One type substituted compressed pellet bursters for F.G. powder in the magazine and perforated pellets for a primer in the central pipe. It had no separate head but a separate screwed-on base; another provided for the bursting charge in the head.

Such intricacies of design, however, belong rather to the present than the past, and handbooks will supply all such details. They do not affect the fundamental principle which has remained unchanged, a principle, moreover, which has rendered the name of its English inventor illustrious throughout the civilized world.

Appendix VII

FUZES

The early history of fuzes is in the main a record of man's endeavour to control the bursting of shell by means of a priming composition, without understanding the rules of combustion.

The igneous missiles flung by machine and hand at the beginning of the 15th century were ignited by some slow-burning mixture inserted on the top of the charge, which filled up the case, flush with its loading hole.

The first incendiary projectiles fired from guns were set in action in a similar manner, and such primitive methods of procedure undoubtedly led to numerous accidents. In fact the gunner of that period was faced with Hobson's choice. If he loaded his piece with the shell's priming contiguous to the charge, the explosion of the latter ignited the former and burst the gun; if he rammed home his projectile in the reversed manner, either a blind occurred on firing, or the composition set back on the shock of discharge and again a burst bore was the result. For this reason shells never became popular during their infancy, as stated in Appendix VI.

The first mention of a fuze, by which is meant some composition enclosed in a separate cover, occurs in 1421 when a "sheet-iron tube enclosing priming" is said to have been used at the siege of St. Boniface in Corsica. An improvement appears to have taken place by 1543, as the fuze then described was a hollow cylindrical iron tube containing a kind of match, screw-threaded externally to fit a corresponding socket in the shell.

Towards the end of the same century a new fuze was invented by the Governor of Jemappe, which consisted of a tube filled with moist powder. Experiments with it were carried out, but proving disastrous to life and to limb, they were promptly brought to a close.

The general form of a fuze employed in the 16th and 17th centuries was of cast iron, cylindrical in form and usually about the size of a finger. In some cases they were larger and their length varied according to their use. They were filled with a mixture of saltpetre, sulphur and mealed powder. Their ignition always proved the great stumbling block; four methods being in use at the time.

(1) Loading the shell, fuze towards the charge with a perforated tampion between the two, the fuze being placed in the hole in the tampion.
(2) Placing a perforated wooden tampion containing mealed powder between the projectile, loaded fuze up bore, and the charge, so that the mealed powder acted as a relay in lighting the fuze.
(3) Using a gun with two vents, one for the fuze and the other for the charge, in order that both may be ignited in quick succession.
(4) Loading the shell fuze towards the muzzle, and lighting the latter with a port-fire thrust down the bore.

It was not realized till half-way through the 18th century that the gun of the period had sufficient windage to allow the flash of discharge to pass up the bore, encircle the shell, and ignite the fuze.

"Single-fire," the ignition of the charge and fuze in one operation, did not, therefore, appear till about the year 1747.

These fuzes had no "time" arrangement and when once constructed their time of burning was fixed. The only attempt at control was by varying the initial lengths of fuzes to suit the average range of the gun from which they would be eventually fired, and by inserting different natures of priming into the fuze-cases.

TIME FUZES

The first suggestion of regulating the burning of fuzes was, so far as is known, put forward by Sebastian Hälle in 1596 ; but no notice seems to have been taken of his foresight for nearly a hundred years. In 1682 the idea was revived ; and the fact that fuzes at this period were manufactured of paper, wood and iron proves that research in this direction was being actively pursued.

The lack of any accurate time-piece added greatly to the difficulties of producing a " time-fuze " in the sense of which we know the term ; for an efficient " time-fuze " depends on an exact method of proof. In the 17th century the repetition of the " Apostles Creed " was one of the Proof-master's favourite measurements of time, and although such a method may have commended itself to the orthodox, it could scarcely be said to have constituted a standard of accuracy.

Later on the use of the pendulum was advocated, but little trust apparently was placed in its performance ; for at this time it was generally conceded that the only sure method of timing a fuze was to fire it in a gun. In other words gun-proof was to be preferred to rest-proof.

The invention of the watch, however, in 1674, was soon to become the solvent of all these difficulties and in the 18th century progress began to be apparent.

By the middle of that century fuzes were made of beechwood and cut to length according to the requisite time of burning. For short ranges, however, the bodies had to be curtailed to such an extent, that either the flash failed to ignite the bursting charge and the shell became blind, or the fuze composition set back and caused a burst in the bore. To obviate this, special fuzes, filled with a quick burning composition, were proposed and adopted, for in 1779 three fuzes burning at the rates of 4, 4½ and 5 seconds an inch were to be found in our service.

The next improvement was that suggested by Captain Mercier during the siege of Gibraltar, in which he proposed to fire shells from guns with calculated fuzes.

By the year 1819 there were 21 different varieties of bored shrapnel fuzes in the service, exclusive of a mealed powder and unbored fuze, and of fuzes for common and mortar shell. In that year a committee was appointed for the purpose of investigating the question of reducing the number of common and mortar fuzes from five to two or three, but although they agreed that such a course in itself would be most desirable, they took no action on the ground of the great difficulty that would be experienced in disposing of existing stocks of shell.

About this time eight types of fuzes were sent on service ; their description and lettering being marked on the tangent scales of the guns, on the cases in which they were packed, and on the fuzes themselves.

A proposition was again submitted on 17th October, 1828, to reduce the number of bored spherical case fuzes from seven to three ; this was apparently approved on the 19th November, although it was never put into effect, as these seven fuzes remained in the service until 1850.

In 1829 metal time fuzes were first adopted for the Navy on the recommendation of General Miller, and the first ship to be thus equipped was H.M.S. *Talavera* which carried a complement of 20 filled shell a gun. On 5th November, 1832, this practice became general for the Navy.

Five reasons were adduced for using metal in naval fuze construction They were as follows :—

(1) The greater proportion of shells were carried filled and fuzed in H.M. ships and metal was considered to be superior to wood, and less dangerous in case of accidents.

(2) Metal fuzes were less liable to deteriorate in store.

(3) Bursting charges in shell fitted with metal fuzes were less susceptible to damp, even in the case of a temporary immersion.
(4) Metal fuzes were less likely to become broken in transport and haulage.
(5) Metal fuzes, acting as more efficient tamping agents, developed the force of explosion.

On 25th November, 1830, the number of spherical fuzes for each shell in the field was reduced from eight to four. In 1845 another committee sat with the object of simplifying the general fuze position and abolishing some of the numbers, but failed to achieve any results.

Until the middle of the 19th century there was little or no improvement in time fuze construction. Then the repeated failures of shrapnel shell forced the pace, and on 27th April, 1849, Colonel Boxer proposed his wooden fuze to surmount this difficulty.

Before this new departure there were 19 time fuzes in the service, 16 wood and three metal. Out of the wooden fuzes, ten were for use with shrapnel, five with common and mortar shell, and one with the hand grenade. All these fuzes were prepared by cutting to length or by boring into the composition from the bottom, thus suffering from the defect of having their filling unsupported.

Colonel Boxer's fuzes arranged for such support by boring a hole in the side during preparation. They were subjected to severe trials, which they passed so satisfactorily that the Ordnance Select Committee recommended their adoption for all gun and howitzer shells. This recommendation was approved on 2nd September, 1850.

They consisted essentially of a wooden straight coned casing with one powder channel and one row of holes reading $\frac{2}{10}$ of an inch. Therefore, two fuzes, one reading odd and the other even tenths, had to be supplied for each shell. The composition bore was concentric, and the fuze was capped with paper, painted white for odd and black for even tenths.

On 21st August, 1852, Colonel Boxer suggested some minor alterations, one being the adoption of a projecting head. In August, 1853, further improvement took place in the method of priming. On 21st December, 1853, the adoption of two channels and 1 inch of composition for all shrapnel fuzes, and two channels and 2 inches of composition for all common and howitzer shell fuzes was recommended. The recommendation was approved, and such fuzes were introduced on 8th February, 1854.

The projecting head, approved in 1852, not only rendered the fuze susceptible to damage, but caused it to be easily capable of being knocked out of its seating on ricochet and on Colonel Boxer's suggestion it was removed on 16th March, 1854.

The Boxer fuze, although considered at the time to be a great stride towards perfection, yet lacked one essential—regularity of burning. Experiments were carried out in the summer of 1854 to investigate the causes of this defect, and these conclusively demonstrated that the grease used with the boring instruments during manufacture entered the wood, and impregnated the composition. Steps were, therefore, taken to institute the use of dry bits for drilling, and the fault then was, to a great extent, eradicated.

Trouble now began to be experienced with these straight coned fuzes after storage. Originally correct to gauge, they either became too swollen, owing to the absorption of moisture, to enter the fuze hole, or so shrunken, owing to the dessication of the wood in a dry atmosphere, that, being deprived of their projecting head, they set back into the shell on firing and caused prematures.

Colonel Boxer, therefore, recommended on 11th May, 1855, that the angle of cone should be increased for all common, shrapnel and mortar fuzes. On the same day the Ordnance Select Committee recommended the use of a metal cap for all fuzes in place of those of paper used hitherto.

Both these alterations were approved on 18th August, 1855, a date which may be taken to indicate the adoption of Boxer's fuzes in their final form.

The first Boxer mortar fuze on the small cone principle was introduced on 27th January, 1855, but it was subsequently brought into line with those for gun and howitzer shells on 18th August of the same year.

The next type of fuze to be recommended was Boxer's metal time fuze for the navy on 21st January, 1857. There were two patterns, the 20 seconds and 7½ seconds respectively, but the latter was not approved for service till 22nd February, 1858. They underwent alterations, and were not finally introduced till 11th August, 1859.

Metal, however, in itself was not considered to be an ideal agent for time fuze bodies at this period, as the improvements since introduced in wooden fuzes were held to stultify the arguments previously advanced for its retention in the navy. Therefore, the introduction on 26th June, 1866, of the 9 seconds M.L. fuze and of the 20 seconds M.L. fuze on 7th June, 1867, were considered to seal for ever the ultimate fate of all metal time fuzes.

Six advantages were claimed for Boxer's fuzes over their predecessors :—

(1) Freedom from premature explosions,
(2) Greater regularity of burning,
(3) Greater safety in preparation,
(4) Simplification in preparation,
(5) Great reduction in the numbers of fuzes,
(6) General all-round improvements,

and the era of wooden time fuzes was definitely and triumphantly established.

A description of the 9 seconds M.L. fuze will give some idea of the general make-up of wooden time fuzes in use at this stage.

The body, about 4 inches long, is made of beechwood, shaped in the form of a truncated cone, and is cupped out to take the priming and match in the head. The interior is bored out to take a lining of paper filled with a composition of saltpetre, sulphur and mealed powder. The side holes, recording tenths of inches, are plugged with rifled powder, the clay of the earlier patterns being abolished. The bottoms of the powder channels are connected with a piece of quick-match, and at the level of the top side hole the composition bore is driven with mealed powder to ensure accuracy at short ranges. In the head is a gunmetal plug, screwed into the upper part of the composition bore so as to be flush with the top of the fuze. From the centre of this plug a copper pin projects downwards round which is looped a piece of quick-match, the ends being passed through two escape holes in the side of the head.

The quick-match is laid in a groove round the head of the fuze, and is protected by a strip of thin sheet copper covered with a tape band, one end of the former remaining exposed.

The copper band served two purposes, protection from moisture and safety from premature ignition.

On preparing the fuze for action the requisite side hole is bored by a hook-borer allowing one-tenth of an inch for each half-second of time. After loading, the tape is torn off.

On the gun being fired the flash from the propellant ignites the fuze, and the composition burns at the rate of one inch in five seconds until it reaches the bored hole. There the flame from the burning composition passes into the powder channel, in which a communication has been opened at the bored side hole, and is then flashed down into the shell. If the fuze be unbored, it will burn its full time.

The introduction of rifled guns led to an alteration of design in the wooden fuze. As windage under these new conditions was reduced to a minimum, the flash from the charge could no longer be relied upon to ignite the fuze. A detonator, cap composition, and a hammer suspended by means of a copper wire, were therefore incorporated, the needle being actuated on the shock of discharge. This design was first provisionally approved on 13th January, 1864.

From then onwards the number of fuzes became legion and no further attempt can be made to describe them. Their details can all be found in the various editions of ammunition text-books or in the early volumes of " List of Changes."

Armstrong's fuzes were introduced on 13th April, 1860, for use with the equipment of R.B.L. guns. The first was known as the " A " pattern, and was made of white metal. Numerous modifications and marks, the nomenclature following the musical scale, succeeded one another in rapid succession, until type " F " was evolved, which was approved on 21st September, 1867, as the first " time and percussion " fuze.

As the percussion arrangement frequently failed, it was withdrawn from the service and the E Mark III remained the only fuze in general use with Armstrong guns.

This fuze was of brass, containing a ring of slow-burning composition, which was ignited by a pellet holding a detonating cap which set back on the shock of discharge, thus bringing the cap against a firing-pin. It was, in fact, the prototype in principle of the modern time fuze. It was certainly superior to the Boxer fuze in regard to its accuracy of setting, and although originally used for naval segment shell only, it was ultimately recognized as being very suitable for shrapnel.

Although the original T. & P. fuze proved unsatisfactory, the type was destined to become the standard for, with the exception of a few time fuzes employed in star shell, anti-aircraft shell, and certain mortar bombs, the modern means of bursting shell during flight are devised on this principle.

On 12th December, 1881, Sir William Armstrong brought out a combined time and concussion fuze, afterwards numbered " 52." It was an amalgamation of his " E " time fuze with a graze percussion arrangement, very similar in action to the modern time and percussion fuze.

In " fuze, time and percussion, short " which followed a little later, it was hoped that a fuze had at last been produced for field artillery which could be relied upon to burn with great regularity, to act by time both at very short intervals and at very short ranges, and to function on graze with certainty. It was introduced to supersede fuze No. 52 in the small natures of B.L. guns. To a certain extent that hope was realized.

About this time, when the quality and range of guns were increasing, the want of a satisfactory time fuze for long ranges was keenly felt. It was realized that to fill the time ring with a slow-burning mixture was folly, as the burning of such compositions was known to be more irregular than that of a quick powder. The difficulty, therefore, was how to lengthen the time of burning without, at the same time, introducing further irregularities. Only two alternatives appeared feasible ; to increase the diameter of the time ring, or to adopt the principle of two rings.

The last arrangement had been actually employed by Sir William Armstrong before 1887, but the " double-banked " fuze was not then considered to have sufficiently advanced in design to allow of its adoption in the service.

The action of all fuzes was, up to that time, initiated by the shock of discharge shearing small pieces of wire retaining needles or pellets, but with certain low-velocity guns the shock of firing was so small that these shearing wires had to be very thin in consequence. In fact, their diameter and strength had to be so reduced that the fuzes became dangerous to load. To circumvent this difficulty, a sensitive fuze was devised which depended for its correct functioning on centrifugal force actuating pellets outward on to needles.

Fuze, T.P., sensitive, middle, No. 24, was, therefore, introduced on 28th May, 1887.

It was not until 1901 that fuze, T. & P., No. 58, the first built up on the double-banked system, was approved. It constituted a big advance in fuze evolution as it went far to solve the troublous problem of long-range shrapnel fire. It was introduced for the 15-pr. Q.F. field equipments.

The last distinctive step to be taken in composition time fuze design was the introduction of the No. 80 type of fuze in 1905. The design was originally Krupp's and was taken over by this country for use with 13-pr. and 18-pr. guns. The original No. 80 was made of aluminium, but the later marks of this and allied fuzes are now manufactured from brass.

It inaugurated the saddle or stirrup spring, the straightening of which armed the fuze, after firing, and eliminated the shearing wire and safety-pin. The gauge was 2-inch and the shape, conforming to the contour of the shell, tended to decrease air resistance.

As the range of modern guns increased, their trajectories became even higher, and as the burning of powder at high altitudes is apt to be so erratic, the meticulous accuracy required with modern shooting suffers in consequence. The only method of dispelling this source of error was to abolish the composition rings and substitute another method of recording time.

For this reason the Germans introduced mechanical time fuzes during the Great War.

PERCUSSION FUZES

The earliest proposal for igniting the bursting charges of shell by percussion appears to have been made in 1596 by Sebastian Hälle. His suggestion with regard to such an arrangement, however, seems to have fared no better than his proposition concerning calculated time fuzes. In 1610, two hand grenades having the property of being able to explode on impact with the ground, were invented ; although one was fitted with a safety arrangement to prevent premature action, both operated on Hälle's principle.

In 1650, the first account of a percussion fuze intended for a gun projectile is recorded. It consisted of an iron tube or barrel, hollow almost up to the top, at which point it was screwed into the shell. The sides of the fuze were perforated, and its internal surface roughened like a file. An iron rod, to the foot of which was screwed an iron plate, carried two flints on its upper end which fitted inside the barrel.

On striking an object, the weight of the projectile forced the bar carrying the flints into the tube, and the ensuing sparks were supposed to pass through the interstices and ignite the charge. Unless the shell fell in certain positions, however, a blind would result, and the old fire-masters must have been moved by a touch of unconscious humour when they named this type of missile " blind shell."

The use of a percussion powder to burst shell was advocated in 1655, but no attempt seems to have been made to put this idea into practice, owing, no doubt, to the difficulties and dangers attendant upon its employment. Berthollet's discovery of fulminate of silver towards the end of the 18th century did not help matters, as it was found to be incapable of utilization owing to the violence of its action. It was not, therefore, till after Howard had isolated fulminate of mercury in 1800 that any progress could be made, and then various percussion mixtures appeared. On 11th April, 1807, the Rev. Alexander Forsyth patented the use of such mixtures for the priming of fire-arms, and 11 years later, Colonel Peter Hankins invented the copper percussion cap.

This principle of percussion priming was generally adopted into the service on 31st March, 1842, and with its introduction, especially with the addition of the copper cap, the solution of the percussion fuze problem became automatically simplified. Several such fuzes were tried before a mixed Naval and Military Committee sitting under the presidency of Sir Thomas Hastings in 1845. None of the five types submitted, however, were considered sufficiently satisfactory to warrant their adoption, and it was not until a year later that Quarter-Master Freeburn, R.A., invented his concussion fuze which was approved for the land service on 12th October, 1846.

This fuze was constructed of beechwood, the head being cup shaped and recessed internally, and the body bored out to take the composition. The total length was about 6 inches. About 2–3 inches from the head, three rectangular holes were bored through the sides into the composition; and each hole was fitted with a gun-metal wedge tending to fall inwards. These wedges were supported internally by the fuze composition, and externally by wedges of wood secured with a thin copper wire covered by a paper band. A long piece of match was coiled in the recess in the head, one end passing into the composition bore. The head was covered by a paper patch.

On the shock of discharge, the propellant flash lighted the fuze, and the composition therein burnt away. On impact the copper wire was broken and the gun-metal wedges, deprived of their rear support, fell inwards allowing the flame from the still burning composition to reach the shell's bursting charge by means of the three uncovered holes. The fuze was active for a period of 12 to 13 seconds.

On 1st December, 1847, minor alterations in the general shape of this fuze were approved.

This fuze, being of wood, was not considered suitable for the navy, and several attempts were made to introduce a metal fuze for that service. All efforts, however, proved fruitless till 1850, when Commander Moorsom, R.N., invented his percussion fuze, which was finally approved for sea service on 16th July, 1851.

This fuze was made of gunmetal, cylindrical in form, and screw-headed to the "Moorsom" gauge which was right-handed and larger in diameter than previous systems. It consisted of three chambers, each containing a patch of detonating composition over which was suspended a gunmetal hammer by means of a wire.

A leaden pillar and a guard wire were incorporated as safety arrangements to prevent premature action.

On firing, the wires were sheared, and on impact, the hammers were thrown forward striking the detonating composition; the flash thus formed passing through the necessary fire-holes and igniting the bursting charge.

This type was the forerunner of the present "graze" percussion principle without its modern refinements.

In 1851, therefore, each service had its own percussion fuze, Freeburn's and Moorsom's. The introduction of Boxer's fuzes in 1850, however, caused a new fuze-hole gauge to appear, which, being adopted in 1852, rendered Freeburn's concussion fuze useless and obsolescent. The land service, therefore, ceased to have a percussion fuze from this date until a new type, designed by Mr. Pettman, was introduced.

In January, 1860, this gentleman, a foreman of the Royal Laboratory, designed a fuze which stood up to departmental trials very well. In April, 1860, his fuze was formally submitted with a view to its adoption by the navy. Experiments were carried out under the *ægis* of the Ordnance Select Committee and H.M.S. *Excellent* which, in the main, proved successful. Several modifications, however, had to be introduced before this fuze was finally adopted for the navy on 2nd August, 1862.

A land service fuze on a similar principle proposed by Mr. Pettman was approved on 3rd October, 1861.

In these early days of what are now termed "percussion fuzes," a distinction was drawn between two principles called, respectively, "concussion" and "percussion."

This distinction was entirely arbitrary and the terms were applied indiscriminately until 1863, although the idea usually conveyed was that "percussion" signified a fuze dependent upon direct impact for its action, and "concussion" one that functioned solely on the shock of discharge. In order to standardize these expressions, the following definition recommended by the Ordnance Select Committee, was adopted on 6th January, 1864.

"A percussion fuze is one which is prepared to act by the shock of discharge, but put in action by the second shock on striking the object. A concussion fuze is one which is put in action by the shock of discharge, but the effect of that action is restrained until it strikes the object."

Under this definition, all the fuzes existing at that date became automatically "percussion" as the only fuze, Freeburn's, which had satisfied the requirements of the other class, was already virtually obsolete.

In modern parlance, the direct action fuze might be compared with the old percussion type, the preparation for action on firing being the gradual opening of the shutter; and the graze fuze with the concussion type, the putting in action on the shock of discharge being the "arming" and the restraining action being the inertia of the creep spring.

Subsequently, Mr. Pettman proposed a modification of his sea service fuze in order to render it also suitable for rifled guns. This general service fuze was introduced on 19th May, 1866, and Pettman's fuzes for sea service gradually superseded Moorsom's pattern, which was declared obsolete on 2nd May, 1865.

In 1867, therefore, there were only three percussion fuzes for spherical shell in the service :—

Pettman's Land Service percussion fuze,
Pettman's Sea Service percussion fuze, and
Pettman's General Service percussion fuze.

The latter fuze, capable of acting with S.B., B.L. or R.M.L. guns on impact only, was a very ingenious design. It consisted essentially of a hollow brass cone, screw-threaded externally, in the centre of which was placed a roughened brass ball coated with detonating composition. This ball was retained in position between brass pellets, the lower one being supported on a leaden collar. The upper brass pellet had detonating composition round its top edge and was bored through in the centre to seat a plain brass ball beneath the closing plug.

On firing, the set-back caused the leaden collar to crush up and released the detonating pellet, which was then free to strike against the wall on impact. With rifled ordnance, the action of the detonating pellet was liable to fail, owing to the striking position of the shell. In this case the action of the fuze was as follows :—The plain brass ball, being released on discharge, was kept against the wall by centrifugal force and was in a position to crush the detonating composition placed round the top edge of the upper pellet when the latter came forward on impact. A small powder magazine was secured in the lower brass pellet to pass the flash on to the bursting charge of the shell.

There was still a school of thought in favour of re-introducing the concussion type of fuze. This type was considered safer to handle and less liable to deteriorate in store than the percussion type. However, as such fuzes were unsuitable for guns without windage such as B.L. guns, which were still felt to have a future before them, they never succeeded in justifying their adoption into the service, although several officers submitted designs in 1870.

At this period there were six percussion fuzes for rifled guns, including Pettman's General Service pattern which was common to all classes of ordnance. One of these, the Armstrong C percussion, was modified into "fuze, percussion, B.L. plain" when smooth bore guns ceased to be considered of primary importance. This type, introduced on 13th October, 1870, was of similar construction to the "fuze, percussion, R.L.," approved on 23rd January, 1872.

In the top of these fuzes was a plate carrying a central needle which projected downwards on to a lead alloy pellet fitted with a detonating cap and a small powder magazine. A pin, withdrawn before firing, was fitted under the needle to ensure safety while travelling. The pellet had feathers projecting from it, and these supported a collar which kept the pellet in position in transit. On firing, the collar set back, shearing off the feathers, and when the flight of the shell became checked on impact the pellet, being freed, moved forward and the needle struck the cap.

Direct action fuzes soon followed, the first being " Fuze, percussion, D.A., No. 3," approved on 17th November, 1880. This type consisted essentially of a copper disc carrying a steel needle suspended over a detonating cap and powder magazine. On the shock of discharge no action takes place, but on impact the copper disc is crushed in and the needle driven on to the detonator.

The general lay-out of D.A. designs has remained the same, the only difference being the inclusion of certain safety devices such as shutters, tapes, etc., to prevent accidents during transport and prematures on firing. The filling of the fuze has naturally varied with requirements, and to-day the greater percentage of these fuzes are filled with C.E. for use with H.E. shell.

The first base fuze to be introduced was " Fuze, percussion, base, Hotchkiss," on 9th April, 1886, a pattern which depended for its action on the plasticity of lead. It is of simple construction and the current marks act on the same principle as their predecessors.

Palliser shell were originally unfuzed, as the friction set up in the bursting charge was deemed sufficient to generate enough heat to cause their explosion. As these projectiles were afterwards converted into shot, base fuzes for the larger types of shell were not introduced till the 'nineties.

On 24th August, 1890, " Fuze, percussion, base, Armstrong, No. 9," was introduced for Q.F. guns. This fuze had a graze pellet retained in the safe position by two bolts screwed into a pressure plate in the base. The pellet was further secured by two centrifugal bolts, one of which was held by a spring bridle. A creep spring was fitted to prevent the pellet moving forward during flight. There was a small powder magazine in the head.

The action was as follows : The pressure of the propellant gases forced in the pressure plate so that the heads of the retaining bolts advanced clear of the pellet. The bridle was also driven forward, releasing the centrifugal bolt, which moved outwards, due to the rotation of the shell. The pellet, being only restrained by the creep spring, was then free to move forward on impact.

In 1894, " Fuze, percussion, base, large, No. 11," appeared for Palliser and common pointed shell. It differed from the Armstrong No. 9 mainly in having a flexible copper pressure plate and one central retaining bolt or spindle. Only one centrifugal bolt was used, and this was carried in the needle pellet and retained in the safe position by the nut of the spindle.

On firing, the pressure plate was crushed in ; and the nut on the central spindle moving clear, released the centrifugal bolt. The latter, flying outwards, left the needle pellet free to move forward on impact by overcoming the weak resistance of a phosphor bronze spring.

Improved base fuzes, such as those used with modern A.P.C. shell, have since appeared, but their difference lies in their refinements, not in their principles.

In fact, the principles in percussion fuzes are simple and have been in existence practically ever since the first reliable patterns were produced. All other improvements such as centrifugal bolts, safety tapes, safety shutters, and creep springs, etc., are incidentals. They are, however, most important, especially where safety and reliability are concerned, and they have been in a large measure instrumental in raising the present fuzes to such a high pitch of efficiency.

Appendix VIII

SMALL ARM AMMUNITION

Although the mention of a composite cartridge, with which the musketeers of the period charged their pieces with powder and ball at the same time, occurs in 1590, the early general history of small arm ammunition resolves itself into the story of the bullet, wad, powder horn, ramrod, and flint.

Undoubtedly the charge was occasionally wrapped in paper to facilitate loading, but such a procedure was sporadic rather than general until the close of the 18th century, when again the name of " cartridge " appeared on documents.

These cartridges were, of course, very elementary in character, and simply consisted of a paper twist containing the correct amount of powder to save time in loading during action. On charging the musket, the choke was torn off the paper receptacle and the contents poured down the muzzle and rammed home with wads, the bullets being inserted afterwards.

Any details, therefore, relevant to loading or igniting the charge previous to the Victorian Era really belong to the weapon rather than to its ammunition, and all particulars previous to about 1842 consist merely of a set of figures relating to weights of powder in drams and of bullets in grains.

In 1800 Baker's rifle was introduced into the British Army and issued to the 95th Regiment, now The Rifle Brigade. It was a 20 bore and fired a bullet weighing 350 grains; being a rifle, however, it was found extremely difficult to load.

The musket of the same date fired a heavier ball of 480 grains with a charge of 6 drams.

It was considered that the rifle bullet should be heavier and of a standard size. On 25th November, 1830, therefore, it was approved that all rifle ball cartridges should have the weight of their bullets raised to 350 grains, *i.e.*, 20 to the lb., instead of 22 to the lb., which had been the practice hitherto. At the same time the word " rifle " was to be marked on the outside of the ammunition packages to distinguish these cartridges from those of the ordinary carbine. On the same date a simple cartridge-filling machine, which ensured uniformity in charging, was also adopted.

The next rifle to appear was the Brunswick in 1836, the ammunition of which consisted of a paper cartridge containing 2½ drams of powder and a belted spherical ball of lead weighing 557 grains.

On 10th July, 1837, the following charges for percussion small arms were laid down by the Board of Ordnance :—

Musket ball	4½ drams
Rifle ball	2½ ,,
Carbine ball	3½ ,,
Blank for all percussion arms	3½ ,,

On 31st March, 1842, the principle of percussion for priming was generally adopted for all small arms, and as the copper percussion cap had been invented previously in 1818 by Colonel Peter Hankins, the ignition of charges from this date proved easier of accomplishment.

The first service weapon to have a distinctive cartridge case in this country, apart from a mere cartouche of paper, appears to have been the rifle musket (sea service) of 1842. This cartridge consisted of many turns of thick paper enclosing a charge of 3 drams F.G. powder and the bullet. The latter, weighing 825 grains, was cylindro-conoidal in shape, and being hollow at the base was strengthened by an iron cup. On 23rd September, 1854, this bullet and charge were introduced for the Royal Navy and Royal Marines. In loading, the top of the cartridge was torn off, the powder poured down the barrel, and the bullet rammed home. Experience showed that the iron cup in the base of the bullet penetrated and blew through the latter on firing, and therefore on 23rd January, 1861, it was decided to substitute a wooden plug, slightly reducing the weight and diameter of the bullet in consequence. A few months later the charge was, not unnaturally, found to be excessive, and it was reduced to 2$\frac{3}{4}$ drams by order on 27th August, 1861. Trouble now began to be experienced in obtaining really reliable wooden plugs for the bullets, and on 15th December, 1863, it was decided to approve of baked clay as a substitute.

The Minie regulation rifle of 1851, the Enfield musket rifle of 1853, and the Lancaster rifle of 1855 successively followed, and were issued to various units of the British infantry. Their cartridges, however, call for no comment, as they were all manufactured to the standard M.L. paper pattern already described.

Cap composition had up to this period consisted of the following mixture :—

Chlorate of potash	6 parts
Fulminate of mercury	4 parts
Powdered glass	2 parts

But in order to obtain an increase of power, a composition of 80 per cent. fulminate of mercury and 20 per cent. chlorate of potash was introduced on 7th May, 1861. A few years' practical experience, however, proved this detonating powder to be far too powerful, as the caps were blown to pieces on firing. On the 10th August, 1864, therefore, the following modified composition was approved :—

Fulminate of mercury	6 parts
Chlorate of potash	6 parts
Antimony sulphide	4 parts

Caps at this period were punched from copper plate in the form of a cross, and when made up had four flanges. Their dimensions were as follows :—Greatest diameter, 0·230 inches ; least diameter, 0·212 inches ; diameter of flange, 0·43 inches ; and internal depth, 0·248 inches.

As the cartridges of these early Victorian days did not contain their own means of ignition, the caps which functioned on being struck by the hammer of the musket or rifle lock, were carried separately. On 17th January, 1848, it was laid down that they should be carried in zinc cylinders placed in the middle of the S.A.A. barrels (either ball or blank), and in zinc cases with ball cartridges when packed in boxes for carrying that ammunition in limber wagons.

The next improvement in cartridge manufacture was incorporated in "Cartridge Ball Sharps B.L. Carbine." This pattern, being for a breech-loading weapon, was no doubt easier to construct, as a composite cartridge capable of being loaded directly into the breech was obtainable.

The bullet was cylindro-conoidal in shape and weighed 530 grains. The charge of 2$\frac{1}{2}$ drams of F.G. powder, was held in pulp bags tapering to a blunt end about 0·28 inches across. The bags were made on a former and charged with powder when dry. The bullet was then placed in the mouth of the bag, base downwards, so that 1/10th inch of the paper pulp overlapped the bullet. A strip of paper 3/10ths inch wide, was then pasted over the junction of bullet and case, and, after banding, the cartridges were lubricated.

On 14th June, 1860, an improved wire-wound ball cartridge for Sharp's carbine was introduced. It consisted of a double roll of thin paper fastened at one end by a wooden bottom and at the other by a piece of thread. At the base of the bullet was another wooden wad containing a fine copper wire which was passed through the paper folds and wound round the outside of the cartridge holding the whole firmly together. Varnish was then applied to the exterior, and the shoulder of the bullet lubricated with wax.

The next improvement was introduced with " Cartridge Terry's B.L. Carbine " on 8th May, 1860. This cartridge was considered an improved ball cartridge. The bullet used was Pritchett's, weighing 530 grains. The cartridge itself was made from a single sheet of strong but thin paper, closed at the base by two thick paper wads to which one of felt was attached. The felt wad was lubricated with wax and tallow, and the cartridge was shellac-varnished and closed at the top by the bullet. On the introduction, however, of waterproof bags for holding rifle cartridges, the treatment of the paper with shellac, which had a tendency to render the paper brittle, was discontinued at the end of 1863.

Wesley Richard cartridges followed about 1861–65 and were built up according to the same general design.

A new departure was introduced in 1863 when the Whitworth rifle was issued to certain units. The bullet was elongated and hexagonal in section to take the six-sided twisted bore of the rifle. It weighed 480 grains and was propelled by a charge of 75 grains of powder. The weapon never came into universal use.

The next development of cartridge construction was known as the " Boxer " type. This cartridge originally introduced on 20th August, 1866, as the " Cartridge, Boxer, Ball, for Snider rifles, 0·577 inch, Pattern I," had its nomenclature changed on 14th December, 1877, to " Cartridge, S.A., Ball, Snider, 0·577-inch, Mark I." This departure marked a great advance on any previous design of cartridge used in the service, and may in reality be considered as the first attempt at modern cartridge construction. It contained its own means of ignition, a great innovation, as the authorities in this country had up to this time regarded such cartridges as a public danger, until the Prussian needle gun showed the groundless nature of such fears.

The case, formed of brass 0·003 inches thick, was rolled round to form a double thickness with an overlap of 2 inches, being covered externally by thin paper cemented on. The case and brass chamber for the anvil and cap were fixed in the brass base known as the " Pôtet " base by a proper wad. An anvil, on which a percussion cap was placed, was inserted in the brass chamber in the centre of the base of the cartridge.

The bullet was made of pure lead with a plug of wood in front and one of clay in rear. It had four cannelures on it, was lubricated with wax, and weighed 52 grains.

The charge, consisting of 68–72 grains of powder rifle F.G., was capped by about half a grain of carded wool. The bullet was fixed to the case by choking the latter into the lowest cannelure of the former.

Experience proved, however, that the Pôtet case was unreliable, and therefore the Mark II cartridge fitted with an improved form of anvil, appeared about fifteen months later. Minor modifications succeeded one another rapidly and finally the Mark IX cartridge, approved on 16th August, 1871, was introduced for the Snider rifle, an arm still current for prison warders and certain frontier police forces.

The next improvement was the " Cartridge, Martini-Henry," rolled case type, introduced on 16th August, 1873. The case, in its essentials, consists of fine brass sheeting 0·004 inches thick rolled round twice with an overlap of ½ inch. The case is lined with paper to prevent interaction between the charge and brass ; it is also strengthened by an inner and an outer brass cup, and supported by an iron disc at its base. A paper pellet is pressed down inside against the bottom of the case. A brass cap chamber, pierced with a firing hole, passes through the paper pellet, the brass cups, and the iron disc riveting all into one solid piece.

The cap chamber contains an anvil and percussion cap, and the neck of the case is crimped into the cannelures on the bullet after filling.

Following on the rolled case appeared the Martini-Henry cartridge on the solid-case system. This brass case approximates so closely to the modern ·303-inch service case in general principles that a description is considered unnecessary. It was introduced on 9th June, 1885.

The present service rifle—the R.S.M.L.E.—has the same bore, ·303-inch, as its predecessor, the Lee-Metford, and consequently the dimensions of their ammunition are the same. A brass cartridge of this size was therefore the last to be introduced into the British service.

The original cartridge, S.A. Ball ·303-inch, introduced on 20th February, 1889, was powder-filled. Its cap, containing the following composition, was adopted on the same date :—

Fulminate of mercury	6 parts
Chlorate of potash	14 ,,
Sulphide of antimony	18 ,,
Sulphur	1 part
Mealed powder	1 ,,

The invention of cordite in 1889 by Abel and Dewar revolutionized the S.A.A. cartridge when the possibilities of the new propellant were realized. It was first adopted as the charge for the ·303-inch rifle on 3rd November, 1891, the Martini-Henry cordite-filled, solid-case cartridge following on 18th December, 1902.

Owing to the greater pressures and heat generated by these new charges, the days of the solid leaden bullet became numbered. Henceforth it was necessary to adopt the principle of encasing the soft core in a hard envelope.

The modern type of bullet, consisting of a cupro-nickel envelope with a lead and antimony core, therefore appeared.

On 1st December, 1896, the present cap composition was introduced for all the latest types of cordite-loaded ammunition ; it differed from its fore-runner in having the addition of two more parts of fulminate of mercury.

The latest service cartridge is made of solid drawn brass, and varies slightly in regard to the cap chamber and anvil from its original prototype. The bullet has also changed to a minor degree in shape, weight and internal construction, and, under pressure of war, armour-piercing, tracer, incendiary, and other types of bullets have been introduced in addition to the original Mark VII, with its soft core of lead and antimony with an aluminium tip.

So far the history of ball cartridges only have been traced out, but the evolution of other cartridges, such as pistol, buckshot and blank, has followed the same course, with the exception that early revolver cartridges were made of skin instead of paper. As progress has thus been maintained on parallel lines, it is not proposed to go into details of such other types, as descriptions of their earlier patterns may be found in paragraphs, List of Changes, or in previous editions of the Treatise on Ammunition.

Appendix IX

PYROTECHNICS

The science of pyrotechny is of considerable antiquity, and doubtless existed prior to that of artillery. It was known in the East, especially in China, in very early times, and the Chinese are always credited with being skilled in the art of firework-making many centuries ago.

The original reason for firework introduction was spectacular, probably for the purpose of display at great festivals, and possibly the realization of their moral effect in war, against fighting men of an uneducated race and beasts, developed from this source. The glowing fiery trails across the heavens would naturally represent the incarnation of the Evil One to the untutored mind, and terrorize his animals unaccustomed to the sight; its value as a military weapon thus began to be established. In course of time, as the primitive methods of manufacture improved, the projected firework or rocket became dangerous as an incendiary missile, and therefore the physical effect gradually outgrew the moral, with the result that pyrotechnics came to be definitely regarded as a type of early weapon. It is only in recent years that their employment has been extended to signalling generally, which, owing to the advance of artillery, has now become their only use from a military point of view.

Ancient writers being as uncertain of facts as they were rich in metaphor and hyperbole, the task of identifying incendiary compositions from fireworks proper becomes difficult in old records, but in the midst of extravagant phraseology the rocket can be clearly recognized.

Caligula was stated to have had rockets, but such incendiary projectiles as he possessed were possibly more in the nature of balls of tow soaked in some inflammable mixture, as a rocket without gunpowder or saltpetre appears to be an impossible proposition.

The Chinese are said to have employed incendiary rockets against the Tartars in 1232, and Timur's rockets were probably used in the battle of Delhi in 1399. History relates that the fall of Bitar Fort in 1657 was caused by a rocket falling upon a pit of explosive material, which blew up to the detriment of the defenders.

In Europe we read of their use in the war of Chioggia in 1380, but in the West they never attained that degree of popularity enjoyed by them in the East. In fact they fell into desuetude about the beginning of the 15th century, and remained forgotten until revived by the efforts of Congreve at the beginning of the 19th century.

Rockets, in reaching their zenith in early centuries, differed from all other projectiles. Modern factory methods have undoubtedly contributed towards producing a better article, but in essentials fireworks have remained the same, and their composition has not appreciably altered for centuries, except in regard to the proportion of their ingredients. In other words, rockets have always been filled with gunpowder in some form, with the addition of various substances for colouring effects.

A curious old book, containing a section entitled " The Schoole of Artificial fire-workes," with a chapter on " Necessary and serviceable fireworks both for Land and Sea execution, and first for the pike," printed in 1653, gives complete instructions for the manufacture of rockets, their filling, stars and priming. The illustrations practically portray the rocket of to-day, the body or " coffin " of which was made of strong paper. It was filled with mealed powder and coaldust, and the stars were made of mealed powder, saltpetre, sulphur vivum, aqua vitæ and oil of spike. This composition was mixed and moulded inside a paper container, one end of which was dusted with powder and the other covered with glue. Another recipe suggests camphor and oil of turpentine, instead of alcohol and oil of spike.

In the time of Tipu Sultan rockets were employed for warlike purposes in India and the one then in use was 8 inches long and 1½ inches in diameter. They were, however, explosive as well as incendiary and must have improved enormously in a short time, for at the siege of Seringapatam in 1799, Colonel Gerrard, the Adjutant-General of the Indian Army, states that he saw three men killed and four badly wounded by one rocket, and that our troops suffered more from that form of missile than from the enemy's shells.

Realizing the value of rockets in war, the Ordnance Office applied to the Royal Laboratory for the services of someone conversant with their manufacture; that Department in turn referred them to the East India Company; but in vain, for no expert was forthcoming. This state of affairs led Sir William Congreve to turn his attention to the subject. This officer—then a colonel—held his commission in the Hanoverian Army, and he might aptly be termed the pioneer of the modern rocket, for it was he who in 1805 first conceived of a systematic construction of pyrotechnical projectiles.

The chief recommendation of such a system appeared to him to be that the projectile force was exerted without any reaction upon the point from which it was discharged, and therefore he visualized boats' crews and individuals discharging with ease missiles equivalent in destructive power to the heaviest cannon of the day. Owing to the limitations of artillery in his day he was right, for the range and accuracy of the rockets he perfected equalled those of any gun of the period.

Sir William Congreve brought a powerful imagination to bear upon this subject, and expanded his original idea of an incendiary rocket into those filled as shell or shrapnel. He conceived the possibility of rocketeers dispersing bodies of troops in the field and enfilading trenches by arming the lightest troops with his new weapon. Owing to his ingenuity, England gained certain advantages in her coming wars.

On 8th October, 1806, 18 boats discharged 200 rockets into Boulogne and set fire to the town in many places without incurring the slightest opposition or loss. In 1807 Copenhagen suffered serious damage from our rockets, and the same result occurred at Walcheren. In 1813 the rocket brigade at the battle of Leipsic vindicated Sir William Congreve's opinion as to the possible value of rockets in the field, and, lastly, they did good service against the French boats at the passage of the Adour.

Sir William Congreve's rockets were first constructed of paper and afterwards of iron, and it was he who initiated the system of naming a rocket according to its total weight. They were made in different sizes, the 32-pr. being the one most commonly in use, was designed to act as explosive, incendiary or shrapnel shell, and attained a range of nearly 3,000 yards. On the 14th September, 1864, the 6-pr., 12-pr., and 24-pr. Congreve rockets were provisionally superseded by those of Boxer's design, which were sealed in August, 1864. These new rockets had two advantages over their predecessors :—

(1) Their bodies were strengthened by altering the position of the vents.

(2) Their accuracy was improved by employing a stronger composition and consequently increasing the initial velocity.

Boxer's 3-pr. rocket was provisionally approved on 1st October, 1866; on 24th April following it was decided to discontinue the use of war rockets as shells.

Finally, Congreve's rockets were declared obsolete on the 14th August, 1866, and in the following year Boxer's pattern was superseded by Hale's, which were considered to be superior in construction.

Four sizes of Hale's rockets were originally made, being introduced in the service as follows :—

3-pr., 6-pr. and 12-pr. on 25th July, 1867 ; and
24-pr. on 31st August, 1867 ;

but the 3-pr. and 12-pr. soon dropped into disuse and only the 6-pr. and 24-pr. continued to be made. As the 6-pr. rocket, however, weighed in reality 9 lb. its nomenclature was correspondingly changed on 27th November, 1867, and it was adopted for general field service at the same time.

War rockets, although considered in the past to possess great moral and incendiary effects, suffered from five serious defects, which owing to the development of modern artillery, have rendered their use nugatory. These disadvantages were :—

(1) Liability to corrosion and rapid deterioration.
(2) The gradual method by which their velocity is imparted to them renders their flight slow and erratic.
(3) They are easily acted on by gravity and wind, and are therefore constantly liable to change their direction during flight.
(4) Rockets carrying a stick (such as Congreve's) increased the inaccuracies introduced by wind and air currents.
(5) As their charge continued to burn during flight, their centre of gravity constantly altered and disturbed their steadiness.

Hale's rockets travelled a considerable distance, some 24-prs. fired in 1868 at 15 degrees elevation attaining a mean range of 1,896 yards.

The case, originally manufactured from the best charcoal iron, was afterwards made of Atlas metal. It was formed into a cylinder with edges lapped, riveted and brazed at the longitudinal joint ; the head, conoidal in shape, was made from cast iron, containing a hollow portion subsequently plugged with oak. The case was corrugated in three places in order to give a better hold on the composition, and to prevent its twisting under rotational velocity.

The composition, consisting of saltpetre, sulphur and charcoal, was separated from the head by a millboard disc and from the inner base by a washer of the same material. It was introduced into the case in successive pellets and then pressed down hydraulically ; afterwards a conically shaped hole was bored out for about two-thirds of its length. The base of the case was closed by a thick iron disc, varying from 0·8 inches in the 9-pr. rocket to 1·25 inches in the 24-pr. fitted in and fixed by screws. This was bored and tapped to receive a tail piece of cast iron containing three vents prolonged in the form of three curved shields by means of which the rocket was kept point first in flight ; the vents themselves were conical, the apex of the cone being towards the rear. These rockets differed from those designed by Congreve in having a turbine rotational principle for flight control instead of a long stick.

The cases of the original marks of Hale's rockets were greased internally with tallow, but this method of treatment was abandoned on 1st September, 1870, in favour of two coats of paint. A machine or trough for firing these rockets was introduced on 17th September, 1867. A sea service machine already existed for this purpose, being introduced on 13th June, 1866, but it was superseded by Hale's machine, which in time gave place to the sea service rocket-tube machine, Mark II, proposed by Lieut. Fisher, R.N., and approved for the Navy on 7th September, 1869.

Hale's original machine did not long survive, few of them being made. A special machine for firing the 9-pr. rocket was provisionally approved for the Abyssinian expedition on 19th November, 1867 ; and finally trough machines for firing the 9-pr. and 24-pr. rockets were introduced on the 8th June and 10th July, 1868, respectively.

Thus the 9-pr. and 24-pr. were the only survivals from the war rockets of the past. As guns improved they fell more and more into disuse as the development of ballistics filched their potency and usefulness. As early as the 'seventies they were practically withdrawn from our service, except when occasions warranted their retention. However, they lingered on and Mark succeeded Mark, until finally the Great War set a seal upon their doom, and the Mark VII patterns of both rockets were declared obsolete on 11th September, 1919.

Although signal lights and coloured fire had been used for display purposes, from early times, no practical use of the illuminating as contrasted with the incendiary properties of fireworks appears to have been made until the middle of last century, when signal or sky rockets were introduced as follows :—

(1) The 2-lb., 1-lb. and ½-lb. on 24th April, 1863, finally sealed as Mark I on 27th January, 1866.
(2) A special ¼-lb. rocket and stick provisionally approved for the Abyssinian expedition on 19th November, 1867.

Rockets, light and sound followed, being approved for service on 25th April, 1878.

There was little development in pyrotechnics from this date until the Great War, when a tremendous impetus was given to this branch of warlike stores owing to the development of trench warfare on the one hand and the growth of the Royal Air Force on the other. Now the number of rockets, lights and signals is very large and pyrotechny has embarked on a new phase of development.

Besides rockets and lights, such for instance as " Lights G.S. long," the introduction of signal pistols has created a whole host of signal cartridges designed for lighting portions of trenches at night or for general signalling purposes.

However all these ramifications in use and design, even including the " rocket life saving " originally introduced by Colonel Boxer and approved on the 15th March, 1865, belong more to a current treatise on pyrotechnics than to the history of their past, and space forbids the inclusion of any details in this appendix. One thing is certain, however, and that is that pyrotechny is once again in the ascendant owing to the complexity of modern problems on land, sea and in the air, and further developments may be expected.

INDEX

All references are to page numbers.

G.

H.

I.

www.ingramcontent.com/pod-product-compliance
Ingram Content Group UK Ltd.
Pitfield, Milton Keynes, MK11 3LW, UK
UKHW050146280726
14058UKWH00007B/867